Contents

Contents

Electrical Installation Work

8th Edition

Brian Scad[dan]

Routledge
Taylor & Francis Group

LONDON AND NEW YORK

Eighth edition published 2016
by Routledge
2 Park Square, Milton Park, Abingdon, Oxon OX14 4RN

and by Routledge
711 Third Avenue, New York, NY 10017

Routledge is an imprint of the Taylor & Francis Group, an informa business

First edition published 1992 by Newnes
Seventh edition published 2011 by Newnes

British Library Cataloguing-in-Publication Data
A catalogue record for this book is available from the British Library

Library of Congress Cataloging in Publication Data
Scaddan, Brian.
Electrical installation work / Brian Scaddan. – 8th edition.
pages cm
1. Electric wiring, Interior. 2. Electrical engineering. 3. Electric apparatus and appliances–Installation. 4. Electric power systems. I. Title.
TK3271.S267 2015
621.319'24--dc23
2015004914

ISBN: 978-1-138-84927-3 (pbk)
ISBN: 978-1-315-72572-7 (ebk)

Typeset in Univers LT by
Servis Filmsetting Ltd, Stockport, Cheshire

Printed in Great Britain by Ashford Colour Press Ltd

Electrical Installation Work

► Mapped closely to the learning outcomes of City & Guilds and EAL exams
► Coverage of Level 2 and Level 3 units in one volume
► Fully aligned to the 3rd Amendment of the 17th Edition of the IET Wiring Regulations

Brian Scaddan's *Electrical Installation Work* explains in detail how and why electrical installations are designed, installed and tested.

You will be guided in a logical, topic by topic progression through all the areas required to complete City & Guilds and EAL courses. Rather than following the order of the syllabus, this approach will make it easy to quickly find and learn all you need to know about individual topics, and makes this title an indispensable resource for electrical trainees of all ability levels, both during their training and once qualified.

With a wealth of colour pictures, clear layout, and numerous diagrams and figures providing visual illustration, mastering difficult concepts will be a breeze.

Brian Scaddan, I Eng, MIET, is a consultant for and an Honorary Member of City & Guilds. He has over 40 years' experience in Further Education and training. He is Director of Brian Scaddan Associates Ltd, an approved City & Guilds training centre offering courses on all aspects of electrical installation contracting including the C&G 2382 series. He is also a leading author of books on electrical installation.

Contents

CHAPTER 16 Earthing and Bonding

CHAPTER 17 Protection

CHAPTER 18 Circuit and Design

Contents

Preface

This book is intended for the trainee electrician who is working towards NVQs, gaining competences in various aspects of installation work.

It covers both installation theory and practice in compliance with the 17th edition of the *IET Wiring Regulations*, and also deals with the electrical contracting industry, the environmental effects of electricity and basic electronics.

The material in this book has been arranged to cater for student-centred learning programmes. Self-assessment questions and answers are provided at the end of chapters.

Since January 1995, the United Kingdom distribution **declared** voltages at consumer supply terminals have changed from 415 V/240 V 6±6% to 400 V/230 V+1 10% 2-6%. As there has been no physical change to the system, it is likely that measurement of voltages will reveal little or no difference to those before, nor will they do so for some considerable time to come. However, I have used only the new values in the examples in this book.

Brian Scaddan

CHAPTER 1

Basic Information and Calculations

Units

A unit is what we use to indicate the measurement of a quantity. For example, a unit of **length** could be an **inch** or a **metre** or a **mile**, etc.

In order to ensure that we all have a common standard, an international system of units exists known as the SI system. There are seven basic SI units from which all other units are derived.

Basic units

Quantity	Symbol	Unit	Symbol
Length	l	Metre	m
Mass	m	Kilogram	kg
Time	s	Second	s
Current	I	Ampere	A
Temperature	t	Kelvin	K
Luminous intensity	I	Candela	cd
Amount of substance		Mole	mol

Conversion of units

Temperature

Kelvin (K) = °C + 273.15

Celsius (°C) = K − 273.15

Celsius (°C) = $\dfrac{5}{9}$ (°F − 32)

Fahrenheit (°F) = $\left(\dfrac{9 \times °C}{5}\right)$ + 32

Boiling point of water at sea level = 100°C or 212°F

Freezing point of water at sea level = 0°C or 32°F

Normal body temperature = 36.8°C or 98.4°F

Length

To Obtain	Multiply	By
mm	cm	10^1
	m	10^3
	km	10^6
cm	mm	10^{-1}
	m	10^2
	km	10^5
m	mm	10^{-3}
	cm	10^{-2}
	km	10^3
km	mm	10^{-6}
	cm	10^{-5}
	m	10^{-3}

mm: millimetre; cm: centimetre; m: metre; km: kilometre.

Area

To Obtain	Multiply	By
mm^2	cm^2	10^2
	m^2	10^6
	km^2	10^{12}
cm^2	mm^2	10^{-2}
	m^2	10^4
	km^2	10^{10}
m^2	mm^2	10^{-6}
	cm^2	10^{-4}
	km^2	10^6
km^2	mm^2	10^{-12}
	cm^2	10^{-10}
	m^2	10^{-6}

mm^2: square millimetre; cm^2: square centimetre; m^2: square metre; km^2: square kilometre; also, 1 km^2 = 100 hectares (ha).

Electrical Installation Work. 978-1-138-84927-3 © Brian Scaddan. Published by Taylor & Francis. All rights reserved.

Volume

To Obtain	Multiply	By
mm³	cm³	10³
	m³	10⁹
cm³	mm³	10⁻³
	m³	10⁶
m³	mm³	10⁻⁹
	cm³	10⁻⁶

mm³: cubic millimetre; cm³: cubic centimetre; m3: cubic metre.

Capacity

To Obtain	Multiply	By
ml	cl	10¹
	l	10³
cl	ml	10⁻¹
	l	10²
l	ml	10⁻³
	cl	10⁻²

ml: millilitre; cl: centilitre; l: litre; also, 1 l of water has a mass of 1 kg.

Mass

To Obtain	Multiply	By
mg	g	10³
	kg	10⁶
	t	10⁹
g	mg	10⁻³
	kg	10³
	t	10⁶
kg	mg	10⁻⁶
	g	10⁻³
	t	10³
t	mg	10⁻⁹
	g	10⁻⁶
	kg	10⁻³

mg: milligram; g: gram; kg: kilogram; t: tonne.

Multiples and submultiples of units

Name	Symbol	Multiplier	Example
tera	T	10¹² (1 000 000 000 000)	terawatt (TW)
giga	G	10⁹ (1 000 000 000)	gigahertz (GHz)
mega	M*	10⁶ (1 000 000)	megawatt (MW)
kilo	k*	10³ (1000)	kilovolt (kV)
hecto	h	10² (100)	hectogram (hg)
deka	da	10¹ (10)	dekahertz (daHz)
deci	d	10⁻¹ (1/10th)	decivolt (dV)
centi	c	10⁻² (1/100th)	centimetre (cm)
milli	m*	10⁻³ (1/1000th)	milliampere (mA)
micro	μ*	10⁻⁶ (1/1 000 000th)	microvolt (mV)
nano	n	10⁻⁹ (1/1 000 000 000th)	nanowatt (nW)
pico	P*	10⁻¹² (1/1 000 000 000 000th)	picofarad (pF)

Multiples most used in this book.

Mass (kg)

This is the amount of material that an object is made of. It remains constant. It is often confused with weight, in physics it is quite different.

Weight or force (newtons, N)

This is the effect of gravity on a mass. It is not constant, the gravity at different places on the Earth is not the same. Snooker balls travel horizontally in exactly the same way on a table on the moon as they do on the Earth but they will fall into the pockets more slowly.

Mass may be converted to force by multiplying by 9.81.

Force (newtons) = Mass (kg × 9.81)

Work (joules, J)

This is the process of moving objects over a distance using force. Work is force times distance.

Work (joules) = Force (newtons) × Distance (m)

Energy (joules)

This is the ability to carry out work. 100 joules of energy is required to do 100 joules of work.

Power (watts)

This is the rate at which energy is used or work is done. Power is energy divided by time.

$$\text{Power (watts)} = \frac{\text{Energy (joules)}}{\text{Time (seconds)}}$$

Indices

It is very important to understand what **indices** are and how they are used. Without such knowledge, calculations and manipulation of formulae are difficult and frustrating.

So, what are **indices**? Well, they are perhaps most easily explained by examples. If we multiply two identical numbers, say 2 and 2, the answer is clearly 4, and this process is usually expressed as

$2 \times 2 = 4$

However, another way of expressing the same condition is

$2^2 = 4$

The superscript 2 simply means that the online 2 is multiplied by itself. The superscript 2 is known as the indice. Sometimes this situation is referred to as 'two **raised to the power of** two'. So, 2^3 means 'two multiplied by itself **three** times',

i.e. $2 \times 2 \times 2 = 8$

Do not be misled by thinking that 2^3 is 2×3.

$2^4 = 2 \times 2 \times 2 \times 2 = 16$ (not $2 \times 4 = 8$)
$24^2 = 24 \times 24 = 576$ (not $24 \times 2 = 48$)

Here are some other examples:

$3^3 = 3 \times 3 \times 3 = 27$
$9^2 = 9 \times 9 = 81$
$4^3 = 4 \times 4 \times 4 = 64$
$10^5 = 10 \times 10 \times 10 \times 10 \times 10 = 100\ 000$

A number by itself, say 3, has an invisible indice, 1, but it is not shown. Now, consider this: $2^2 \times 2^2$ may be rewritten as $2 \times 2 \times 2 \times 2$, or as 2^4 which means that the indices 2 and 2 or the invisible indices 1 have been added together. So the rule is, when multiplying, **add** the indices.

Examples

$4 \times 4^2 = 4^1 \times 4^2 = 4^3 = 4 \times 4 \times 4 = 64$
$3^2 \times 3^3 = 3^5 = 3 \times 3 \times 3 \times 3 \times 3 = 243$
$10 \times 10^3 = 10^4 = 10 \times 10 \times 10 \times 10 = 10\ 000$

Let us now advance to the following situation:

$10^4 \times \dfrac{1}{10^2}$ is the same as $\dfrac{10^4}{10^2} = \dfrac{10 \times 10 \times 10 \times 10}{10 \times 10}$

Cancelling out the tens

$$\frac{\cancel{10} \times \cancel{10} \times 10 \times 10}{\cancel{10} \times \cancel{10}}$$

we get

$10 \times 10 = 10^2$

which means that the indices have been **subtracted**, that is $4 - 2$. So the rule is, when dividing, **subtract** the indices.

How about this though: $4 - 2$ is either 4 subtract 2 or 4 add −2, and remember, the addition of indices goes with multiplication, so from this we should see that 10^4 divided by 10^2 is the same as 10^4 multiplied by 10^{-2}.

So,

$\dfrac{1}{10^2}$ is the same as 10^{-2}

Example

$\dfrac{1}{3^4} = 3^{-4} \qquad \dfrac{1}{2^6} = 2^{-6}$

and conversely,

$\dfrac{1}{10^{-2}} = 10^2$

Hence we can see that indices may be moved above or below the line provided the sign is changed.

Example

1 $\dfrac{10^6 \times 10^7 \times 10^{-3}}{10^4 \times 10^2} = \dfrac{10^{13} \times 10^{-3}}{10^6}$

  $= \dfrac{10^{10}}{10^6} = 10^{10} \times 10^{-6} = 10^4 = 10\ 000$

2 $\dfrac{10^4 \times 10^{-6}}{10} = 10^4 \times 10^{-6} \times 10^{-1}$

  $= 10^4 \times 10^{-7} = 10^{-3} = \dfrac{1}{10^3}$

  $= \dfrac{1}{1000} = 0.001$

Simple algebra

Algebra is a means of solving mathematical problems using letters or symbols to represent unknown quantities. The same laws apply to algebraic symbols as to real numbers.

Hence: if **one ten times one ten** = 10^2, then **one X times one X** = X^2. That is,

$$X \times X = X^2$$

In algebra the multiplication sign is usually left out. So, for example $A \times B$ is shown as AB and $2 \times Y$ is shown as $2Y$. This avoids the confusion of the multiplication sign being mistaken for an X. Sometimes a dot ($\cdot$) is used to replace the multiplication sign. Hence $3 \cdot X$ means 3 times X, and $2F \cdot P$ means 2 times F times P.

The laws of indices also apply to algebraic symbols. For example,

$$X \cdot X = X^2 \quad Y^2 \cdot Y^2 = Y^4 \quad \frac{1}{X} = X^{-1} \quad \frac{1}{Y^3} = Y^{-3}, \text{ etc.}$$

Addition and subtraction are approached in the same way. For example,

$$X + X = 2X$$
$$3X - X = 2X$$
$$10P - 5P = 5P$$
$$4M + 6M + 2F = 10M + 2F$$

Also with multiplication and division. For example,

$$X \cdot X = X^2$$
$$3M \cdot 2M = 6M^2$$
$$\frac{14P}{7} = 2P$$
$$\frac{10Y}{2Y} = 5$$

Formulae or equations

A formula or equation is an algebraic means of showing how a law or rule is applied. For example, we all know that the money we get in our wage packets is our gross pay less deductions. If we represent each

of these quantities by a letter say W for wages, G for gross pay and D for deductions, we can show our pay situation as

$$W = G - D$$

Similarly, we know that if we travel a distance of 60 km at a speed of 30 km per hour, it will take us 2 h. We have simply divided distance (D) by speed (S) to get time (T), which gives us the formula

$$T = \frac{D}{S}$$

Manipulation or transposition of formulae

The equals sign (=) in a formula or equation is similar to the pivot point on a pair of scales (Fig. 1.1).

Pivot

Figure 1.1 Balancing an equation.

If an item is added to one side of the scales, they become unbalanced, so an identical weight needs to be added to the other side to return the scales to a balanced condition. The same applies to a formula or equation, in that whatever is done on one side of the equals sign must be done to the other side.

Consider the formula $X + Y = Z$.

If we were to multiply the left-hand side (LHS) by, say, 2, we would get $2X + 2Y$, but in order to ensure that the formula remains correct, we must also multiply the right-hand side (RHS) by 2, hence $2X + 2Y = 2Z$.

Formulae may be rearranged (transposed) such that any symbol can be shown in terms of the other symbols. For example, we know that our pay formula is $W = G - D$, but if we know our wages and our gross pay how do we find the deductions? Clearly we need to transpose the formula to show D in terms of W and G. However, before we do this, let us consider the types of formula that exist.

There are three types:

1. Pure addition/subtraction
2. Pure multiplication/division
3. Combination of (1) and (2).

Points to Note

1. A symbol on its own with no sign is taken as being positive (i.e. K is $+K$).
2. Symbols or groups of symbols will be on either the top or the bottom of each side of an equation, for example

   $$\frac{A}{B} = \frac{M}{P}$$

   A and M are on the top, B and P are on the bottom. In the case of, say,

   $$X = \frac{R}{S}$$

   X and R are on the top line and S is on the bottom.
   (Imagine X to be divided by 1, i.e. $X/1$.)
3. Formulae are usually expressed with a single symbol on the LHS, for example $Y = P - Q$, but it is still correct to show it as $P - Q = Y$.
4. Symbols enclosed in brackets are treated as one symbol. For example, $(A + C + D)$ may, if necessary, be transposed as if it were a single symbol.

Let us now look at the simple rules of transposition.

(a) Pure addition/subtraction

Move the symbol required to the LHS of the equation and move all others to the RHS. **Any move needs a change in sign.**

Example

If $A - B = Y - X$, what does X equal?

Move the $-X$ to the LHS and change its sign. Hence,

   $$X + A - B = Y$$

Then move the A and the $-B$ to the RHS and change signs. Hence,

   $$X = Y - A + B$$

Example

If $M + P = R - S$, what does R equal?

We have

   $$-R + M + P = -S$$
   $$\therefore -R = -S - M - P$$

However, we need R, not $-R$, so simply change its sign, but remember to do the same to the RHS of the equation. Hence,

   $$R = S + M + P$$

So we can now transpose our wages formula $W = G - D$ to find D:

   $$W = G - D$$
   $$D + W = G$$
   $$D = G - W$$

(b) Pure multiplication/division

Move the symbol required **across** the equals sign so that it is on the **top** of the equation and move all other symbols away from it, **across** the equals sign but in the opposite position (i.e. from top to bottom or vice versa). Signs are not changed with this type of transposition.

Example

If

   $$\frac{A}{B} = \frac{C}{D}$$

what does D equal?

Move the D from bottom RHS to top LHS. Thus,

   $$\frac{A \cdot D}{B} = \frac{C}{1}$$

Now move A and B across to the RHS in opposite positions. Thus,

   $$\frac{D}{1} = \frac{C \cdot B}{A} \text{ or simply } D = \frac{C \cdot B}{A}$$

Example

If

   $$\frac{X \cdot Y \cdot Z}{T} = \frac{M \cdot P}{R}$$

what does P equal?

As P is already on the top line, leave it where it is and simply move the M and R. Hence,

   $$\frac{X \cdot Y \cdot Z \cdot R}{T \cdot M} = P$$

which is the same as

$$P = \frac{X \cdot Y \cdot Z \cdot R}{T \cdot M}$$

(c) Combination transposition

This is best explained by examples.

Example

If

$$\frac{A(P+R)}{X \cdot Y} = \frac{D}{S}$$

what does S equal?

We have

$$\frac{S \cdot A(P+R)}{X \cdot Y} = D$$

Hence,

$$S = \frac{D \cdot X \cdot Y}{A(P+R)}$$

Example

If

$$\frac{A(P+R)}{X \cdot Y} = \frac{D}{S}$$

what does R equal?

Treat $(P + R)$ as a single term and leave it on the top line, as R is part of that term. Hence,

$$(P+R) = \frac{D \cdot X \cdot Y}{A \cdot S}$$

Remove the brackets and treat the RHS as a single term. Hence,

$$P + R = \left(\frac{D \cdot X \cdot Y}{A \cdot S} \right)$$

$$R = \left(\frac{D \cdot X \cdot Y}{A \cdot S} \right) - P$$

Self-Assessment Questions

1. Write down the answers to the following:
   (a) $X + 3X =$
   (b) $9F - 4F =$
   (c) $10Y + 3X - 2Y + X =$
   (d) $M \cdot 2M =$
   (e) $P \cdot 3P \cdot 2P =$
   (f) $\dfrac{12D}{D} =$
   (g) $\dfrac{30A}{15} =$
   (h) $\dfrac{X^3}{X^2} =$

2. Transpose the following to show X in terms of the other symbols:
   (a) $X + Y = P + Q$
   (b) $F - D = A - X$
   (c) $L - X - Q = P + W$
   (d) $2X = 4$
   (e) $XM = PD$
   (f) $\dfrac{A}{X} = W$
   (g) $\dfrac{B}{K} = \dfrac{H}{2X}$
   (h) $A \cdot B \cdot C = \dfrac{MY}{X}$
   (i) $X(A + B) = W$
   (j) $\dfrac{M+N}{2X} = \dfrac{P}{R}$

The theorem of Pythagoras

Pythagoras showed that if a square is constructed on each side of a right-angled triangle (Fig. 1.2), then the area of the large square equals the sum of the areas of the other two squares.

Hence: 'The square on the hypotenuse of a right-angled triangle is equal to the sum of the squares on the other two sides.' That is,

$$H^2 = B^2 + P^2$$

Or, taking the square root of **both** sides of the equation

$$H = \sqrt{B^2 + P^2}$$

Or, transposing

$$B = \sqrt{H^2 - P^2}$$

or

$$P = \sqrt{H^2 - B^2}$$

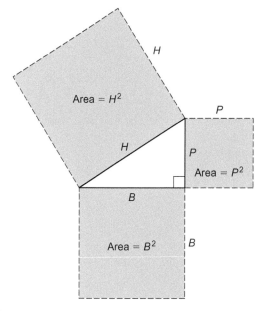

Figure 1.2 Pythagoras's theorem.

Example

1. From Fig. 1.3 calculate the value of H:

$$H = \sqrt{B^2 + P^2}$$
$$= \sqrt{3^2 + 9^2}$$
$$= \sqrt{9 + 81}$$
$$= \sqrt{90}$$
$$= 9.487$$

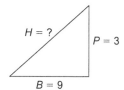

Figure 1.3　　　　　Figure 1.4

2. From Fig. 1.4 calculate the value of B:

$$B = \sqrt{H^2 - P^2}$$
$$= \sqrt{15^2 - 12^2}$$
$$= \sqrt{225 - 144}$$
$$= \sqrt{81}$$
$$= 9$$

Basic trigonometry

This subject deals with the relationship between the sides and angles of triangles. In this section we will deal with only the very basic concepts.

Consider the right-angled triangle shown in Fig. 1.5. *Note*: Unknown angles are usually represented by

Figure 1.5

Greek letters, such as alpha (α), beta (β), phi (ϕ), theta (θ), etc.

There are three relationships between the sides H (hypotenuse), P (perpendicular), and B (base), and the base angle θ. These relationships are known as the **sine**, the **cosine** and the **tangent** of the angle θ, and are usually abbreviated to sin, cos and tan.

The **sine** of the base angle θ,

$$\sin \theta = \frac{P}{H}$$

The **cosine** of the base angle θ,

$$\cos \theta = \frac{B}{H}$$

The **tangent** of the base angle θ,

$$\tan \theta = \frac{P}{B}$$

The values of sin, cos and tan for all angles between 0° and 360° are available either in tables or, more commonly now, by the use of a calculator.

How then do we use trigonometry for the purposes of calculation? Examples are probably the best means of explanation.

Example

1. From the values shown in Fig. 1.5, calculate P and H:

$$\cos \theta = \frac{B}{H}$$

Transposing,

$$H = \frac{B}{\cos \theta} = \frac{3}{\cos 53.13°}$$

From tables or calculator, cos 53.13° = 0.6

$$\therefore H = \frac{3}{0.6} = 5$$

Now we can use sin or tan to find P:

$$\tan \theta = \frac{P}{B}$$

Transposing,

$P = B \cdot \tan \theta$

$\tan \theta = \tan 53.13° = 1.333$

$\therefore P = 3 \times 1.333 = 4(3.999)$

2. From the values shown in Fig. 1.6, calculate α and P:

$$\cos \alpha = \frac{B}{H} = \frac{6}{12} = 0.5$$

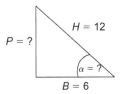

Figure 1.6

We now have to find the angle whose cosine is 0.5. This is usually written as $\cos^{-1} 0.5$. The superscript -1 is not an **indice**; it simply means 'the angle whose sin, cos or tan is …'

So the angle $\alpha = \cos^{-1} 0.5$.

We now look up the tables for 0.5 or use the INV cos or ARC cos, etc., function on a calculator. Hence,

$\alpha = 60°$

$$\sin \alpha = \frac{P}{H}$$

Transposing,

$P = H \cdot \sin \alpha$

$= 12 \cdot \sin 60°$

$= 12 \times 0.866$

$= 10.4$

Self-Assessment Questions

1. What kind of triangle enables the use of Pythagoras' theorem?
2. Write down the formula for Pythagoras' theorem.
3. Calculate the hypotenuse of a right-angled triangle if the base is 11 and the perpendicular is 16.
4. Calculate the base of a right-angled triangle if the hypotenuse is 10 and the perpendicular is 2.
5. Calculate the perpendicular of a right-angled triangle if the hypotenuse is 20 and the base is 8.
6. What is the relationship between the sides and angles of a triangle called?

7. For a right-angled triangle, write down a formula for:
   (a) The sine of an angle.
   (b) The cosine of an angle.
   (c) The tangent of an angle.
8. A right-angled triangle of base angle 25° has a perpendicular of 4. What is the hypotenuse and the base?
9. A right-angled triangle of hypotenuse 16 has a base of 10. What is the base angle and the perpendicular?
10. A right-angled triangle of base 6 has a perpendicular of 14. What is the base angle and the hypotenuse?

Areas and volumes

Areas and volumes are shown in Fig. 1.7.

Areas

Square $= a \times a = a^2$

Rectangle $= h \times l$

Parallelogram $= h \times l$

Cube $= 6a^2$

Cuboid $= 2hb + 2hl + 2bl$
or $2(hb + hl + bl)$

Cone:
open ended $= \pi rl$
solid $= \pi rl + \pi r^2$
or $\pi r(l + r)$

Cylinder:
hollow $= 2\pi rh$
one ended $= 2\pi rh + \pi r^2$
or $\pi r(2h + r)$
solid $= 2\pi rh + 2\pi r^2$
or $2\pi r(h + r)$

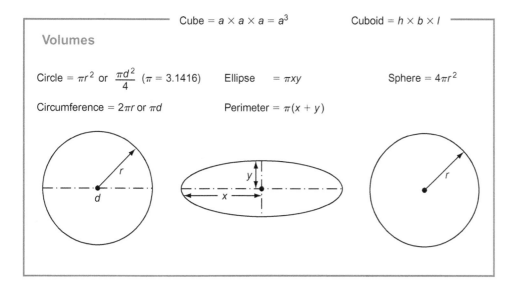

Cube $= a \times a \times a = a^3$ Cuboid $= h \times b \times l$

Volumes

Circle $= \pi r^2$ or $\dfrac{\pi d^2}{4}$ ($\pi = 3.1416$)   Ellipse $= \pi xy$ Sphere $= 4\pi r^2$

Circumference $= 2\pi r$ or πd Perimeter $= \pi(x + y)$

Figure 1.7 Areas and volumes.

CHAPTER 2

Electricity

What is electricity? Where does it come from? How fast does it travel? In order to answer such questions, it is necessary to understand the nature of substances. The following paragraphs give a very simple explanation of the relationship between atomic particles as this area of physics and chemistry is extremely complex.

Molecules and atoms

Every known substance is composed of molecules which in turn are made up of atoms. Substances whose molecules are formed by atoms of the same type are called **elements**, of which there are known to be, at present, 118 (Table 2.1). Some of these are synthetic.

Table 2.1 Elements

Atomic Number	Name	Symbol	Atomic Number	Name	Symbol
1	Hydrogen	H	52	Tellurium	Te
2	Helium	He	53	Iodine	I
3	Lithium	Li	54	Xenon	Xe
4	Beryllium	Be	55	Caesium	Cs
5	Boron	B	56	Barium	Ba
6	Carbon	C	57	Lanthanum	La
7	Nitrogen	N	58	Cerium	Ce
8	Oxygen	O	59	Praseodymium	Pr
9	Fluorine	F	60	Neodymium	Na
10	Neon	Ne	61	Promethium	Pm
11	Sodium	Na	62	Samarium	Sm
12	Magnesium	Mg	63	Europium	Eu
13	Aluminium	Al	64	Gadolinium	Gd
14	Silicon	Si	65	Terbium	Tb
15	Phosphorus	P	66	Dysprosium	Dy
16	Sulphur	S	67	Holmium	Ho
17	Chlorine	Cl	68	Erbium	Er
18	Argon	A	69	Thulium	Tm
19	Potassium	K	70	Ytterbium	Yb
20	Calcium	Ca	71	Lutecium	Lu
21	Scandium	Sc	72	Hafnium	Hf

Electrical Installation Work. 978-1-138-84927-3 © Brian Scaddan.
Published by Taylor & Francis. All rights reserved.

Table 2.1 (Continued)

Atomic Number	Name	Symbol	Atomic Number	Name	Symbol
22	Titanium	Ti	73	Tantalum	Ta
23	Vanadium	V	74	Tungsten	W
24	Chromium	Cr	75	Rhenium	Re
25	Manganese	Mn	76	Osmium	Os
26	Iron	Fe	77	Iridium	Ir
27	Cobalt	Co	78	Platinum	Pt
28	Nickel	Ni	79	Gold	Au
29	Copper	Cu	80	Mercury	Hg
30	Zinc	Zn	81	Thallium	Tl
31	Gallium	Ga	82	Lead	Pb
32	Germanium	Ge	83	Bismuth	Bi
33	Arsenic	As	84	Polonium	Po
34	Selenium	Se	85	Astatine	At
35	Bromine	Br	86	Radon	Rn
36	Krypton	Kr	87	Francium	Fr
37	Rubidium	Rb	88	Radium	Ra
38	Strontium	Sr	89	Actinium	Ac
39	Yttrium	Y	90	Thorium	Th
40	Zirconium	Zr	91	Protoactinium	Pa
41	Niobium	Nb	92	Uranium	U
42	Molybdenum	Mo	93	Neptunium	Np
43	Technetium	Tc	94	Plutonium	Pu
44	Ruthenium	Ru	95	Americium	Am
45	Rhodium	Rh	96	Curiam	Cm
46	Palladium	Pd	97	Berkelium	Bk
47	Silver	Ag	98	Californium	Cf
48	Cadmium	Cd	99	Einsteinium	Es
49	Indium	In	100	Fermium	Fm
50	Tin	Sn	101	Mendelevium	Md
51	Antimony	Sb	102	Nobelium	No
103	Lawrencium	Lr	111	Roentgenium	Rg
104	Rutherfordium	Rf	112	Ununbium	Uub
105	Dubnium	Db	113	Ununtrium	Uut
106	Seaborgium	Sg	114	Ununquadium	Uuq
107	Bohrium	Bh	115	Ununpentium	Uup
108	Hassium	Hs	116	Ununhexium	Uuh
109	Meitnerium	Mt	117	Ununseptium	Uus
110	Darmstadtium	Ds	118	Ununoctium	Uuo

Substances whose molecules are made up of atoms of different types are known as **compounds**. Hence, water, which is a compound, comprises two hydrogen atoms (H) and one oxygen atom (O), that is H_2O. Similarly, sulphuric acid has two hydrogen, one sulphur and four oxygen atoms: hence, H_2SO_4.

Molecules are always in a state of rapid motion, but when they are densely packed together this movement is restricted and the substance formed by these molecules is stable (i.e. a **solid**). When the molecules of a substance are less tightly bound there is much free movement, and such a substance is known as a **liquid**. When the molecule movement is almost unrestricted the substance can expand and contract in any direction and, of course, is known as a **gas**.

The atoms which form a molecule are themselves made up of particles known as protons, neutrons and electrons. Protons are said to have a positive (+ve) charge, electrons a negative (−ve) charge and neutrons no charge. Since neutrons play no part in electricity at this level of study, they will be ignored from now on.

So what is the relationship between protons and electrons; how do they form an atom? The simplest explanation is to liken an atom to our Solar System, where we have a central star, the Sun, around which are the orbiting planets. In the tiny atom, the protons form the central nucleus and the electrons are the orbiting particles. The simplest atom is that of hydrogen which has one proton and one electron (Fig. 2.1).

Figure 2.1 The hydrogen atom.

The atomic number (Table 2.1) gives an indication of the number of electrons surrounding the nucleus for each of the known elements. Hence, copper has an atomic number of 29, indicating that it has 29 orbiting electrons.

Electrons are arranged in layers or clouds at varying distances from the nucleus (like the rings around Saturn); those nearest the nucleus are more strongly held in place than those farthest away. These distant electrons are easily dislodged from their orbits and hence are free to join those of another atom whose own distant electrons in turn may leave to join other atoms, and so on. These wandering or **random** electrons that move about the molecular structure of the material are what makes up electricity.

So, then, how do electrons form electricity? If we take two dissimilar metal plates and place them in a chemical solution (known as an electrolyte) a reaction takes place in which electrons from one plate travel across the electrolyte and collect on the other plate. So one plate has an excess of electrons which makes it more −ve than +ve, and the other an excess of protons which makes it more +ve than −ve. What we are describing here, of course, is a simple cell or battery (Fig. 2.2).

Now then, consider a length of wire in which, as we have already seen, there are electrons in random movement (Fig. 2.3).

Figure 2.2 Battery.

Figure 2.3 Conductor with random electrons.

If we now join the ends of the wire to the plates of a cell, the excess electrons on the −ve plate will tend to leave and return to the +ve plate, encouraging the random electrons in the wire to **drift** in the same direction (Fig. 2.4). This drift is what we know as electricity. The process will continue until the chemical action of the cell is exhausted and there is no longer a difference, +ve or −ve, between the plates.

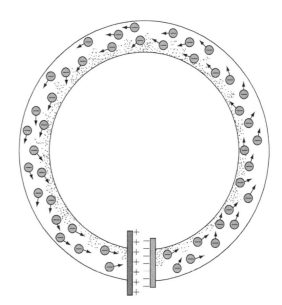

Figure 2.4 Conductor connected to battery (electron drift).

Potential difference

Anything that is in a state whereby it may give rise to the release of energy is said to have **potential**. For example, a ball held above the ground has potential in that, if it were let go, it would fall and hit the ground.

So, a cell or battery with its +ve and −ve plates has potential to cause electron drift. As there is a difference in the number of electrons on each of the plates, this potential is called the **potential difference** (p.d.).

Electron flow and conventional current flow

As we have seen, if we apply a p.d. across the ends of a length of wire, electrons will drift from −ve to +ve. In the early pioneering days, it was incorrectly thought that electricity was the movement of +ve protons and, therefore, any flow was from +ve to −ve. However, as the number of proton charges is the same as the number of electron charges, the convention of electric current flow from +ve to −ve has been maintained.

Conductors and insulators

Having shown that electricity is the general drift of random electrons, it follows that materials with large numbers of such electrons give rise to a greater drift than those with few random electrons. The two different types are known as conductors and insulators. Materials such as PVC, rubber, mica, etc., have few random electrons and therefore make good **insulators**, whereas metals such as aluminium, copper, silver, etc., with large numbers, make good **conductors**.

Electrical quantities

The units in which we measure electrical quantities have been assigned the names of famous scientific pioneers, brief details of whom are as follows. (Others will be detailed as the book progresses.)

André Marie Ampère (1775–1836)

French physicist who showed that a mechanical force exists between two conductors carrying a current.

Charles Augustin de Coulomb (1746–1806)

French military engineer and physicist famous for his work on electric charge.

Georg Simon Ohm (1789–1854)

German physicist who demonstrated the relationship between current, voltage and resistance.

Allessandro Volta (1745–1827)

Italian scientist who developed the electric cell, called the 'voltaic pile', which comprised a series of copper and zinc discs separated by a brine-soaked cloth.

Electric current: symbol, *I*; unit, ampere (A)

This is the flow or drift of random electrons in a conductor.

Electric charge or quantity: symbol, *Q*; unit, coulomb (C)

This is the quantity of electricity that passes a point in a circuit in a certain time. One coulomb is said to have passed when one ampere flows for one second:

$$Q = I \times t$$

Electromotive force (e.m.f.): symbol, *E*; unit, volt (V)

This is the total potential force available from a source to drive electric current around a circuit.

Potential difference (p.d.): symbol, *V*; unit, volt (V)

Often referred to as 'voltage' or 'voltage across', this is the actual force available to drive current around a circuit.

The difference between e.m.f. and p.d. may be illustrated by the **pay** analogy used in Chapter 1. Our gross wage (e.m.f.) is the total available to use. Our net wage (p.d.) is what we actually have to spend after deductions.

Resistance: symbol, *R*; unit, ohm (Ω)

This is the opposition to the flow of current in a circuit.

When electrons flow around a circuit, they do not do so unimpeded. There are many collisions and deflections as they make their way through the complex molecular structure of the conductor, and the extent to which they are impeded will depend on the material from which the conductor is made and its dimensions.

Resistivity: symbol, ρ; unit, μΩ mm

If we take a sample of material in the form of a cube of side 1 mm and measure the resistance between opposite faces (Fig. 2.5), the resulting value is called the **resistivity** of that material.

This means that we can now determine the resistance of a sample of material of any dimension. Let us suppose that we have a 1 mm cube of material of resistivity, say 1 Ω (Fig. 2.6a). If we double the length of that sample, leaving the face area the same (Fig. 2.6b), the resistance now measured would be 2 Ω (i.e.

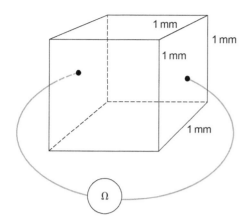

Figure 2.5 Resistivity of a unit cube.

Example

Calculate the resistance of a 50 m length of copper conductor of c.s.a. 2.5 mm², if the resistivity of the copper used is 17.6 μΩ mm.

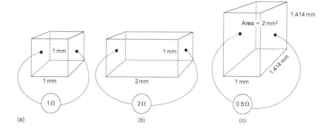

Figure 2.6 Change of resistivity with length and csa.

Note

All measurements should be of the same type, that is resistivity, microohm millimetres; length, millimetres; c.s.a., square millimetres. Hence 10^3 to convert metres to millimetres.

$$R = \frac{\rho \times l}{a}$$

$$\therefore R = \frac{17.6 \times 10^{-6} \times 50 \times 10^3}{2.5}$$

$$= \frac{17.6 \times 10^{-3} \times 50}{2.5}$$

$$= 17.6 \times 10^{-3} \times 20$$

$$= 352 \times 10^{-3}$$

$$= 0.352 \, \Omega$$

the resistance has doubled). If, however, we leave the length the same but double the face area (Fig. 2.6c), the measured value would now be 0.5 | (i.e. the resistance has halved).

Hence, we can now state that whatever happens to the length of a conductor also happens to its resistance (i.e. **resistance is proportional to length**) and whatever happens to the cross-sectional area (c.s.a.) has the opposite effect on the resistance (i.e. **resistance is inversely proportional to area**).

So,

$$\text{Resistance } R = \frac{\text{resistivity } \rho \times \text{length } l}{\text{area } a}$$

$$R = \frac{\rho \times l}{a}$$

In practice, the resistance across a 1 mm cube of a material is extremely small, in the order of millionths of an ohm (μΩ) as shown in Table 2.2.

Table 2.2 Examples of resistivity

Material	Resistivity, ρ (μΩ mm at 20°C)
Copper, International Standard	17.2
Copper, annealed	16.9–17.4
Copper, hard drawn	17.4–18.1
Aluminium, hard drawn	28
Silver, annealed	15.8
Silver, hard drawn	17.5
Platinum	117
Tungsten	56
Eureka (constantan)	480
German silver (platinoid)	344
Manganin	480

Example

Calculate the resistivity of aluminium if a 100 m length of conductor of c.s.a. 4 mm² has a measured resistance of 0.7 Ω.

$$R = \frac{\rho \times l}{a}$$

$$\therefore \rho = \frac{R \times a}{l}$$

$$= \frac{0.7 \times 4}{100 \times 10^3}$$

$$= 7 \times 10^{-1} \times 4 \times 10^{-5}$$

$$= 28 \times 10^{-6}$$

$$= 28 \, \mu\Omega \, mm$$

Reference to Table 2.2 will show that values of resistivity are based on a conductor temperature of 20°C which clearly suggests that other temperatures would give different values. This is quite correct. If a conductor is heated, the molecules vibrate more vigorously making the passage of random electrons more difficult (i.e. the conductor resistance increases). On the other hand, a reduction in temperature has the opposite effect, and hence a decrease in conductor resistance occurs. The amount by which the resistance of a conductor changes with a change in temperature is known as the temperature coefficient of resistance.

Temperature coefficient: symbol, α; unit, ohms per ohm per °C ($\Omega/\Omega/°C$)

If we were to take a sample of conductor that has a resistance of $1\,\Omega$ at a temperature of 0°C, and then increase its temperature by 1°C, the resulting increase in resistance is its temperature coefficient. An increase of 2°C would result in twice the increase, and so on. Therefore the new value of a $1\,\Omega$ resistance which has had its temperature raised from 0°C to t°C is given by $(1 + \alpha t)$. For a $2\,\Omega$ resistance the new value would be $2 \times (1 + \alpha t)$, and for a $3\,\Omega$ resistance, $3 \times (1 + \alpha t)$, etc. Hence we can now write the formula:

$$R_f = R_0(1 + \alpha t)$$

where R_f is the final resistance, R_0 is the resistance at 0°C, α is the temperature coefficient and t°C is the change in temperature.

For a change in temperature between any two values, the formula is:

$$R_2 = \frac{R_1(1 + \alpha t_2)}{(1 + \alpha t_1)}$$

where R_1 is the initial resistance, R_2 is the final resistance, t_1 is the initial temperature and t_2 is the final temperature.

The value of temperature coefficient for most of the common conducting materials is broadly similar, ranging from 0.0039 to $0.0045\,\Omega/\Omega/°C$, that of copper being taken as $0.004\,\Omega/\Omega/°C$.

Example

A sample of copper has a resistance of $10\,\Omega$ at a temperature of 0°C. What will be its resistance at 50°C?

$$R_f = R_0(1 + \alpha t)$$
$$R_f = ?$$
$$R_0 = 10$$
$$t = 50$$
$$\alpha = 0.004\,\Omega/\Omega/°C$$
$$\therefore R_f = 10(1 + 0.004 \times 50)$$
$$= 10(1 + 0.2)$$
$$= 10 \times 1.2$$
$$= 12\,\Omega$$

Example

A length of tungsten filament wire has a resistance of $200\,\Omega$ at 20°C. What will be its resistance at 600°C ($\alpha = 0.0045\,\Omega/\Omega/°C$)?

$$R_2 = \frac{R_1(1 + \alpha t_2)}{(1 + \alpha t_1)}$$
$$R_2 = ?$$
$$R_1 = 200$$
$$t_1 = 20°C$$
$$t_2 = 600°C$$
$$\alpha = 0.0045\,\Omega/\Omega/°C$$

$$\therefore R_2 = \frac{200(1 + 0.0045 \times 600)}{(1 + 0.0045 \times 20)}$$
$$= \frac{200(1 + 2.7)}{(1 + 0.09)}$$
$$= \frac{200 \times 3.7}{1.09}$$
$$= 679\ \Omega$$

There are certain conducting materials such as carbon and electrolytes whose resistances display an **inverse** relationship with temperature, that is their resistances **decrease** with a rise in temperature, and vice versa. These conductors are said to have **negative** temperature coefficients. Carbon, for example, is used for the brushes in some types of motor, where friction causes the brushes to become very hot. In this way current flow to the motor is not impeded.

We have already learned that random electrons moving in the same direction (electric current) through the molecular structure of a conductor experience many collisions and deflections. The energy given off when this happens is in the form of heat; hence the more electrons the more heat and thus the greater the

2

(a)

(b)

(c)

(d)

Figure 2.7 The water analogy.

resistance. So current flow can, itself, cause a change in conductor resistance.

The water analogy

Consider a tap and a length of hose. With the tap just turned on, only a trickle of water will issue from the hose (Fig. 2.7a). Turn the tap further and more water will flow (Fig. 2.7b). Hence pressure and flow are **proportional**. Leave the tap in this position and squeeze the pipe: less water will flow (Fig. 2.7c). Increase the opposition by squeezing more and even less water will flow (Fig. 2.7d). Hence opposition and flow are **inversely proportional**.

Now, for an electric circuit, replace the tap with some source of electricity supply, change the hose to a conductor and the constriction in the hose into added resistance. The flow of water becomes the current. We will now have the same effect, in that a small voltage will only give rise to a small current (Fig. 2.8a), an increase in voltage produces a greater current (Fig. 2.8b) and a constant voltage but with an increase in resistance results in reduced current flow (Fig. 2.8c and d).

Ohm's law

Georg Simon Ohm demonstrated the relationships we have just seen, and stated them in his famous law which is: 'The current in a circuit is proportional to the circuit voltage and inversely proportional to the circuit resistance, at constant temperature'.

So, we can show Ohm's law by means of the formula

$$I = \frac{V}{R}$$

Also, transposing,

$$R = \frac{V}{I} \quad \text{and} \quad V = I \times R$$

Example

A 230 V electric heating element has a measured resistance of 19.2 Ω (Fig. 2.9). Calculate the current that will flow. (*Note*: Whenever possible, draw a diagram, no matter how simple; this will help to ensure that correct values are assigned to the various circuit quantities.)

Figure 2.8 Ohm's Law.

Figure 2.9

$$I = \frac{V}{R}$$

$$= \frac{230}{19.2}$$

$$= 11.9 \text{ A}$$

Example

What is the resistance of an electric lamp filament, if it draws a current of 0.625 A from a 230 V supply? (Fig. 2.10).

Figure 2.10

$$R = \frac{V}{I}$$

$$= \frac{230}{0.625}$$

$$= 368 \ \Omega$$

Example

What voltage would be required to cause a current of 2 A to flow through a resistance of 55 Ω? (Fig. 2.11).

Figure 2.11

$$V = I \times R$$

$$= 2 \times 55$$

$$= 110 \text{ V}$$

Electricity and the human body

Water is a conductor of electricity and since the human body is made up of a high proportion of water, it follows that it also is a conductor. However, unlike other materials we have dealt with so far, there is no exact value for body resistivity, and therefore body resistance can vary, not only between individuals but between values for each person. Depending on whether the body is dry, moist or wet, the value measured between hands or between hands and feet can be anywhere between 1000 and 10 000 Ω. As we have just seen from Ohm's law, the current flowing through a body will depend on the voltage and the body resistance. Different levels of current will have different effects, the worst occurring when the heart goes out of rhythm and will not return to normal. This condition is known as ventricular fibrillation, and will

Table 2.3 Shock levels

Current (mA)	Effect
1–2	Perception level, no harmful effects
5	Throw-off level, painful sensation
10–15	Muscular contraction begins, cannot let go
20–30	Impaired breathing
50 and above	Ventricular fibrillation and death

often result in death. Table 2.3 is a guide to the various levels of shock current and their effects on the body.

It will be seen that 50 mA (0.05 A) is considered to be the minimum **lethal** level of shock current, so, if a person's body resistance was as low as 1000 Ω, the voltage required to cause this current to flow would be

$$V = I \times R$$
$$= 50 \times 10^{-3} \times 1000$$
$$= 50 \text{ V}$$

Note this voltage, it is important.

Types and sources of supply

There are only two types of electricity supply, **direct current** (d.c.) and **alternating current** (a.c.); d.c. is obtained from cells and batteries, d.c. generators or electronically derived from a.c. (rectification); a.c. is obtained from a.c. generators.

The methods of producing a.c. and d.c. supplies are discussed in later chapters.

Voltage bands

Extra low   Not exceeding 50 V a.c. or 120 V d.c. between conductors or conductors and earth (Fig. 2.12a).

Low   Exceeding extra low, but not exceeding 1000 V a.c. or 1500 V d.c. between conductors; or 600 V a.c. or 900 V d.c. between conductors and earth (Fig. 2.12b).

Figure 2.12   Voltage bands.

Measuring current and voltage

As current flows **through** a conductor, it seems logical to expect that any instrument used to measure current would need to have that current flowing **through** it. This is known as a **series** connection (Fig. 2.13).

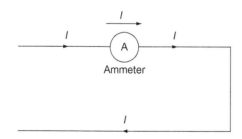

Figure 2.13   Ammeter connections.

Voltage, on the other hand, is a measure of potential difference **between** or **across** two points, and hence a voltage measuring instrument would need to be

connected **between** or **across** two points in a circuit. This is known as a **parallel** connection (Fig. 2.14).

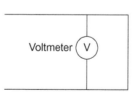

Figure 2.14   Voltmeter connections.

So, the arrangement of instruments to measure the current and voltage associated with a circuit supplying, say, a lamp would be as shown in Fig. 2.15.

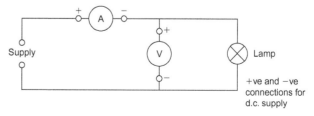

Figure 2.15   Ammeter and voltmeter connections.

In order to allow all the current needed by the lamp to flow, the ammeter should not impede that flow, and hence should have a very **low** resistance. To ensure that very little of the current needed is diverted away from the lamp via the voltmeter, it should have a very **high** resistance.

Components of a circuit

An electric circuit (Fig. 2.16) comprises the following main components:

1. A source of supply, a.c. or d.c.
2. A fuse or circuit breaker which will cut off supply if too much current flows. This is called circuit **protection**.
3. Conductors through which the current will flow. Two or more conductors embedded in a protective sheathing is called a cable.

Figure 2.16   Circuit components.

4. A switch with which to turn the supply on or off. This is called circuit **control**.
5. A device which needs current to make it work. This is known as the **load**.

Self-Assessment Questions

1. Protons and electrons have equal and opposite charges. What are they and which is which?
2. What does an atomic number give an indication of?
3. What are random electrons?
4. (a) What is the difference between a conductor and an insulator?
   (b) Give three examples of each.
5. In what units are the following quantities measured?
   (a) Current
   (b) Potential difference
   (c) Resistance
   (d) Charge
   (e) Resistivity.
6. What length of 4.0 mm² copper conductor, having a resistivity of 17 μΩmm, would have a measured resistance of 0.68 Ω?
7. A 100 m length of aluminium wire ($\rho = 28\,\mu\Omega$ mm) has a resistance of 0.7 Ω. What is its c.s.a.?
8. A length of copper wire has a resistance of 0.5 Ω at 0°C. What would be its resistance at 50°C if the temperature coefficient is 0.004 Ω/Ω/°C?
9. A coil of aluminium wire has a resistance of 54 Ω at 20°C. What would be its resistance at 70°C ($\alpha = 0.004\,\Omega/\Omega/°C$)?

10. What would happen to the current in a circuit if:
    (a) The voltage is constant and the resistance is doubled?
    (b) The voltage is doubled and the resistance is constant?
    (c) The voltage and the resistance were both trebled?
11. Ignoring any effects of temperature, what would happen to circuit resistance if the current was increased?
12. State Ohm's law.
13. Solve the following circuit problems:
    (a) $I = 10\,A$, $V = 230\,V$, $R = ?$
    (b) $R = 38.4\,\Omega$, $V = 230\,V$, $I = ?$
    (c) $R = 60\,\Omega$, $I = 0.4\,A$, $V = ?$
14. At 110 V, what body resistance would allow the accepted lethal level of shock current to flow?
15. What do the letters a.c. and d.c. stand for?
16. What are the limits of a.c. extra-low and low voltage?
17. Draw a labelled diagram of the connections of instruments used to measure the current and voltage in an a.c. circuit supplying a lamp. Include on the diagram circuit protection and control.
18. In such a circuit as described in question 17 above, the instrument readings were 0.5 and 12 A. What is the resistance of the lamp?

CHAPTER 3

Resistance, Current and Voltage, Power and Energy

Resistance

We should know by now what resistance is and how it affects current flow. However, a circuit may contain many resistances connected in various ways, and it is these connections we are to consider now.

Resistance in series

Remember how we connected an ammeter in **series** in a circuit so that the current could flow through it? Two or more resistances or resistors connected in the same way are said to be connected in **series**. It is like squeezing a hosepipe in several places (Fig. 3.1).

The more depressions we make in the pipe, the less water will flow – similarly with resistance (Fig. 3.2).

Less current will flow with, for example, R_3 connected than with R_1 and R_2. Hence we can see that the more resistances that are connected in series, the smaller

Figure 3.1 Resistances in series.

Figure 3.2 Resistances in series.

the current flow. So, the total resistance R_T of a number of resistances in *series* is

$$R_T = R_1 + R_2 + R_3 + \cdots$$

As we have seen in Chapter 1, a conductor will have a resistance to the flow of current. Hence if we take a length of conductor and add another length to it (series connection) the resistance will increase. So resistance is proportional to conductor's length. This is important to remember, because if we supply a load with lengthycable, the current flow may be reduced to such an extent that the load may not work properly.

A load has a fixed resistance, so a reduced current due to cable resistance means that there must be less voltage across the load. This voltage loss is called **voltage drop** and is a very important topic.

Consider the following circuit (Fig. 3.3). The conductor resistances and the lamp resistance are all in series:

$$\therefore \text{Cable resistance} = 0.5 + 0.5 = 1\,\Omega$$

$$\therefore \text{Total series resistance} = R_C + R_L$$
$$= 1 + 24$$
$$= 25\,\Omega$$

$$\therefore \text{Circuit current } I = \frac{V}{R}$$
$$= \frac{12}{25}$$
$$= 0.48\,\text{A}$$

Figure 3.3

This current flows **through all the series** components, and where we have current flow we must have potential difference (p.d.). So, each of these components will have a p.d. across them. Hence, the p.d. across each conductor is

$$V_C = I \times R_C \text{ (Ohm's law)}$$
$$= 0.48 \times 0.5$$
$$= 0.24\,V$$

$$\therefore \text{ Total p.d. across cable} = 2 \times 0.24$$
$$= 0.48\,V$$

The p.d. across the lamp is
$$V_L = I \times R_L$$
$$= 0.48 \times 24$$
$$= 11.52\,V$$

Now, if we add these p.ds together, we get

$$0.48 + 11.52 = 12\,V \text{ (the supply voltage)}$$

Hence we have lost 0.48 V due to the cable resistance, and the lamp has to work on 11.52 V not on 12 V. The 0.48 V lost is the **voltage drop** due to the cable.

If we were to make the cable, say, six times longer, its resistance would become $6 \times 1 = 6\,\Omega$.

$$\therefore \text{ Circuit resistance} = 6 + 24$$
$$= 30\,\Omega$$

$$\therefore \text{ Total current } I = \frac{V}{R}$$
$$= \frac{12}{30}$$
$$= 0.4\,A \text{ (less than before)}$$

$$\therefore \text{ Cable voltage drop} = I \times R_C$$
$$= 0.4 \times 6$$
$$= 2.4\,V$$

The p.d. across the lamp is

$$V_L = I \times R_L$$
$$= 0.4 \times 24$$
$$= 9.6\,V$$

Check: 2.4 V + 9.6 V = 12 V.

So, we can now state that, for a series circuit, the voltages across all the components add up to the supply voltage. That is,

$$V_T = V_1 + V_2 + V_3 + \cdots$$

Voltage drop

In order to ensure that loads are not deprived of too much of their operating voltage due to cable resistance, the IET Wiring Regulations recommend that the voltage drop in a circuit should not exceed a certain percentage of the voltage at the origin of the circuit. This percentage is 3% for lighting and 5% for power. Hence, for a 230 V single-phase supply, the voltage drop on a lighting circuit should not exceed 3% of 230:

$$\frac{3 \times 230}{100} = 6.9\,V$$

and on a power circuit 5% of 230 V

$$\frac{5 \times 230}{100} = 11.5\,V$$

Example

A circuit (Fig. 3.4) comprises two heating elements each of 28.8 Ω resistance. These are connected in series across a 230 V supply. If the supply cable has a resistance of 2.4 Ω calculate:

(a) the total circuit current
(b) the total circuit resistance
(c) the cable voltage drop
(d) the voltage available across each heater.

(a) Total resistance:
$$R = R_C + R_E + R_E$$
$$= 2.4 + 28.8 + 28.8$$
$$= 60\,\Omega$$

(b) Total current:
$$I = \frac{V}{R}$$
$$= \frac{230}{60}$$
$$= 3.833\,A$$

Figure 3.4

(c) Cable voltage drop:

$$V_C = I \times R_C$$
$$= 3.833 \times 2.4$$
$$= 9.2\,V$$

(d) Voltage across each element:

$$V_E = I \times R_E$$
$$= 3.833 \times 28.8$$
$$= 110.4\,V$$

Check:

$$V = V_C + V_E + V_E$$
$$= 9.2 + 110.4 + 110.4$$
$$= 230\,V$$

Resistance in parallel

Remember the connection of a voltmeter **across** the ends of a load? This was said to be connected in **parallel**.

Similarly, if we connect one or more conductors across the ends of another conductor (Fig. 3.5), these are said to be wired in parallel.

If each of these conductors has the same cross-sectional area (c.s.a.), then the effect of three wired in parallel is that of having one conductor three times as large as any one. As we have seen in Chapter 2, an increase in c.s.a. results in a corresponding decrease in resistance, so parallel connections reduce resistance.

> **Note**
>
> Wiring two cables in parallel, or, more simply, changing one for a larger size, will reduce resistance and hence lessen voltage drop.

In the IET Regulations, there are tables giving the values of voltage drop for various types and sizes of conductor. These values are given in millivolts (mV) for every ampere (A) that flows along a length of 1 metre (m) (i.e. mV/A/m). So, we should be able to check on our original comments that resistance, and hence voltage drop, reduces with an increase of c.s.a. For example, a 10.0 mm² conductor should have ten times less of a voltage drop than a 1.0 mm² conductor. Reference to table 4D1B, column 3 (IET Regs) confirms this: the millivolt drop for 1.0 mm² being 44 mV and that for 10.0 mm² being 4.4 mV.

Addition of resistances in parallel

Consider, say, three resistances connected in parallel across a supply (Fig. 3.6).

It will be seen from the figure that, unlike a series circuit, the same supply voltage is being applied across each resistance. In this case, it is the current flowing through each resistance that is different (unless, of course, all the resistances are of the same

Figure 3.5 Resistances in parallel.

Figure 3.6 Resistances in parallel.

23

value). The total current, therefore, is the sum of each of the individual currents:

$$I = I_1 + I_2 + I_3$$

If we now use Ohm's law to convert I to V/R we get

$$\frac{V}{R} = \frac{V}{R_1} + \frac{V}{R_2} + \frac{V}{R_3}$$

Now, dividing right through by V, we get

$$\frac{V}{V \cdot R} = \frac{V}{V \cdot R_1} = \frac{V}{V \cdot R_2} + \frac{V}{V \cdot R_3}$$

The V cancels out, leaving

$$\frac{1}{R} = \frac{1}{R_1} + \frac{1}{R_2} + \frac{1}{R_3} + \cdots$$

Example

Referring to Fig. 3.7, calculate the total resistance of the circuit.

$$\frac{1}{R} = \frac{1}{R_1} + \frac{1}{R_2} + \frac{1}{R_3}$$

$$\frac{1}{R} = \frac{1}{4} + \frac{1}{8} + \frac{1}{2}$$

$$\frac{1}{R} = 0.25 + 0.125 + 0.5$$

$$\frac{1}{R} = 0.875$$

Transposing,

$$R = \frac{1}{0.875}$$
$$= 1.143\,\Omega$$

Note

The total resistance in parallel is less than the smallest resistance in the circuit.

Figure 3.7

There are some short-cuts to the calculation of resistance in parallel. These are as follows:

1. For a number of resistances identical in value, the total resistance is the value of any one divided by the number of resistances. That is, twelve $24\,\Omega$ resistances in parallel will have a total resistance of $24/12 = 2\,\Omega$.
2. For any two resistances in parallel, the total may be found by dividing their product by their sum. Hence a 6 and a $3\,\Omega$ resistance in parallel would have a total of

$$\frac{\text{Product}}{\text{Sum}} = \frac{6 \times 3}{6 + 3} = \frac{18}{9} = 2\,\Omega$$

This method can be used for more than two resistances, by simply doing two at a time.

Example

Determine the total resistance of the circuit shown in Fig. 3.8.

Use product/sum for R_1 and R_2, then for R_3 and R_4 and then again for the two totals.

Hence,

$$\text{Total for } R_1 \text{ and } R_2 = \frac{3 \times 6}{3 + 6} = 2\,\Omega$$

$$\text{Total for } R_3 \text{ and } R_4 = \frac{8 \times 2}{8 + 2} = 1.6\,\Omega$$

$$\therefore \text{Overall total} = \frac{2 \times 1.6}{2 + 1.6} = 0.89\,\Omega$$

Figure 3.8

Insulation resistance

As already known, insulation is a very poor conductor and hence presents a very high resistance to the flow of current. Consider then two short lengths of conductor, A and B, separated by insulation (Fig. 3.9).

Current is inhibited from flowing from A to B due to the insulation which could be said to comprise an infinite number of very high resistances in parallel (Fig. 3.10).

If we now extend the length of these conductors and insulation, the effect is that of adding extra parallel resistance to that insulation, and as we have just seen, the greater the number of parallel resistances, the smaller the total resistance. So an increase in cable length results not only in an increase in the conductor resistance, but in a **decrease** in insulation resistance.

The insulation resistance of an installation must, of course, be very high, usually in the order of many millions of ohms (i.e. megohms, MΩ). The IET Regulations require that the minimum acceptable value is 1.0 MΩ for circuits up to 500 V.

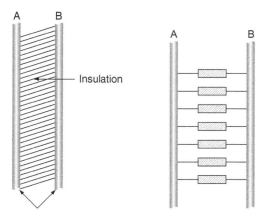

Figure 3.9 Insulation resistance.

Figure 3.10 Insulation resistance.

Current distribution

Example

From Fig. 3.11 determine (a) the total resistance, (b) the total current and (c) the current in each resistance.

(a) Using the product/sum method:

$$\text{Total of } R_1 \text{ and } R_2 = \frac{6 \times 3}{6+3} = 2\,\Omega$$

Using this value with R_3:

$$\text{Total } R = \frac{2 \times 8}{2+8}$$
$$= 1.6\,\Omega$$

(b) Total current:

$$I = \frac{V}{R} = \frac{24}{1.6} = 15\,\text{A}$$

Figure 3.11

(c) Current in R_1:

$$I_1 = \frac{V}{R_1} = \frac{24}{6} = 4\,\text{A}$$

Current in R_2:

$$I_2 = \frac{V}{R_2} = \frac{24}{3} = 8\,\text{A}$$

Current in R_3:

$$I_3 = \frac{V}{R_3} = \frac{24}{8} = 3\,\text{A}$$

Check: Total current I should be

$$I = I_1 + I_2 + I_3$$
$$= 4 + 8 + 3 = 15\,\text{A}$$

Combined series – parallel connections

Current-using pieces of equipment in installation circuits are connected across the supply and are therefore wired in parallel. The cable supplying such equipment, however, is connected in series with the parallel arrangement (Fig. 3.12).

To calculate the total resistance of the circuit, the parallel network must be worked out first and this total is added to the series cable resistance.

Example

Three lamps A, B and C, having resistances of 1440, 960 and 576 Ω, are connected to a 230 V supply by a cable of resistance 2 Ω (Fig. 3.13). Calculate (a) the total circuit resistance, (b) the total current, (c) the cable voltage drop, (d) the voltage across the lamps and (e) the current drawn by each lamp.

(a) Total resistance of A, B and C in parallel:

$$\text{A and B product/sum} = \frac{1440 \times 960}{1440 + 960}$$
$$= 576\,\Omega$$

Figure 3.12 Series/parallel combination.

Figure 3.13

This is the same value as lamp C.

$$\therefore \text{ Total with C} = \frac{576}{2} = 288 \ \Omega$$

So the resistance of the lamp circuit is $288\,\Omega$.

$$\text{Total resistance of circuit } R = R_c + R_L$$
$$= 2 + 288$$
$$= 290 \ \Omega$$

(b) Total current:

$$I = \frac{V}{R} = \frac{230}{290} = 0.793 \,\text{A}$$

(c) Cable voltage drop:

$$V_c = I \times R_c = 0.793 \times 2 = 1.586 \,\text{V}$$

(d) Voltage across lamps:

$$V_L = V - V_c$$
$$= 230 - 1.586 = 228.4 \,\text{V}$$

Check: V_L also equals
$I \times R_L = 0.793 \times 288 = 228.4 \,\text{V}$.

(e) Current through A:

$$I_A = \frac{V_L}{R_A} = \frac{228.4}{1440} = 0.1586 \,\text{A}$$

Current through B:

$$I_B = \frac{V_L}{R_B} = \frac{228.4}{960} = 0.2379 \,\text{A}$$

Current through C:

$$I_C = \frac{V_L}{R_C} = \frac{228.4}{576} = 0.3965 \,\text{A}$$

Check: $I_A + I_B + I_C$ should equal total current *I*.
Hence, $0.1586 + 0.2379 + 0.3965 = 0.793 \,\text{A}$,
correct.

The following examples show other resistance combinations not usually encountered in ordinary installation work, but which, nevertheless, need to be understood.

Example

Calculate the total resistance of the circuit shown in Fig. 3.14.

It is important to recognize which resistances are either in series or in parallel with other resistances. That is,

1. Two or more resistances can only be in series if they are connected end to end with no joins from other resistances between them.
2. Two or more resistances can only be in parallel if the ends of each are directly connected to the ends of any other.

So, in Fig. 3.14 there is no single resistance in parallel with any other, and only C and D are directly in series:

$$\therefore \text{ Total of C and D in series} = 4 + 2 = 6\,\Omega$$

Figure 3.14

Figure 3.17

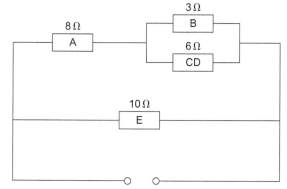

Figure 3.15

So these two could be replaced by a single resistance of 6 Ω (Fig. 3.15). Now we should see that B and CD are in parallel:

∴ Total of B and CD = $\dfrac{\text{product}}{\text{sum}} = \dfrac{3 \times 6}{3+6} = 2\,\Omega$

So B and CD could be replaced by a 2 Ω resistance (Fig. 3.16). Now, A and BCD are in series:

∴ Total of A and BCD = 8 + 2 = 10 Ω

Once again these can be replaced by a single 10 Ω resistance (Fig. 3.17):

∴ Total of E and ABCD in parallel is clearly $\dfrac{10}{2} = 5\,\Omega$

So, total resistance of original circuit is 5 Ω.

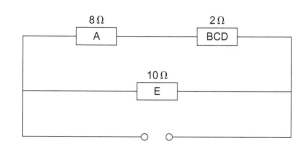

Figure 3.16

Example

Calculate the total resistance in Fig. 3.18, the voltage at point X, and the current through resistance F.

Total of A and B in series = 1.6 + 1 = 2.6 Ω

Total of C and D in parallel $= (7 \times 3)/(7+3) = 2.1\,\Omega$

So, the resultant circuit is shown in Fig. 3.19.

Total of CD and E in series = 2.1 + 1.9 = 4 Ω

So, circuit is now as shown in Fig. 3.20.

Total of CDE and F in parallel $= \dfrac{4 \times 6}{4+6} = 2.4\,\Omega$

So, the resultant circuit is shown in Fig. 3.21.

Total of AB and CDEF in series = 2.6 + 2.4 = 5 Ω

As shown in Fig. 3.22.

Total circuit resistance 5 Ω

Total current $I = \dfrac{V}{R} = \dfrac{10}{5} = 2\,A$

The voltage V_A dropped across $A = I \times R_A$

$$= 2 \times 1.6$$
$$= 3.2\,V$$

Since the supply voltage is 10 V, then the voltage at X is:

$$10 - 3.2 = 6.8\,V$$

To find the current through F, we need to know the p.d. across it (i.e. across YZ). From Fig. 3.21, the total resistance between Y and Z is CDEF which is 2.4 Ω, and the current flowing through it is a total current of 2 A.

∴ The p.d. across YZ = $I \times$ CDEF

$$= 2 \times 2.4$$
$$= 4.8\,V$$

∴ $I_F = \dfrac{V_F}{R_F} = \dfrac{4.8}{6} = 0.8\,A$

Figure 3.18

Figure 3.19

Figure 3.20

Figure 3.21

Figure 3.22

Power: symbol, *P*; unit, watt (W)

Someone with plenty of energy has the potential to convert that energy into work. The rate at which that energy is converted is called **power**. Hence power is the rate of energy conversion, that is:

$$P = \frac{\text{energy}}{\text{time}}$$

Two engineering pioneers, Watt and Joule, gave their names to the units of power and energy.

James Watt (1736–1819)

British engineer who invented the improved steam engine and introduced horsepower (hp) as a means of measuring power.

James Prescott Joule (1818–1889)

British scientist and engineer best known for his mechanical equivalent of heat and his work on the heating effect of an electric current.

Energy is measured in joules and of course the SI unit of time is the second:

$$\therefore \text{ Power in watts} = \frac{\text{joules}}{\text{seconds}}$$

Transposing, we get

$$\text{joules} = \text{watts} \times \text{seconds}$$
$$\therefore 1\,J = 1\,Ws$$

Joule showed by experiment that heat energy was produced when a current flowed through a resistance for a certain time, and from these experiments it was shown that

$$\text{joules} = I^2 \times R \times \text{seconds}$$

But we have already seen that

$$\text{joules} = \text{watts} \times \text{seconds}$$
$$\therefore \text{Watts} = I^2 \times R$$

Hence electrical power

$$P = I^2 \times R$$

By using Ohm's law, we can develop two other equations for *P*. Hence,

$$I = \frac{V}{R}$$

Therefore, by replacing *I* with *V/R*, we get

$$P = \frac{V^2}{R^2} \times R$$
$$\therefore P = \frac{V^2}{R}$$

Similarly,

$$R = \frac{V}{I}$$

So,

$$P = I^2 \times \frac{V}{I}$$
$$\therefore P = I \times V$$

Hence power is

$$P = I^2 R \quad \text{or} \quad P = \frac{V^2}{R} \quad \text{or} \quad P = IV$$

Nearly all of us are familiar with power. It is all around us in our homes, for example a 60 W lamp or a 3 kW fire; in fact all appliances should have ratings of power and voltage marked on them.

The rated values are extremely important. A 60 W 230 V lamp will dissipate 60 W only if connected to a 230 V supply. The resistance of the lamp when working is of course unaffected by the voltage and we can calculate this resistance from the lamp's rated values. For a 60 W 230 V lamp:

$P = 60\,W$; $V = 230\,V$; R_L = lamp resistance

$$P = \frac{V^2}{R_L}$$
$$\therefore R_L = \frac{V^2}{P}$$
$$= \frac{230^2}{60}$$
$$= \frac{52\,900}{60}$$
$$R_L = 881.7\,\Omega$$

If we therefore supply the lamp from a 200 V source, the power dissipated by the lamp would be

$$P = \frac{V^2}{R}$$
$$= \frac{200^2}{881.7}$$
$$= \frac{40\,000}{881.7}$$
$$P = 45.37\,W$$

Example

Two 100 W 230 V lamps are wired in series across a 230 V supply. Calculate the power dissipated by each lamp and the total power dissipated.

Both lamps have the same resistance:

$$P = \frac{V^2}{R_L}$$

Figure 3.23

Figure 3.24

$$R_L = \frac{V^2}{P}$$
$$= \frac{52\,900}{100}$$
$$R_L = 529\,\Omega$$

It is clear from Fig. 3.23 that as both resistances are the same, then the voltage across each is the same:

$$V_A = V_B = \frac{230}{2}$$
$$= 115\text{ V}$$

$$P_A = \frac{V_A^2}{R_A}$$
$$= \frac{13\,225}{529}$$
$$P_A = 25\text{ W}$$

$$\therefore\ P_B = 25\text{ W}$$

Total resistance $R_{total} = R_A + R_B$
$$= 529 + 529$$
$$= 1058\,\Omega$$

$$\therefore \text{Total power } P_{total} = \frac{V^2}{R}$$
$$= \frac{52\,900}{1058}$$
$$P = 50\text{ W}$$

which is $P_A + P_B$.

Thus the individual powers dissipated in a series circuit may be added to find the total power dissipated.

Also notice that when two identical lamps are connected in series the power dissipated by each is a quarter of its original value (i.e. 100 W down to 25 W).

Two lamps A and B are connected in series across a 230 V supply. Lamp A is rated at 40 W 230 V and lamp B at 60 W 230 V. Calculate the power dissipated by each lamp and the total power dissipated (Fig. 3.24).

Lamp A	Lamp B
$R_A = \dfrac{V^2}{P_A}$	$R_B = \dfrac{V^2}{P_B}$
$= \dfrac{52\,900}{40}$	$= \dfrac{52\,900}{60}$
$R_A = 1322.5\,\Omega$	$R_B = 881.7\,\Omega$

In this case, we need not calculate V_A and V_B. As I is common to A and B, we could calculate power from I^2R. To find I:

$$R = R_A + R_B$$
$$= 1322.5 + 881.7$$
$$R = 2204.2\,\Omega$$

and

$$I = \frac{V}{R}$$
$$= \frac{230}{2204.2}$$
$$I = 0.1\text{A}$$
$$P_A = I^2 R_A$$
$$= 0.01 \times 1322.5$$
$$P_A = 13.225\text{ W}$$

$$P_B = I^2 R_B$$
$$= 0.01 \times 881.7$$
$$P_B = 8.817\text{ W}$$
$$P_{total} = P_A + P_B$$
$$= 13.225 + 8.817$$
$$P_{total} = 22\text{ W}$$

Check: P_{total} is also equal to I^2R:
$$P_{total} = 0.01 \times 2204.2$$
$$= 22\text{ W}$$

It can be seen from this example that a considerable amount of power can be lost by series connections. It is also obvious that incorrect selection of cables can cause a loss of power to the equipment they are supplying.

However, these power losses can be put to some useful purpose, for example in an electric grill that has three heat settings (Fig. 3.25).

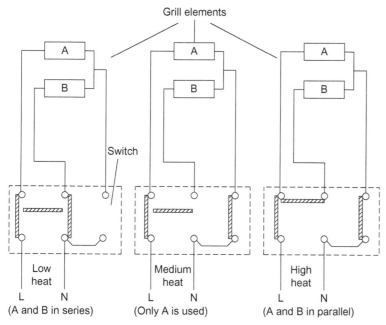

Figure 3.25 Three heat switch.

If each element were rated at 1 kW 230 V, then on high heat A and B would both have their rated voltage and would both dissipate their rated power, that is 1 kW each, giving a 2 kW total. On medium heat only one element is used, giving 1 kW total. On low heat, as A and B are in series each will dissipate 250 W (a quarter of 1 kW), giving a total of 500 W.

Electrical energy: symbol, *W*; unit, kilowatt-hour, (kWh)

Energy is the ability to do work and is measured in joules (J). As we have already seen,

$$1 \text{ joule} = 1 \text{ watt second}$$
$$= \text{power} \times \text{time}$$

Generally, in considering domestic loads we measure the power in kilowatts and time in hours:

$$\therefore \text{ Energy} = \text{kW} \times \text{hours}$$
$$= \text{kWh (kilowatt hours)}$$

To convert watt seconds to kilowatt hour:

$$\text{kWh} = \frac{\text{watt seconds}}{1000 \times 3600}$$

Also

$$\text{kWh} = \frac{\text{joules}}{3\,600\,000}$$

The kilowatt hour is often referred to as a **unit** of electricity, that is

6 kWh = 6 units

which could be

6 kW for 1 h
or   3 kW for 2 h
or   2 kW for 3 h
or   1 kW for 6 h, etc.

As previously mentioned, domestic and industrial appliances are connected in parallel to the supply and therefore the energy each consumes may simply be added together to determine the total energy used.

Example

A domestic consumer has the following daily loads connected to the supply:

Five 60 W lights for 4 h
Two 3 kW electric fires for 2 h
One 3 kW water heater for 3 h
One 2 kW kettle for $\frac{1}{2}$ h

Calculate the energy consumed in 1 week.

Light	$E = 5 \times 60 \times 4 = 1200 =$	1.2 kWh
Fires	$E = 2 \times 3 \times 2 =$	12.0 kWh
Water heater	$E = 1 \times 3 \times 3 =$	9.0 kWh
Kettles	$E = 1 \times 2 \times 0.5 =$	1.0 kWh
	Total for 1 day =	23.2 kWh

$$\therefore \text{ Energy expended in 1 week} = 7 \times 23.2$$
$$= 162.4 \text{ kWh}$$

Tariffs

Tariffs are charges made by the Distribution Network Operator (DNO) for the use of electricity. There are several types of tariff available depending on the kind of installation (i.e. domestic, commercial, etc.). However, the basis of tariff calculation is a charge per unit (kWh) consumed.

Example

If the cost per unit of electrical energy in the previous example is 10 p per unit, calculate the cost to the consumer over a 13-week period (one quarter).

Total energy consumed per week = 162.4 units (kWh)

$$\therefore \text{Cost at 10 p per unit} = 10 \times 162.4$$
$$= 1624 \text{ p}$$
$$= 16.24$$
$$\therefore \text{Cost for 13 weeks} = 13 \times £16.24$$
$$= £211.12$$

Measuring power and energy

A **wattmeter** is connected as shown in Fig. 3.26. It is basically a combination of an ammeter and a voltmeter, and it measures the product of current and voltage:

$$P \text{ (watts)} = I \times V$$

An **energy meter** is similar to a wattmeter and its connections are the same. However, it is designed to show the number of kilowatt hours of energy used. It is familiar to most of us as our electricity meter (Fig. 3.27).

Water heating

We have just discussed electricity energy and showed that

$$\text{kWh} = \frac{\text{joules}}{3\,600\,000}$$

Joule in his experiments showed that 4.2 J of electrical energy = 1 calorie of heat energy. Hence it required 4.2 J of electrical energy to raise the temperature of 1 g of water through 1°C or 4.2 J/g/°C or 4200 J/kg/°C.

This value is called the **specific heat** of water (SH), that is if 2 kg of water was raised through 2°C then the amount of electrical energy required to do this would be

$$2 \times 2 \times 4200 = 16\,800 \text{ J}$$

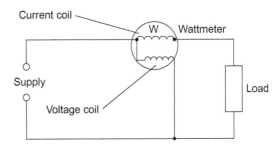

Figure 3.26 Wattmeter connections.

Hence,

Joules = mass (kg) × change in temperature (°C)
× SH of water (4200)

But

$$\text{kWh} = \frac{\text{joules}}{3\,600\,000}$$

$$\therefore \text{Heat output in kWh} =$$
$$\frac{\text{mass} \times \text{change in temperature} \times \text{SH}}{3\,600\,000}$$

If a system is 100% efficient then the output of the system should be equal to its input (i.e. there are no losses).

Example

How long will it take for a 2 kW 230 V kettle to raise the temperature of 2 l of water from 8°C to boiling point? Assume 100% efficiency. (SH of water = 4200 J/kg/°C and 1 l of water has a mass of 1 kg.)

$$\text{kWh}_{\text{output}} = \frac{\text{mass} \times \text{change in temperature} \times \text{SH}}{3\,600\,000}$$
$$= \frac{2 \times 92 \times 4200}{3\,600\,000}$$
$$= 0.215 \text{ kWh}$$

As the system is 100% efficient,

$$\text{kWh}_{\text{input}} = \text{kWh}_{\text{output}}$$

and, as 2 kW is the input power dissipated by the element,

$$2 \times \text{hours} = 0.215$$
$$\therefore \text{Hours} = \frac{0.215}{2}$$
$$= 0.1075 \text{ h}$$
$$= 6.45 \text{ min}$$

Digital readout or dials

0 2 9 5 . 4 1
kwh

Double-insulated 'tails'

To fuse board

Brown

Brown

Grey sheath

Blue

Blue

From mains

Double-insulated 'tails'

Figure 3.27 Energy meter.

Efficiency

The efficiency of a system is the ratio of the output to the input:

$$\therefore \text{ Percentage efficiency} = \frac{\text{output} \times 100}{\text{input}}$$

Example

Calculate the efficiency of a water heater if the output in kilowatt hours is 24 kWh and the input energy is 30 kWh.

$$\text{Efficiency (\%)} = \frac{\text{output}}{\text{input}} \times 100$$
$$= \frac{24}{30} \times 100$$
$$= 80\%$$

Example

Calculate the time taken for a 3 kW immersion heater to be energized to heat 137 l of water from 10°C to 70°C. The efficiency of the system is 80%. (SH of water = 4200 J/kg/°C and 1 l has a mass of 1 kg.)

$$\text{kWh}_{\text{output}} = \frac{\text{mass} \times \text{change in temperature} \times \text{SH}}{3\,600\,000}$$
$$= \frac{137 \times 60 \times 4200}{3\,600\,000}$$
$$= 9.59 \text{ kWh}$$

But

$$\text{Efficiency} = \frac{\text{output}}{\text{input}} \times 100$$
$$\therefore 80 = \frac{9.59 \times 100}{80}$$
$$\therefore \text{kWh}_{\text{input}} = \frac{9.59 \times 100}{\text{kWh}_{\text{input}}}$$
$$= 12 \text{ kWh}$$

As the input power is 3 kW,

$$\text{kWh}_{\text{input}} = \text{kW}_{\text{input}} \times \text{hours}$$
$$\therefore 12 = 3 \times \text{hours}$$
$$\therefore \text{Hours} = \frac{12}{3}$$
$$\therefore \text{Time} = 4 \text{ hours}$$

If the tank were lagged to prevent as many losses as possible, the efficiency would improve, reducing the time taken and hence ensuring greater economy.

Self-Assessment Questions

1. A 230V electric iron has a resistance of 96Ω and is connected to a socket outlet by a twin cable, each conductor of which has a resistance of 0.1Ω. If the total resistance of the cable from the fuse board to the socket is 0.8Ω, calculate the total resistance of the whole circuit.

2. If the total resistance of the three lamps in the diagram is 4128Ω, calculate the resistance of lamp A.

3. An electric kettle, an iron and a food mixer have element resistances of 28.8, 96 and 576Ω, respectively. If they are all connected in parallel, calculate the total resistance.

4. A 2.5mm² 20m long twin copper cable supplies a heating appliance having three elements each of resistance 57.6Ω. If the elements are arranged such that two are in series and the third is in parallel with these two, calculate: (a) the resistance of the cable and (b) the total resistance of the whole circuit. (ρ for copper is 17μΩ mm.)

5. (a) What are the maximum permissible voltage drops in an installation, as recommended by the IET Regulations?
   (b) A length of cable supplying a cooker at 230 V has a resistance of 0.24Ω. If the cooker has a total resistance of 5.76Ω, calculate the voltage drop along the cable. Is this value permissible?

6. If an electric fire of resistance 28.8Ω, an immersion heater of resistance 19.2Ω, a small electric kettle of resistance 57.6Ω and a toaster of resistance 115.2Ω are connected to a 230 V domestic power circuit, calculate the current taken by each appliance and the total current drawn from the supply.

7. What would be the resistance of, and the current drawn by, the following when connected to a 230V supply: (a) a 3kW 230V immersion heater; (b) a 600W 230V food mixer; (c) a 1kW 230V electric fire and (d) a 40W 230V filament lamp?

8. Three lamps have the following rated values: 60W 230V; 100W 200V and 40W 100V. Calculate the power dissipated by each if they are connected in series across a 230V supply.

9. From the diagram, calculate the total resistance, the current in each branch, the voltage at X and Y, and the p.d. across XY.

10. A domestic consumer has the following loads connected each day (supply voltage 230V).
    Five 100W lamps for 4h
    One 3kW immersion heater for 2h
    One 10kW cooker for $1\frac{1}{2}$h
    Two 3kW electric fires for 3h
    Sundry appliances taking 12A for $\frac{3}{4}$h
    If electricity costs 7p per unit, calculate the cost for one quarter (13 weeks).

11. It is required to raise the temperature of 1.5 l of water in a kettle from 5°C to 100°C in 3.5min. Assuming 100% efficiency, calculate the nearest size of element required to do this. (SH of water = 4200 J/kg/°C.)

12. An immersion tank contains 110 l of water at 15°C. An immersion element supplied from a 230 V source takes 12.5 A when energized and heats the water to 78°C. If the system is 90% efficient, calculate for how long the element is energized. (SH of water = 4200 J/kg/°C.)

13. A small 25 l boiler is completely filled with water at 12°C. The heating element has a resistance of 28.8Ω and is connected to a 230 V supply. If the boiler raises the temperature of the water to boiling point in 97min, calculate the efficiency of the system.

CHAPTER 4

Electromagnetism

Before we deal with this major subject, it would be sensible to discuss basic magnetism briefly.

Magnetism

We are all familiar with simple magnets and have probably seen the lines of force traced on paper with iron filings. The quantity of lines of force that come out from a magnet is called the **flux** and is measured in **webers** (Wb).

Wilhelm Edward Weber (1804–1891)

German scientist famous for his work in the measurement of electrical quantities.

Flux density: symbol *B*; unit tesla (T)

Just as population densities are measured in people per km², flux density is measured in flux per m² or Wb/m². This unit, however, is known as the **tesla** (T).

$$\therefore B \ (\text{T}) = \frac{\Phi \ (\text{Wb})}{a \ (\text{m}^2)}$$

Nickola Tesla (1856–1943)

Yugoslavian electrical engineer renowned for his work on a.c. generation and distribution.

Example

A motor field pole has an area of 60 cm². If the pole carries a flux of 0.3 Wb, calculate the flux density.

$B = ?; \ \Phi = 0.3 \, \text{Wb}; \ a = 60 \, \text{cm}^2 = 0.006 \, \text{m}^2$

$$B = \frac{\Phi}{a}$$
$$B = \frac{0.3}{0.006}$$
$$= 50 \, \text{T}$$

Electromagnetism

Field around a conductor carrying a current

When a conductor carries a current, a magnetic field is produced around that conductor (Fig. 4.1). This field is in the form of concentric circles along the whole length of the conductor. The direction of the field depends on the direction of the current – clockwise for a current flowing away from the observer and anti-clockwise for a current flowing towards the observer. In order to show these directions, certain signs are used (Fig. 4.2).

The screw rule

In order to quickly determine the direction of the magnetic field around a current-carrying conductor, the screw rule may be applied (Fig. 4.3).

Figure 4.1 Magnetic field around a conductor.

Electrical Installation Work. 978-1-138-84927-3 © Brian Scaddan.
Published by Taylor & Francis.

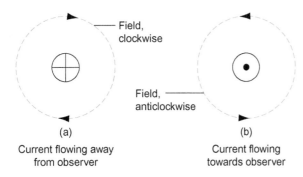

Figure 4.2 Symbols for current direction.

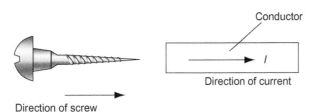

Figure 4.3 The Screw Rule.

Imagine a screw being twisted into or out of the end of a conductor in the same direction as the current. The direction of rotation of the screw will indicate the direction of the magnetic field.

Force between current-carrying conductors

If we place two current-carrying conductors side by side, there will exist a force between them due to the flux. The direction of this force will depend on the directions of the current flow (Fig. 4.4).

In Fig. 4.4a, there is more flux between the conductors than on either side of them, and they will be forced apart.

In Fig. 4.4b, the flux between the conductors is in opposite directions and tends to cancel out leaving more flux on the outside of the conductors than in between them, so they will be forced together.

The direction of movement can be found using Fleming's left-hand rule.

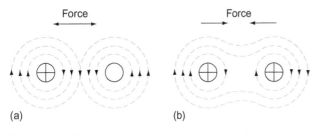

Figure 4.4 Force on a current carrying-conductor.

Fleming's left-hand rule

If the thumb, first and second fingers of the left hand are placed at right angles to one another (Fig. 4.5), they indicate:

First finger Field

seCond finger Current

thuMb Motion

Figure 4.5 Flemings left-hand rule.

Force on a conductor carrying a current in a magnetic field

If a current-carrying conductor is placed at right angles to a magnetic field, a force will be exerted on that conductor (Fig. 4.6). This force is measured in newtons.

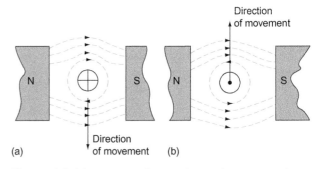

Figure 4.6 Movement of a conductor in a magnetic field.

Sir Isaac Newton (1642–1727)

English scientist of considerable fame, known especially for his work on force, mass, motion and momentum.

In Fig. 4.6a, the flux above the conductor is greater than the flux below, and the conductor is forced downwards. In Fig. 4.6b, the current and hence the field around the conductor is opposite to that in Fig. 4.6a and the conductor is forced upwards.

The magnitude of this force is dependent on three things:

1. the current flowing in the conductor (I).
2. the density of the magnetic field (B).
3. the length of the conductor in the magnetic field (l).

$$\therefore F \text{ (newtons)} = B \text{ (teslas)} \times l \text{ (metres)} \times I \text{ (amperes)}$$

Example

Calculate the force exerted on a conductor 40 cm long carrying a current of 100 A at right angles to a magnetic field of flux density 0.25 T.

$$F = ?; B = 0.25\,T; l = 40\,cm = 0.4\,m; I = 100\,A$$
$$F = B \times l \times I$$
$$= 0.25 \times 0.4 \times 100$$
$$F = 10\,N$$

Example

A circular magnetic field has a diameter of 20 cm and a flux of 149.6 mWb. Calculate the force exerted in a conductor 21 cm long lying at right angles to this field if the current flowing is 15 A.

$$B = \frac{\Phi}{a}$$

But

$$a = \frac{\pi d^2}{4}$$

$\pi = 3.1416$; $d = 20\,cm = 0.2\,m$; $B = ?$;
$\Phi = 149.6\,mWb = 149.6 \times 10^{-3}\,Wb$

$$\therefore a = \frac{3.1416 \times 0.2 \times 0.2}{4}$$
$$= 0.031\,416\,m^2$$

$$B = \frac{\Phi}{a}$$
$$= \frac{149.6 \times 10^{-3}}{0.031\,416} = 4.762\,T$$

$$F = ?; B = 4.762\,T; l = 21\,cm = 0.21\,m; I = 15\,A$$
$$F = B \times l \times I$$
$$= 4.762 \times 0.21 \times 15$$
$$F = 15\,N$$

The e.m.f. induced in a moving conductor

We have seen that passing a current through a conductor in a magnetic field produced a movement

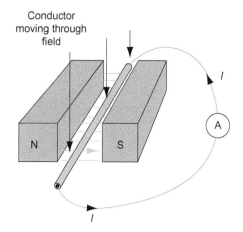

Figure 4.7 E.m.f. induced in a moving conductor.

of that conductor. If we were to reverse the process and physically move the conductor through a magnetic field, such that it cuts across the flux, then a current would flow in that conductor (Fig. 4.7).

As we have already seen in Chapter 2, a pressure is required for a current to flow. Therefore, if a current flows (Fig. 4.7) then an e.m.f. must be producing it. This e.m.f. is called an **induced e.m.f.** and its direction is the same as that of the current flow. This direction can be determined by using Fleming's right-hand rule.

Fleming's right-hand rule

If the thumb, first and second fingers of the right hand are arranged at right angles to one another (Fig. 4.8) they indicate:

First finger, Field (north to south)

seCond finger, Current (and e.m.f.)

thuMb, Motion

The magnitude of the induced e.m.f. depends upon:

1. the flux density of the field (B).
2. the length of the conductor (l).
3. the velocity at which the conductor cuts across the flux (v).

$$\therefore E \text{ (volts)} = B \text{ (teslas)} \times l \text{ (metres)} \times v \text{ (metres/second)}$$

Figure 4.8 Flemings right-hand rule.

Example

A conductor 15 cm long is moved at 20 m/s perpendicularly through a magnetic field of flux density 2 T. Calculate the induced e.m.f.

$E = ?$; $B = 2$ T; $l = 15$ cm $= 0.15$ m; $v = 20$ m/s

$E = B \times l \times v$

$\quad = 2 \times 0.15 \times 20$

$E = 6$ V

Application of magnetic effects

There are major areas in which the magnetic effects of electric current are used. These are measuring instruments and motors (discussed in later chapters), solenoids, electromagnets, inductors and transformers, etc.

Solenoid

If we wind a conductor on to a hollow cardboard cylinder or former and pass a current through it, the whole assembly will act like a magnet, having a north and south pole, and an iron rod will be drawn inside the solenoid when it is energized. This effect may be used in various ways.

One example is a chime bell (Fig. 4.9). When the bell push is depressed, the solenoid is energized, the soft iron rod, with plastic end inserts, is attracted by the magnetic field in the direction shown, and the chime (ding) will sound. When the bell push is released, the spring will return the rod with enough force to sound the chime (dong) again.

Electromagnet

If the coil of the solenoid is wound on an iron core, when energized, the coil will cause the core to act as

Figure 4.9 'Ding-dong' chimes for d.c. or a.c. supplies.

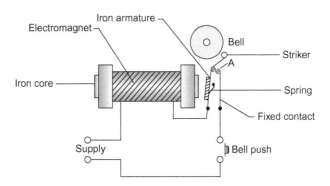

Figure 4.10 Trembler bell.

a magnet. This effect is used in many different ways, for example in bells, relays, contractors, telephones or circuit-tripping mechanisms.

Figure 4.10 shows how a simple trembler bell works. When the bell push is operated, the electromagnet is energized and the iron armature is attracted to it, the striker hitting the bell. This action, however, breaks the circuit at A and the electromagnet de-energizes. The spring returns the striker and the armature to their original position, completing the circuit, and the electromagnet is energized again.

Another device making use of the same principle is the electromagnetic trip. Its latching mechanism is similar to that of the thermal trip, but the bimetal is replaced by an iron armature (Fig. 4.11). The current supplying the load flows through the coil and the closed contacts. The coil is designed to allow a certain value of current to flow, and beyond this amount the core of the electromagnet is magnetized sufficiently to attract the armature, causing the circuit to be interrupted.

Direct-current generator

We have just seen how an e.m.f. is induced in a conductor when it is moved through a magnetic field.

This effect is the principle which enables a simple generator to work (Fig. 4.12). A single-loop conductor arranged as shown in this figure has its ends connected to a simple commutator, which comprises two copper segments insulated from each other. The commutator and the loop are fixed to a central shaft that enables the whole assembly to rotate freely. Two fixed carbon brushes bear on the surface of the commutator, enabling an external load to be connected to the generator.

When the loop is rotated in the direction shown, side A will travel downwards and side B upwards. From Fleming's right-hand rule (p. 64), the current and hence the induced e.m.f. will be in the direction

Figure 4.11 Tripping mechanism.

(a)   (b)

Coil

Armature

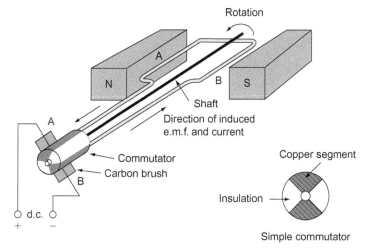

Figure 4.12 Simple single-loop d.c. generator.

shown, making brush A +ve and B −ve. After 180°
revolution, side A will be travelling upwards and side
B downwards, but the induced e.m.f. stays in the
same direction (right-hand rule) and so the polarity at
the brushes remains unchanged. We have therefore
generated a d.c. source of supply.

In large generators the magnetic field is provided
by electromagnets rather than permanent magnets.
The single loop is replaced by many such loops held
in slots in an iron core. This arrangement is called an
armature. The commutator fixed to the armature
shaft has of course many insulated segments to which
the ends of all the loops are connected.

The armature is laminated, that is made up of many
thin sheets insulated from one another. This is done to
reduce eddy currents.

Eddy currents

As we have seen, cutting magnetic flux with a
conductor induces an e.m.f. in it, and if the conductor
is a part of a complete circuit, a current will flow.

The iron of an armature as well as its conductors cuts
the flux and small currents are induced in the armature
core, which, when circulating together, can cause it to
heat up. This effect is overcome to a large extent by

laminating the core. This confines the currents to each
lamination and prevents a large circulating current
from building up.

Alternating-current generator

The principle of a.c. generation is the same as that of
d.c. The ends of the loop in this case, however, are
terminated in slip rings, not a commutator (Fig. 4.13).

As the loop rotates, side A will have an e.m.f. induced
first in one direction and then in the other. Therefore,
as side A is permanently connected to a slip ring A,

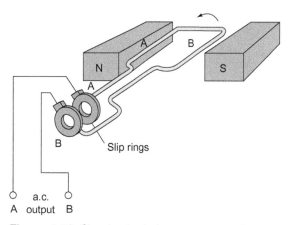

Figure 4.13 Simple single-loop a.c. generator.

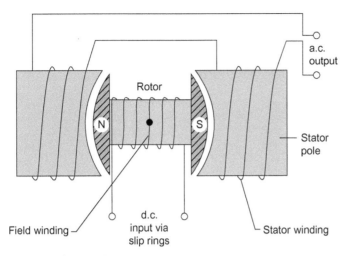

Figure 4.14 Single-phase a.c. generator.

this ring will be alternatively +ve and −ve. The same process applies to ring B. The generated supply output is therefore alternating.

In practice, large generators are arranged such that the magnetic field rotates, its flux cutting across the armature conductors (Fig. 4.14). In a generator of this type, the armature is made stationary and called the **stator**, while the rotating magnetic field, in the form of an electromagnet, is called the **rotor**. This arrangement is preferred to that in Fig. 4.14 because of the excessive cost of providing slip rings and brushes capable of handling the large output currents.

The single-phase a.c. waveform

Figure 4.15 shows a cross-section through a single-loop generator. It will be seen that in the vertical position AB, the loop sides are cutting no flux and hence no e.m.f. is induced. However, as the loop rotates, more and more flux is cut and hence more and more e.m.f. is being induced, up to a maximum in the horizontal position. Further rotation causes the e.m.f. to fall to zero again. This rise and fall of the e.m.f. can be traced graphically.

Note

Since the current will flow in the same direction as the induced e.m.f., it also will rise and fall with time and the e.m.f. The current and the e.m.f. are said to be in **phase** with one another.

Figure 4.16a shows a single conductor rotating in a magnetic field; after each 30° of revolution the conductor is at positions 1, 2, 3, 4, etc. The horizontal axis of Fig. 4.16a is the circular path taken by the conductor simply opened out to form a straight line, each 30° linear space corresponding to each 30° angle of movement. The vertical axis represents the magnitude of the induced e.m.f. As the induced e.m.f. depends on the amount of flux being cut, which itself depends on the position of the conductor, the magnitude of the e.m.f. can be represented by the conductor position.

Hence each 30° position of conductor rotation (Fig. 4.16a) can be represented by an e.m.f. at each 30° space of linear movement (Fig. 4.16b).

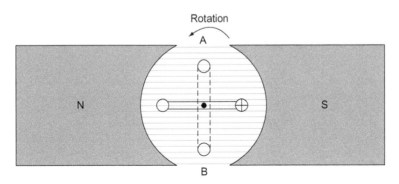

Figure 4.15 Cross-section through a single loop generator.

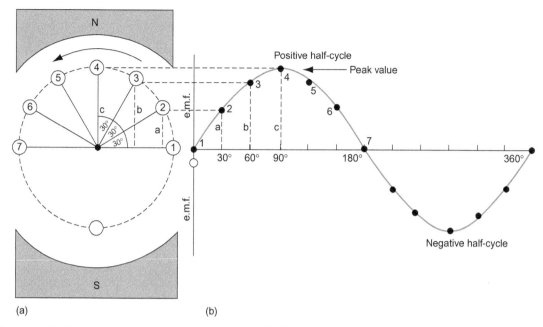

Figure 4.16 Single conductor rotating in a magnetic field: one complete cycle.

The resulting graph indicates the e.m.f. induced in one complete revolution of the conductor. This waveform is called a **sine wave**, and any quantity that has a wave of that nature is called a **sinusoidal quantity**.

Figure 4.16b shows the variation of the e.m.f. during one revolution of the conductor and is termed **one cycle**.

Frequency: symbol, f; unit, hertz (Hz) (=cycles per second)

The number of complete cycles which occur in one second is called the **frequency** (Fig. 4.17). In the British Isles, the frequency of the supply is 50 cycles per second or 50 Hz.

Heinrich Rudolf Hertz (1857–1894)

German physicist who demonstrated the transmission of electromagnetic waves.

Drawing the waveform of an alternating quantity

The following example shows how the waveform of an alternating quantity can be drawn.

Example

Draw the waveform of a sinusoidal e.m.f. having a maximum or peak value of 90 V, and from it determine the value of the e.m.f. after one-third of a cycle.

First, a suitable scale must be chosen and then a circle with a radius representing 90 V is drawn (Fig. 4.18). (This radius representing 90 V is called a **phasor**, phasors simply being straight lines drawn to scale to represent electrical quantities.) We will choose a scale in which 1 cm = 30 V so that 90 V will be represented by 3 cm.

Now the circle is divided up into 30° segments. If the scale permits, it can be divided into 15° segments: the more the points on the waveform the easier to draw and the more accurate the result.

After one-third of a cycle the conductor has moved 1/3 of 360° (i.e. 120°) and the vertical measurement l, to scale, is the value of induced e.m.f. at that point. By measurement,

$$l = 2.6 \, \text{cm}$$
$$= 2.6 \times 30$$
$$= 78 \, \text{V}$$

Figure 4.17 Frequency.

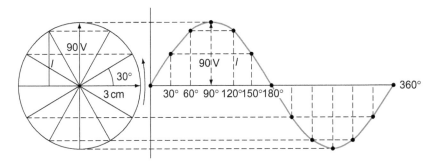

Figure 4.18 Sinusoidal waveform.

Addition of waveforms

Sometimes two or more voltages or currents are acting simultaneously in a.c. circuits, and they may not act together. Under these circumstances they are referred to as being **out of phase**.

Whether the voltages or currents are in or out of phase, their combined effect or **resultant** can be shown.

Example

Two sinusoidal voltages A and B of peak values 90 and 60 V, respectively, act together in a circuit. If voltage B lags behind voltage A by 70°, draw the two waveforms on the same axis and show the resultant voltage. What is its peak value?

The construction is carried out as follows (Fig. 4.19):

1. Choose a suitable scale.
2. Draw two circles, one inside the other, of radius corresponding to 90 and 60 V.
3. Starting at A, divide the larger circle into 30° segments.
4. Starting at B, divide the smaller circle into 30° segments.
5. Draw each waveform.

The resultant waveform is drawn by adding the value of waves A and B at each 30° interval (e.g. at 60°, B has a value of −4 mm and A has +25 mm).

$$\therefore A + B \text{ at } 60° = -4 + 25 = 21 \text{ mm}$$

This is the value of the resultant at 60°. By measurement, the peak value of the resultant is 4 cm.

$$\therefore \text{ Peak voltage of resultant} = 4 \times 30 = 120 \text{ V}$$

Root-mean-square (r.m.s.) value

As discussed in Chapter 3, a current passing through a resistance has a heating effect, the magnitude of which is measured in watts, and we have also seen that

$$P = I^2 \times R$$

and

$$P = \frac{V^2}{R}$$

so that the power (or heat) dissipated in a resistance is proportional to either the square of the current or the voltage. Take, say, an a.c. current of peak value I amperes, draw its waveform over half a cycle and

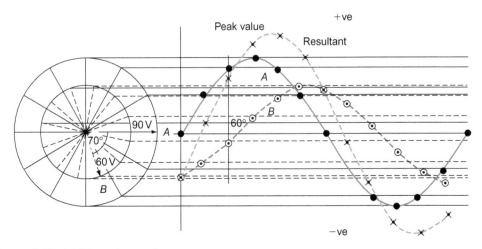

Figure 4.19 Addition of waveforms.

then square the value of the current at each 30° or 15° spacing (instantaneous values), and then draw a waveform using these squared values. The resulting wave will represent the power dissipated. The average or mean heating effect will therefore be the sum of all the instantaneous values divided by the number of instantaneous values,

$$\therefore P = I^2 = \frac{i_1^2 + i_2^2 + i_3^2 + i_4^2 + \cdots}{\text{number of values}}$$

$$\therefore I = \sqrt{\frac{i_1^2 + i_2^2 + i_3^2 + i_4^2 + \cdots}{\text{number of values}}}$$

(in other words, the **root** of the **mean** of the **squares**). This can be shown mathematically to be 0.7071 of the peak value,

$$\therefore I_{\text{r.m.s.}} = I_{\text{max}} \times 0.7071$$

where $I_{\text{r.m.s.}}$ is the root-mean-square current and I_{max} is the peak value of the current.

If we connect a resistance to a d.c. supply and draw a d.c. current equal to the value of the a.c. r.m.s. value, the heating effect will be the same in both cases. We can therefore define the r.m.s. value of alternating current or voltage as 'that value of alternating current or voltage which will give the same heating effect as the same value of direct current or voltage', that is

10 A (r.m.s.) = 10 A (d.c.)

Note

Unless otherwise stated, all values of voltage and current quoted on a.c. equipment are given as r.m.s. values.

It is interesting to note that the peak value of our 230 V domestic supply is

$$V_{\text{r.m.s.}} = V_{\text{max}} \times 0.7071$$

$$\therefore V_{\text{max}} = \frac{V_{\text{r.m.s.}}}{0.7071}$$

$$= \frac{230}{0.7071}$$

$$= 325.2\,\text{V}$$

It is also interesting that the r.m.s. value of an alternating quantity is achieved when the conductor has rotated through an angle of 45° (Fig. 4.20). It will be seen that

$$v = V \times \sin 45°$$

$$= V \times 0.7071$$

Figure 4.20 R.m.s. value of a sine wave.

Average value

The average value of all the instantaneous values that make up a sine wave is given by

$$I_{\text{average}} = I_{\text{max}} \times 0.637$$

Three-phase a.c. generator

The principle of generating a three-phase supply is the same as that for a single phase. In this case, however, the stator poles are arranged 120° apart (Fig. 4.21). With the rotor in the position shown in the figure, the induced e.m.f. is at a maximum in the brown phase, is increasing in the black phase and is decreasing in the grey phase. There are therefore three waveforms, each 120° apart (Fig. 4.22). If we were to find the resultant of, say, the brown and black phases first and then add this to the grey phase, we would find that the total resultant was zero.

Hence, in a balanced (all phases equal) three-phase system the resultant voltage and current is zero. As discussed later, this fact is important in distribution systems and in the design of three-phase motors.

Inductance: symbol, *L*; unit, henry (H)

Let us consider the effect of forming a coil from a length of wire and connecting it to a d.c. source of supply. Figure 4.23 shows the distribution of the magnetic lines of force, or flux, produced by such a circuit. We know that if we wind the same coil on to an iron core, the lines of force tend to be confined to that core and the flux is much greater (Fig. 4.24), and that when a conductor is cut by magnetic lines of force, a current, and hence an e.m.f., is produced in that conductor. Consider what happens when the switch S is first closed (Fig. 4.24).

As the current increases from zero to a maximum, the flux in the core also increases, and this growing magnetic field cuts the conductors of the coil, inducing an e.m.f. in them. This e.m.f., called the **back e.m.f.**, operates in the reverse direction to the supply voltage

Figure 4.21 Simple three-phase star-connected a.c. generator.

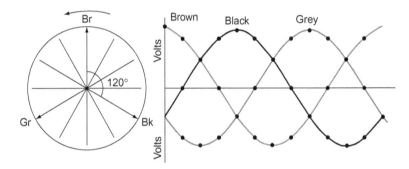

Figure 4.22 Balanced three phase waveform.

Figure 4.23 Fiux distribution for a coil.

Figure 4.24 Self-inductance.

and opposes the change in the circuit current that is producing it. The effect of this opposition is to slow down the rate of change of current in the circuit.

When the switch S is opened, the current falls to zero and the magnetic field collapses. Again, lines of force cut the conductors of the coil inducing an e.m.f. in them. In this case, the e.m.f. appears across the switch contacts in the form of an arc.

Induced e.m.f. due to change in flux

The average value of the induced e.m.f. in a circuit such as the one shown in Fig. 4.24 is dependent on the rate of change of flux and the number of turns of the coil. Hence the average induced e.m.f.

$$E = -\frac{\Phi_2 - \Phi_1}{t} \times N \text{ volts}$$

The minus sign indicates that the e.m.f. is a back e.m.f. and is opposing the rate of change of current.

Example

The magnetic flux linking the 1800 turns of an electromagnet changes from 0.6 to 0.5 mWb in 50 ms. Calculate the average value of the induced e.m.f. (E = induced e.m.f.).

$\Phi_2 = 0.6\,\text{mWb};\ \Phi_1 = 0.5\,\text{mWb};\ t = 50\,\text{ms};\ N = 1800$

$$E = -\frac{(\Phi_2 - \Phi_1)}{t} \times N$$

$$= -\frac{(0.6 - 0.5)}{50 \times 10^{-3}} \times 10^{-3} \times 1800$$

$$= -\frac{0.1 \times 1800}{50} = \frac{18}{5}$$

$$= -3.6\,\text{V}$$

Self-inductance

Self-inductance is the property of a coil in which a change of current, and hence a change of flux, produces an e.m.f. in that coil. The average induced e.m.f. in such a circuit is given by

$$E = \frac{-L(I_2 - I_1)}{t}\,\text{volts}$$

The inductance L can be calculated from

$$L = \frac{N\Phi}{I}$$

where N = number of turns, Φ = flux in webers and I = current.

The unit of inductance

The unit of inductance is the **henry** (symbol H) and is defined as follows: 'A circuit is said to possess an inductance of 1 H when an e.m.f. of 1 V is induced in that circuit by a current changing at the rate of 1 A/s'.

Joseph Henry (1797–1878)

American electrophysicist whose work with magnetism led him to discover self-inductance.

Mutual inductance: symbol, *M*; unit, henry (H)

Let us consider the effect of winding two coils on the same iron core (Fig. 4.25).

A change of current in coil 1 produces a change of flux which links with coil 2, thus inducing an e.m.f. in that coil. These two coils are said to possess the property

Figure 4.25 Mutual inductance.

of **mutual inductance**, which is defined as: 'A mutual inductance of 1 H exists between two coils when a uniformly varying current of 1 A/s in one coil produces an e.m.f. of 1 V in the other coil'.

If a change of current ($I_2 - I_1$), in the first coil induces an average e.m.f. E in the second coil, then

$$E = -\frac{M(I_2 - I_1)}{t}\,\text{volts}$$

But E can also be expressed as

$$E = -\frac{(\Phi_2 - \Phi_1)N}{t}\,\text{volts}$$

Then

$$\frac{M(I_2 - I_1)}{t} = \frac{(\Phi_2 - \Phi_1)}{t} \times N$$

$$\therefore M = \frac{(\Phi_2 - \Phi_1)}{(I_2 - I_1)} \times N\,\text{henrys}$$

Example

Two coils A and B have a mutual inductance of 0.5 H. If the current in coil A is varied from 6 to 2 A, calculate the change in flux if coil B is wound with 500 turns.

$M = 0.5\,\text{H};\ I_2 = 6\,\text{A};\ I_1 = 2\,\text{A};\ N = 500;\ (\Phi_2 - \Phi_1) = ?$

$$M = \frac{(\Phi_2 - \Phi_1)}{(I_2 - I_1)} \times N$$

$$(\Phi_2 - \Phi_1) = \frac{M \times (I_2 - I_1)}{N} = \frac{0.5 \times (6 - 2)}{500}$$

$$= \frac{0.5 \times 4}{500} = \frac{2}{500} = 4\,\text{mWb}$$

Time constant: symbol, *T*

When considering inductive circuits, it is useful to represent the inductance and resistance of a coil as separate entities on a circuit diagram. A typical inductive circuit is shown in Fig. 4.26.

Figure 4.26 Typical inductive circuit.

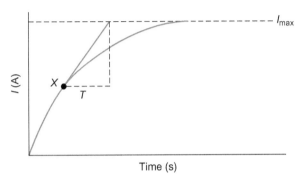

Figure 4.27 Growth of current in an inductive circuit.

When switch S is closed, the current increases from zero to a steady maximum, given by

$$I_{max} = \frac{V}{R} \text{ amperes}$$

Figure 4.27 shows the growth of current in an inductive circuit. At any instant, say X, on the growth curve, if the rate of growth of current at that instant is such that if it continues to increase at that rate it would reach its maximum value in L/R seconds, then this period of time is called the time constant and is given by

$$T = \frac{L}{R} \text{ seconds}$$

Graphical derivation of current growth curve

Construction

1. Select suitable scales for the two axes.
2. Draw the dotted line corresponding to the value of the maximum current (from $I = V/R$).
3. Along the time axis, mark off OA corresponding to the time constant T (from $T = L/R$).
4. Draw the perpendicular AB.
5. Join OB.
6. Select a point C close to O along OB.
7. Draw CD = T horizontally.
8. Draw the perpendicular DE.
9. Join CE.
10. Repeat procedures 6–9 for the line CE, and continue in the same manner as shown in the figure.

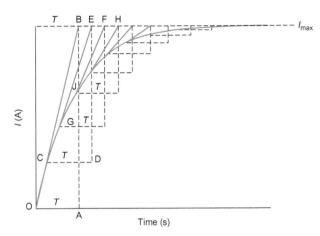

Figure 4.28 Graphical representation of current growth curve.

11. Join all the points O, C, G, J, etc., to form the growth curve (Fig. 4.28).

The more points that are taken, the more accurate the final curve will be.

Example

A coil having a resistance of $25\,\Omega$ and an inductance of $2.5\,H$ is connected across a $50\,V$ d.c. supply. Derive the curve of the current growth graphically (Fig. 4.29).

$$I_{max} = \frac{V}{R} = \frac{50}{25} = 2\,A; \text{ time constant, } T = \frac{L}{R} = \frac{2.5}{25} = 0.1\,s$$

Scales: $10\,cm = 1\,A$ and $10\,cm = 0.2\,s$

Derivation of curve of current decay

The curve of current decay is constructed in the same manner as the growth curve, but in reverse as is shown in Fig. 4.30.

Figure 4.29 Current growth.

Figure 4.30 Current decay.

Energy stored in a magnetic field

As we have already seen, opening an inductive circuit produces an arc across the switch contacts. This arc is the dissipation of the magnetic energy which was stored in the coil; the value of this energy can be calculated from

$$W = \tfrac{1}{2} \times L \times I^2 \text{ joules}$$

Example

When carrying a current of 1.2 A, each field coil of a generator has an inductance of 2.5 H. Calculate the value of the energy stored in each coil.

W = energy stores; I = 1.2 A; L = 2.5 H

$$W = \tfrac{1}{2} \times L \times I^2$$

$$= \frac{2.5 \times 1.2 \times 1.2}{2}$$

$$= 1.8 \text{ J}$$

Inductance in a.c. circuits

Inductive reactance: symbol, X_L; unit, ohm (Ω)

Let us now consider the effect of supplying an iron-cored coil of negligible resistance with an alternating current and voltage.

In this instance the current, and therefore the magnetic field, is building up and collapsing (in the case of a 50 Hz supply) 50 times every second and hence a continual alternating back e.m.f. is produced. As we have seen at the beginning of the chapter, the back e.m.f. opposes the change in circuit current

which is producing that e.m.f. Therefore, under a.c. conditions the e.m.f. produces a **continual** opposition to the current (much in the same way as resistance does in a resistive circuit). This opposition is called the **inductive reactance** (symbol X_L and is measured in ohms). X_L is given by

$$X_L = 2\pi f L \, \Omega$$

where f = frequency in hertz and L = inductance in henrys.

Example

Calculate the inductive reactance of a coil of inductance 0.5 H when connected to a 50 Hz supply.

$X_L = ?; f = 50 \text{ Hz}; L = 0.5 \text{ H}$

$$X_L = 2\pi f L$$

$$= 2\pi \times 50 \times 0.5$$

$$= 2\pi \times 25$$

$$= 50\pi$$

$$= 157.1 \, \Omega$$

When an a.c. supply is given to a **pure** inductance, the principles of Ohm's law may be applied (i.e. $V = I \cdot X_L$).

Example

Calculate the current taken by a coil of inductance 0.8 H when connected to a 100 V, 50 Hz supply.

$X_L = ?; V = 100 \text{ V}; f = 50 \text{ Hz}; L = 0.8 \text{ H and } I = ?$

In order to find the current, the formula $V = I \times X_L$ must be used; therefore the value of X_L must be calculated first.

$$X_L = 2\pi f L$$

$$= 2\pi \times 50 \times 0.8$$

$$= 80\pi$$

$$= 251.36 \, \Omega$$

$$V = I \times X_L$$

$$\therefore I = \frac{V}{X_L}$$

$$= \frac{100}{251.36}$$

$$= 0.398 \text{ A}$$

Representation of current by a phasor diagram

In a purely resistive circuit (Fig. 4.31a), only the magnitude of the current is opposed by the resistance,

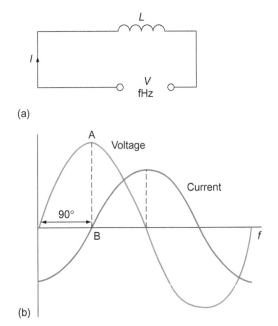

Figure 4.31 (a) Circuit diagram, (b) waveform.

Figure 4.33

Current lags voltage by 90° (out of phase)

Figure 4.34 Voltage and current in phase.

Resistance and inductance in series (*R–L* circuits)

Consider a coil which has inductance and resistance as shown in Fig. 4.35.

It is clear that the applied voltage, *V*, comprises the voltage across *L*, V_L, and the voltage across *R*, V_R, the current remaining. However, unlike purely resistive circuits, we cannot merely add V_L to V_R to obtain *V*. The reason for this is that in the inductive part of the circuit the common current *I* is out of phase with V_L and in the resistive part *I* is in phase with V_R, V_L and V_R can only, therefore, be added graphically (or by phasors) as in Fig. 4.36.

By construction, we can see that V_R is in phase with *I* and V_L is 90° out of phase with *I* and that the resultant is the applied voltage *V*. θ is the number of degrees that *I* lags behind the applied voltage *V*.

Figure 4.32 (a) Circuit diagram, (b) waveform.

and as the current and voltage alternate at the same time they are said to be **in phase**. Figure 4.31b shows the waveforms of current and voltage in a resistive circuit.

In a purely inductive circuit (Fig. 4.32a), the rate of change of current is opposed by the reactance of the coil, and the effect of this opposition is to make the current lag behind the applied voltage or be **out of phase** by 90°. The waveforms of current and voltage in a purely inductive circuit are shown in Fig. 4.32b.

The current lags the voltage by 90°, as *V* has reached its maximum at point A when current is zero at point B.

We can represent this effect by means of **phasors** (scaled lines representing electrical quantities). Figure 4.33 shows the phasor representation of current and voltage in a purely resistive circuit. Figure 4.34 shows the phasor representation of current and voltage in a purely inductive circuit.

Figure 4.35 R & L in series.

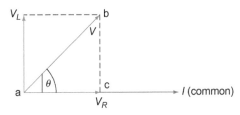

Figure 4.36 Phasor diagram.

From Fig. 4.36 it can be seen that the triangle abc is right-angled, having ab = V, bc = V_L and ac = V_R. Hence by Pythagoras' theorem:

$$ab = \sqrt{ac^2 + bc^2}$$
$$\therefore V = \sqrt{V_R^2 + V_L^2}$$

Note

This formula need not be remembered; it is simply a useful check after a value has been obtained with the aid of a phasor diagram.

Impedance: symbol, *Z*; unit, ohm (Ω)

It is clear that there are two separate oppositions to the flow of current in an R–L circuit, one due to resistance and the other due to reactance.

The combination of these oppositions is called the **impedance** of the circuit: its symbol is *Z* and it is measured in ohms.

Impedance may be defined as the **total** opposition offered by the components in that circuit. Ohm's law may once again be applied:

$$Z = \frac{V}{I}$$

where *V* is the applied voltage of the whole circuit.

Impedance triangle

From Fig. 4.37, it can be seen that a triangle can represent all the voltages in the circuit.

If we now divide these voltages by the current *I* which is common to all components in the circuit the triangle can be shown as in Fig. 4.38.

$$\frac{V}{I} = Z \quad \frac{V_L}{I} = X_L \quad \frac{V_R}{I} = R$$

Therefore the triangle can be shown as in Fig. 4.39. This triangle is called the **impedance triangle**. Applying Pythagoras' theorem:

$$Z = \sqrt{R^2 + X_L^2}$$

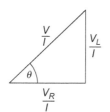

Figure 4.37 Voltage triangle.   **Figure 4.38**

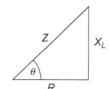

Figure 4.39 Impedance triangle.

Example

A choke coil has a resistance of 6 Ω and an inductance of 25.5 mH. If the current flowing in the coil is 10 A when connected to a 50 Hz supply (Fig. 4.40), find the supply voltage *V*.

In order to solve the problem by the use of phasors it is necessary to know the values of V_L and V_R. Hence,

$$V_R = I \times R = 10 \times 6 = 60 \text{ V}$$
$$V_L = I \times X_L$$
$$X_L = 2\pi f L$$
$$\quad = 2\pi \times 50 \times 25.5 \times 10^{-3}$$
$$\quad = 8\,\Omega$$
$$\therefore V_L = I \times X_L$$
$$\quad = 10 \times 8$$
$$\quad = 80 \text{ V}$$

By phasors we can draw Fig. 4.41 (scale 1 cm = 10 V).

By measurement *V* = 100 V. Check by Pythagoras' theorem:

$$Z = \sqrt{R^2 + X_L^2}$$
$$\quad = \sqrt{6^2 + 8^2}$$
$$\quad = \sqrt{100}$$
$$\quad = 10\,\Omega$$
$$V = I \times Z$$
$$\therefore V = 10 \times 10 = 100 \text{ V}$$

Figure 4.40

Figure 4.41

Resistance and inductance in parallel

Again, unlike resistive circuits, currents in parallel branches of an *R–L* circuit cannot simply be added to find the total current. It will be seen from Fig. 4.42a that the common quantity in the circuit is the voltage. This is used as the reference phasor (as the current was in the series circuit).

$$Z = \frac{V}{I} \, \Omega$$

The current in the resistive branch I_R is in phase with the applied voltage and the current in the inductive branch I_L lags the applied voltage by 90°. The resultant of these two currents is the supply current I (Fig. 4.42b). The impedance Z is given by

$$Z = \frac{V}{I} \, \Omega$$

Example

A non-resistive inductor of 51 mH is connected to a non-inductive resistor of $10\,\Omega$ across a 200 V, 50 Hz supply. What is the value of the supply current and the impedance of the circuit (Fig. 4.43)?

In order to find I, the values of I_R and I_L must be found:

$$I_L = \frac{V}{X_L} \quad \text{and} \quad I_R = \frac{V}{R}$$

$$X_L = 2\pi fL$$
$$= 2\pi \times 50 \times 51 \times 10^{-3}$$
$$= 16\,\Omega$$

$$\therefore I_L = \frac{V}{X_L} = \frac{200}{16}$$
$$= 12.5\,\text{A}$$

$$I_R = \frac{V}{R} = \frac{200}{10}$$
$$= 20\,\text{A}$$

By phasor diagram (Fig. 4.44), $I = 23.7\,\text{A}$:

$$Z = \frac{V}{I}$$
$$= \frac{200}{23.7}$$
$$= 8.45\,\Omega$$

Power in a.c. circuits

All the power (watts) in a circuit is dissipated in the circuit resistance (Chapter 3). The purely inductive part of the circuit consumes **no** power; it only provides a magnetic field.

Consider the circuit shown in Fig. 4.45. We have already seen the phasor diagram of the voltage in the circuit and developed a voltage triangle from it. Now we **multiply** the voltage by the common current (Fig. 4.46).

But $V \times I = P$. Therefore the triangle becomes as shown in Fig. 4.47.

This is called the **power triangle** but as the inductive part of the circuit consumes no power, it is called the **wattless component of power (VAr)**. The resistive part is called the **wattful component** or **true power** and the combination of the two is known as the

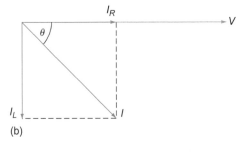

Figure 4.42 R & L in parallel.

Figure 4.43

Figure 4.44

Figure 4.45 R & L in series.

Figure 4.46 Voltage triangle.

Figure 4.47 Power triangle.

Figure 4.48 Power triangle.

apparent power (VA). The power triangle is usually shown in terms of kVA$_r$, kW and kVA (Fig. 4.48).

The relationship between the true power and the apparent power is very important.

Power factor

The ratio of the kW (true power) to the kVA (apparent power) is called the power factor (PF):

$$PF = \frac{kW}{kVA}$$

By trigonometry

$$\frac{kW}{kVA} = \cos\theta$$
$$\therefore PF = \cos\theta$$

As the voltage, impedance and power triangles have the same angles, cos θ is either

$$\frac{V_R}{V} \quad \text{or} \quad \frac{R}{Z} \quad \text{or} \quad \frac{kW}{kVA}$$

all of which equal the PF.

As the original triangle was formed from the phasor diagram (Fig. 4.36), θ is the angle between the current and the supply voltage, and therefore PF may be defined as: 'the cosine of the angle of phase difference between the current and the applied voltage'.

PFs in inductive circuits are termed **lagging** as the current lags the voltage.

When the true power equals the apparent power, the PF of 1 is usually referred to as **unity**. Under these circumstances there would be no wattless power (kVA$_r$) and the current taken by the circuit would be at a minimum. This is clearly an ideal situation.

The beer analogy

This is a useful way to explain the PF. Figure 4.49 shows a pint beer glass with the main body of beer and the head. Although the glass is full, part of it is useless (remember this is only an analogy) and the true amount of beer is less than a pint. A ratio of true to apparent beer would indicate how much head there was. So, if this ratio (pint factor) were 1 or unity there would be no head, and a PF of 0.5 would mean half beer and half head. Clearly, it is better to have a PF close to unity.

All large plants (motors, transformers, etc.) are rated in kVA, unlike most domestic appliances which are rated in kW. The reason for this is best explained by an example.

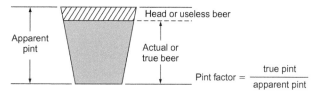

Figure 4.49 Beer analogy.

Example

If a heating appliance has a power rating of 1 kW at 230 V it will take a current of

$$I = \frac{P}{V}$$
$$= \frac{1000}{230}$$
$$= 4.34\,A$$

But if a motor has a power rating of 1 kW at 230 V and the motor windings cause a PF of 0.6, then as

$$PF = \frac{kW}{kVA}$$
$$kVA = \frac{kW}{PF}$$
$$= \frac{1}{0.6}$$
$$= 1.667\,kVA$$
$$= 1667\,VA$$

and since

$$\text{Current} = \frac{\text{volt amperes}}{\text{volts}}$$
$$I = \frac{1667}{230}$$
$$= 7.25\,A$$

had the cable supplying the motor been rated on the kW value it would clearly have been undersized.

Applications of inductance

Inductors, or chokes as they are more popularly called, are used in many areas of modern technology. In electrical installation work the main application is in fluorescent lighting, where the choke is open-circuited across the ends of the tube to cause it to strike. This effect is discussed further in Chapter 10. Motor windings are also inductances.

Transformers

As we have seen earlier, two coils that are wound on the same iron core have the property of mutual inductance, because a change in flux, and hence in e.m.f., in one coil produces, via the iron core, a corresponding change in the other coil.

If we take the same arrangement and apply an alternating voltage to one coil, it will induce an alternating e.m.f. in the other coil; this is called the **transformer effect**. The coil or winding to which the supply is connected is called the **primary** and the winding from which the induced voltage is taken is called the **secondary** (Fig. 4.50).

The relationship between the voltage, current and number of turns for each winding is as follows:

$$\frac{V_p}{V_s} = \frac{N_p}{N_s} = \frac{I_s}{I_p}$$

where V_p = primary voltage, I_p = primary current, N_p = primary turns, V_s = secondary voltage, I_s = secondary current and N_s = secondary turns.

Transformers which have a greater secondary voltage are called **step-up** transformers, while those with a smaller secondary voltage are called **step-down** transformers.

Laminated steel core

I_p I_s

V_p N_p N_s V_s

Primary winding   Secondary winding

Figure 4.50 Simple transformer.

Example

A single-phase step-down transformer has 763 turns on the primary and 365 turns on the secondary winding. If the primary voltage is 230 V calculate the secondary voltage. Also calculate the secondary current if the primary current is 10 A.

$$\frac{V_p}{V_s} = \frac{N_p}{N_s}$$

$$\therefore V_s = \frac{V_p \times N_s}{N_p}$$

$$= \frac{230 \times 365}{763}$$

$$= 110 \, V$$

Also

$$\frac{V_p}{V_s} = \frac{I_s}{I_p}$$

$$\therefore I_s = \frac{I_p \times V_p}{V_s}$$

$$= \frac{10 \times 230}{110}$$

$$= 21 \, A$$

Note the larger secondary current. The secondary winding would need to have a larger conductor size than the primary winding to carry this current. If the transformer were of the step-up type, the secondary current would be smaller.

Types of transformer

Double wound

This type is constructed as shown in Fig. 4.50. Two electrically separate coils are wound on to a common silicon steel core.

The core is laminated to lessen the effects of eddy currents and silicon steel is preferred, as there are few losses due to hysteresis. These losses are dealt with later in this section.

The double-wound transformer is the commonest form of transformer and has a wide range of applications.

Auto-transformer

In this type of transformer a single coil is wound on to a steel core, the primary and secondary windings being part of one winding (Fig. 4.51).

The main use of this type of transformer is in the grid system. When 400 000 V (400 kV) has to

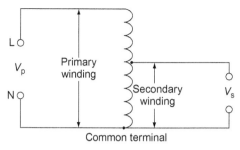

Figure 4.51 Auto-transformer.

be transformed (stepped down) to 132 kV, huge transformers are required. Auto-transformers are used mainly because, a great saving in copper and hence expense is achieved as there is only one winding.

The main disadvantage in using auto-transformers for applications such as bells or train sets, etc., is that the primary and secondary windings are **not** electrically separate and a short circuit on the upper part of the winding (see Fig. 4.51) would result in the whole of the primary voltage appearing across the secondary terminals.

The same ratio applies between the voltages, currents and number of turns.

The current transformer

The action of this transformer is the same as those previously discussed. It is a step-up (voltage) transformer and is used extensively for taking measurements. The most common form is the bar primary type (Fig. 4.52).

It is clearly impracticable to construct an instrument to measure currents as high as, say, 200–300 A, so a current transformer is used to step down the secondary current to a value which can be measured on a standard instrument.

Example

A current of 300 A flowing in a bus-bar needs to be measured. The ammeter available has a maximum rating of 0.5 A. How many turns on the secondary of a current transformer would be required to measure

Figure 4.52 Bar primary transformer.

the primary current, where $N_p = 1$ (single bus-bar), $I_p = 300$ A and $I_s = 0.5$ A (instrument rating).

$$\frac{I_s}{I_p} = \frac{N_p}{N_s}$$

$$N_s = \frac{N_p \times I_p}{I_s}$$

$$N = \frac{1 \times 300}{0.5}$$

$$= 600 \text{ turns}$$

Transformer losses

Ideally, the power input to a transformer ($I_p \times V_p$) should equal the power output ($I_s \times V_s$). However, there are power losses which reduce the efficiency. These losses are copper, eddy current and hysteresis losses.

Copper (I^2R) loss

Current flowing in the copper windings causes a heating loss.

Eddy current loss

This loss is caused by alternating currents which are induced magnetically in the core. They are reduced by laminating.

Hysteresis loss

This is an energy loss due to the changing magnetism in the core.

If we take a sample of unmagnetized iron (Fig. 4.53), wind a coil on it and pass a current through the winding, the core will become magnetized. The density of the flux will depend on the current and the number of turns. The product of the current and the turns is called the **magnetizing force (H)**.

Clearly, once the coil has been wound it is only the current that is variable, and if it is increased the core becomes magnetized. A graph of this effect is shown in Fig. 4.54a–e. The current in the coil is increased from zero to saturation point. Beyond this, an increase in the current does not increase the magnetism (Fig. 4.54a). If the current is now decreased to zero

Figure 4.53

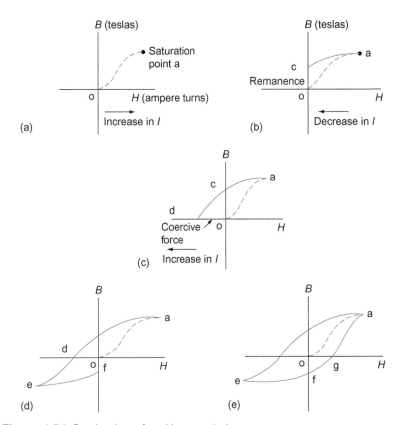

Figure 4.54 Production of an Hysteresis loop.

(see Fig. 4.54b), its path is along (a,c) and not along (a,o), leaving the core slightly magnetized. This remaining magnetism (o,c) is called the residual magnetism or **remanence**.

If the polarity of the supply is now reversed and the current increased again, the current follows (c,d) (see Fig. 4.54c). Clearly some force has been used to reduce the remanence to zero (o,d). This force is called the **coercive force**. This would not have occurred if the current had followed the original route (o,a). Hence, energy has been used≈to overcome the remanence. This is an **energy loss**.

If the current is further increased in the same direction, saturation point will be reached again (e) and the core will have reversed polarity. Decreasing the current to zero will result in a remanence (o,f) which has to be overcome (Fig. 4.54d).

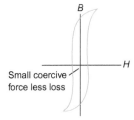

Figure 4.55 Hysteresis loop for transformer core.

Another reversal of polarity and an increase in current will result in a coercive force (o,g) being used (energy loss). A further increase in current will bring the curve back to (a) (Fig. 4.54). This complete curve is called the **hysteresis loop** (Fig. 4.55).

If the supply to the coil were alternating current, the polarity would be changing constantly and there would be a continual energy loss. It can be minimized to some extent by using a silicon steel core, the remanence of which is easily overcome.

Self-Assessment Questions

1. A magnetic flux density of a circular field is 27 T. If the flux is 108 mWb, calculate the area of the field.
2. A magnetic field has a flux of 54 mWb and a cross-sectional area of 9 cm². Calculate the field's flux density.
3. A conductor 11 cm long is lying at right angles to a magnetic field of flux density 10 T; if the conductor carries a current of 10.91 A, calculate the force on the conductor.
4. A conductor 8 cm long connected to a 50 V d.c. supply is situated at right angles to a magnetic field of flux 30 mWb and a cross-sectional area 15 cm². If the force exerted on the conductor is 16 N, calculate the resistance of the conductor.
5. With the aid of diagrams, explain why a force is exerted on a conductor carrying a current in a magnetic field.
6. Give three examples of how the magnetic effect of an electric current may be used, using explanatory diagrams.
7. With the aid of sketches, explain the difference between simple d.c. and a.c. generators.
8. Draw a diagram showing the arrangement of a typical single-phase a.c. generator and explain its action.
9. An alternating current has a peak value of 50 A. Draw to scale the sine wave of this current over half a cycle and from it determine the value of the current after 70°. What is the r.m.s. value of the current?
10. Explain the meaning of the term 'root-mean-square value'.
11. What is the meaning of the term 'frequency'? Illustrate your answer with sketches.
12. What is meant by three-phase generation?
13. If a coil has an e.m.f. of 6 V induced in it by a flux changing from 0 to 36 mWb in 0.18 s, calculate the number of turns on the coil.
14. (a) What is self-inductance? (b) A relay coil of 300 turns produces a flux of 5 mWb when carrying a current of 1.5 A. Calculate the inductance of the coil.
15. An iron-cored coil having an inductance of 0.1 H and a resistance of 1.25 Ω is connected to a 25 V d.c. supply. Calculate the circuit time constant and the maximum current. Draw to scale the

curve of the current decay when the supply is switched off. What will be the value of current after 0.15 s?
16. (a) What is the effect of opening an inductive circuit? How can this effect be used? (b) The energy stored in a coil is 2 J; if the inductance of the coil is 160 mH, calculate the coil current.
17. An inductor of negligible resistance has an inductance of 100 mH and an inductive reactance of 31.42 Ω when connected to an a.c. supply. Calculate the supply frequency.
18. A coil of inductance 1528 mH and negligible resistance takes a current of 0.5 A when connected to a 50 Hz supply. Calculate the value of the supply voltage.
19. An inductor has a resistance of 16 Ω and an inductive reactance of 12 Ω. If the current flowing in the circuit is 12 A, find, by means of a phasor diagram, the value of the supply voltage.
20. A pure inductance which has a resistance of 12 Ω is wired in parallel with a resistance of 8 Ω across a 230 V supply. Calculate the current in each component and determine the value of the supply current by means of a phasor diagram.
21. (a) Define the term power factor. (b) A circuit consists of a resistance and an inductance in series. The voltage across the resistance is 192 V and the PF is 0.8. Determine the value of the supply voltage and the voltage across the inductance.
22. An ammeter, a voltmeter and a wattmeter are to be connected into the circuit supply of a single-phase motor. Draw a diagram showing how these instruments would be connected. If the readings obtained were 230 V, 1.3 A and 180 W, respectively, calculate the PF of the motor.
23. (a) Explain the action of a transformer. (b) A double-wound transformer has a primary voltage of 230 V and a secondary voltage of 110 V. If there are 690 primary turns, calculate the number of turns on the secondary.
24. With the aid of a diagram explain what an **auto-transformer** is.
25. List the losses which occur in a transformer. What steps can be taken to overcome them?

CHAPTER 5

Capacitors and Capacitance

Capacitors

A capacitor consists of two metal plates separated by an insulator, called a **dielectric**; the whole assembly is able to store electricity. This store is in the form of an excess of electrons on one plate and a deficiency on the other. In this state the capacitor is said to be charged. The charge is achieved by applying a voltage across the plates.

The use of water is once again excellent for an analogy. Consider Fig. 5.1 which shows a water-filled system where pressure on the plunger P causes the flexible diaphragm D to distort. In this way, energy is stored in the diaphragm, because when the plunger pressure is removed, the diaphragm will cause the plunger to return to normal.

The type of capacitor commonly used in installation work is the electrolytic capacitor. This consists of plates of metal foil placed on either side of a waxed paper dielectric like a sandwich (Fig. 5.2). It is manufactured in a long strip, rolled up and sealed into a metal container.

Capacitance: symbol, *C*; unit, farad (F)

The unit of capacitance is the **farad** and may be defined as: 'the capacitance of a capacitor which requires a potential difference of 1 V to maintain a charge of 1 C on that capacitor'.

Movement of
diaphragm
(stored charge)

D

P

Pressure
(voltage)

Figure 5.1 Water analogy for capacitor.

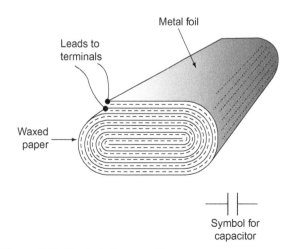

Metal foil

Leads to
terminals

Waxed
paper

Symbol for
capacitor

Figure 5.2 Electrolytic capacitor.

Hence,

Charge = capacitance × voltage

$$Q \text{ (C)} = C \text{ (F)} \times V \text{ (V)}$$

Michael Faraday (1791–1867)

British physicist and chemist known as the 'Father of electricity'. He experimented in many different areas of physics but is probably best known for his discovery of electromagnetic induction and hence the transformer.

Example

Calculate the charge on a 50 µF capacitor when it is connected across a 200 V d.c. supply:

$$Q = C \times V$$
$$= 50 \times 10^{-6} \times 200$$
$$= 0.01 \text{ C}$$

Dimensions of capacitors

If we take a simple parallel plate capacitor with an air dielectric, measure its capaci-tance, and then move the plates further apart, we find that the capacitance is smaller when measured a second time. We can therefore state that an increase in dielectric thickness (d) causes a decrease in capacitance. **Capacitance is inversely proportional to dielectric thickness:**

$$C \propto \frac{1}{d}$$

If, however, we were to keep the dielectric thickness constant and to vary the area of the plates (a), we would find that a change in plate area would cause a corresponding change in capacitance. The larger the plate area the larger the capacitance. **Capacitance is directly proportional to plate area:**

$$C \propto a$$

Combining these two effects we can see that

$$C \propto \frac{a}{d}$$

Capacitors in series

Consider the effect of connecting three similar capacitors in Fig. 5.3.

Figure 5.3 Capacitors in series.

We know that

$$C \propto \frac{1}{d}$$
$$\therefore d \propto \frac{1}{C}$$

So

$$d_1 \propto \frac{1}{C_1} \quad d_2 \propto \frac{1}{C_2} \quad d_3 \propto \frac{1}{C_3}$$

If we were to combine all the dielectrics, we would have one capacitor of dielectric thickness d_t and capacitance C:

$$d_t \propto \frac{1}{C}$$

But

$$d_t = d_1 + d_2 + d_3$$
$$= \frac{1}{C_1} + \frac{1}{C_2} + \frac{1}{C_3}$$

Also

$$d_t = \frac{1}{C}$$
$$\therefore \frac{1}{C} = \frac{1}{C_1} + \frac{1}{C_2} + \frac{1}{C_3} + \cdots$$

Just as the current is common to all parts of a series resistive circuit, so is charge common in a series capacitive circuit. Therefore

$$Q = CV \quad Q = C_1 V_1 \quad Q = C_2 V_2 \quad Q = C_3 V_3 \ldots$$

Example

Three capacitors of 60, 40 and 24 µF are connected in series across a 500 V d.c. supply. Calculate the total capacitance and the charge on each capacitor.

$$\frac{1}{C} = \frac{1}{C_1} + \frac{1}{C_2} + \frac{1}{C_3}$$
$$= \frac{1}{60} + \frac{1}{40} + \frac{1}{24}$$
$$\frac{1}{C} = \frac{1}{12}$$

$$\therefore C = 12 \text{ µF}$$

Q is common to each capacitor.

$$\therefore Q = CV$$
$$= 12 \times 10^{-6} \times 500$$
$$= 6 \times 10^{-3}$$
$$= 6 \text{ mC}$$

Figure 5.4 Capacitors in parallel.

Capacitors in parallel

Let us arrange three similar capacitors in parallel (see Fig. 5.4). We know that $C \propto a$. Therefore $C_1 \propto a_1$, $C_2 \propto a_2$ and $C_3 \propto a_3$.

As the plates connected to either side of the supply are common, we could replace the arrangement with one capacitor C of plate area a_T.

$$\therefore a_T = a_1 + a_2 + a_3$$
$$\therefore a_T = C_1 + C_2 + C_3$$

But $C \propto a_T$.

$$\therefore C = C_1 + C_2 + C_3$$

In this case it is the voltage that is common and the charge Q behaves like the current in a parallel resistive circuit. So,

$$Q = CV \quad Q_1 = C_1V \quad Q_2 = C_2V \quad Q_3 = C_3V$$

Example

Three capacitors of 60, 40 and 24 μF are connected in parallel across a 500 V supply. Calculate the total capacitance, the total charge and the charge on each capacitor.

$$C = C_1 + C_2 + C_3$$
$$= 60 + 40 + 24$$
$$= 124 \, \mu F$$

Total charge $Q = C \times V$
$$= 124 \times 10^{-6} \times 500$$
$$= 62 \, mC$$

$$Q_1 = C_1V$$
$$= 60 \times 10^{-6} \times 500$$
$$= 30 \, mC$$

$$Q_2 = C_2V$$
$$= 40 \times 10^{-6} \times 500$$
$$= 20 \, mC$$

$$Q_3 = C_3V$$
$$= 24 \times 10^{-6} \times 500$$
$$= 12 mC$$

Energy stored in a capacitor

The amount of energy stored in a capacitor is expressed in joules and is given by

$$W = \tfrac{1}{2}CV^2$$

Capacitors in d.c. circuits

A capacitor connected across a d.c. supply is shown in Fig. 5.5a. The curves of the current and the voltage in the circuit are shown in Fig. 5.5b.

As the capacitor begins to charge, its voltage increases until it is equal to the supply voltage. At the same time the charging current decreases. When the supply voltage and the capacitor voltage are equal, the current in the circuit will be zero.

Figure 5.6a shows the charged capacitor connected across a resistor. Figure 5.6b shows the curves of the discharge voltage and current.

Curves of current and voltage change

These curves are plotted in the same manner as those in inductive circuits (Chapter 4).

Maximum charging or discharging current:

$$I = \frac{V}{R}$$

Time constant:

$$T = CR$$

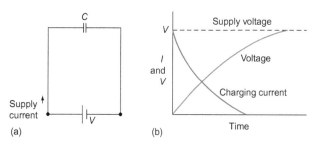

Figure 5.5 Charging: (a) capacitor connected across d.c. supply; (b) curves of discharge voltage and current.

Figure 5.6 Discharging: (a) charged capacitor connected across a resistor; (b) curves of discharge voltage and current.

5 Capacitors and Capacitance

Capacitance in a.c. circuits

In an a.c. circuit, a capacitance has the effect of opposing the voltage, thus causing the circuit current to **lead**. In a purely capacitive circuit, the current leads the voltage by 90°. The waveforms and phasors of such a circuit are shown in Fig. 5.7.

(a)   (b)   (c)

Figure 5.7 (a) Circuit diagram; (b) phasor diagram; (c) waveforms.

Capacitive reactance: symbol, X_c; unit, ohm (Ω)

The opposition offered by a capacitor in an a.c. circuit is called the **capacitive reactance** and is given by

$$X_c = \frac{1}{2\pi f C}$$

where X_c = capacitive reactance (Ω), f = frequency of supply (Hz) and C = capacitance (F).

As in the case of inductive reactance, Ohm's law may be applied, that is

$$V = I \times X_c$$

Example

A purely capacitive circuit of 31.8 μF is connected to a 230 V, 50 Hz supply. Calculate the capacitive reactance and the circuit current.

$$X_c = \frac{1}{2\pi f C}$$

$$= \frac{1}{2\pi \times 50 \times 31.8 \times 10^{-6}}$$

$$= \frac{10^6}{100\pi \times 31.8}$$

$$= 100\,\Omega$$

$$V = I \times X_c$$

$$\therefore I = \frac{V}{X_c}$$

$$= \frac{230}{100}$$

$$= 2.3\,\text{A}$$

Resistance and capacitance in series

Figure 5.8 shows a capacitor in series with a resistor.

As the current leads the voltage across the capacitor and is in line with the voltage across the resistor, the phasor diagram may be drawn as shown in Fig. 5.9.

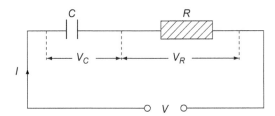

Figure 5.8 Capacitor in series with resistor.

Figure 5.9 Phasor diagram.

Example

A capacitor of 159 μF is connected in series with a non-inductive resistor of 15 Ω across a 50 Hz supply. If the current drawn is 9.2 A calculate X_c, the voltage across each component and find by means of a phasor diagram the value of the supply voltage.

$$X_c = \frac{1}{2\pi f C}$$

$$= \frac{1}{2\pi \times 50 \times 159 \times 10^{-6}} = \frac{10^6}{100\pi \times 159}$$

$$X_c = 20\,\Omega$$

$$V_c = I \times X_c$$

$$= 9.2 \times 20$$

$$V_c = 184\,\text{V}$$

$$V_R = I \times R$$

$$= 9.2 \times 15$$

$$V_R = 138\,\text{V}$$

By measurement (Fig. 5.10) V will be found to be 230 V.

Figure 5.10

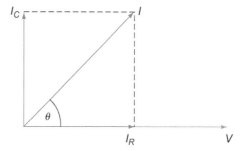

Figure 5.12 Phasor diagram.

Resistance and capacitance in parallel

In this case it is the voltage that is common and the currents that are added by phasors (see Figs 5.11 and 5.12).

Figure 5.11 R & C in parallel.

Working voltage

Every capacitor has the value of its working voltage marked on it. Beyond this value the dielectric would break down and the capacitor would be useless.

Referring to Fig. 5.1, if the pressure on the plunger is too high, the diaphragm would puncture and the system is rendered useless.

Applications of capacitors

Capacitors are used extensively in electrical engineering. In the field of installation work, they are mainly used for motor starting, power factor correction and radio interference suppression and to minimize the stroboscopic effects in fluorescent lighting circuits. Their use in power factor correction and fluorescent lighting is dealt with in Chapter 10.

5

Self-Assessment Questions

1. A capacitor has a value of 73 µF and is connected across a 100 V supply. Calculate the charge on the capacitor.

2. Three capacitors of 20, 80 and 16 µF are connected in series across a 230 V supply. Calculate the charge.

3. A variable capacitor has a capacitance of 100 µF when the distance between the plates is 1 mm. What will be the capacitance if the plates are adjusted to be 10 mm apart?

4. Four capacitors of 10, 20, 15 and 5 µF are connected in parallel across a 230 V supply. Calculate the total capacitance, the total charge and the charge across the 20 µF capacitor.

5. A capacitor has a value of 150 µF and a plate area of 60 cm². What will be the plate area of a similar type of capacitor of 200 µF and the same dielectric thickness?

6. A parallel plate capacitor has a value of 636 µF. At what value of supply frequency will the reactance be 5 Ω?

7. A resistor of 6 Ω is connected in series with a capacitor of 398 µF, the current drawn being 24 A. Calculate the voltage across each component and find the supply voltage by means of a phasor diagram (f = 50 Hz).

8. Draw the waveforms and phasor diagram for an a.c. current and voltage in a purely capacitive circuit.

9. A 127.3 µF capacitor is connected in parallel with a 50 Ω resistor across a 230 V supply. Calculate the current taken by each component. Determine the value of the supply current (f = 50 Hz).

10. Explain with the aid of a sketch the construction of an electrolytic capacitor. What is meant by the term working voltage?

CHAPTER 6

Resistance, Inductance and Capacitance in Installation Work

In Chapters 4 and 5, we have discussed separately the effects of inductance and capacitance in an a.c. circuit. Here we consider how these effects may be applied, and in some cases combined, in a.c. circuits.

Let us first refresh our memories regarding the phasor diagrams for R, L and C.

Pure resistance (Fig. 6.1)

$$R = \frac{V}{I}$$

Pure inductance (Fig. 6.2)

$$X_L = \frac{V}{I}$$

and

$$X_L = 2\pi f L$$

Pure capacitance (Fig. 6.3)

$$X_C = \frac{V}{I}$$

and

$$X_C = \frac{1}{2\pi f C}$$

R and L in series (Fig. 6.4)

$$Z = \frac{V}{I}$$

Circuit diagram
(a)

Waveform
(b)

Phasor diagram
(c)

Figure 6.1

Circuit diagram
(a)

Waveform
(b)

Phasor diagram
(c)

Figure 6.2

Circuit diagram
(a)

Waveform
(b)

Phasor diagram
(c)

Figure 6.3

From the phasor diagram of voltages (Fig. 6.4), an impedance triangle may be formed (Fig. 6.5).

By Pythagoras' theorem:

$$Z = \sqrt{R^2 + X_L^2}$$

Electrical Installation Work. 978-1-138-84927-3 © Brian Scaddan.
Published by Taylor & Francis. All rights reserved.

Circuit diagram
(a)

Phasor diagram
(b)

Figure 6.4 R & L in series.

Also

$$\cos\theta = \frac{R}{Z} = \text{power factor (PF)}$$

Figure 6.5 Development of impedance triangle.

R and *C* in series (Fig. 6.6)

Circuit diagram
(a)

Phasor diagram
(b)

Figure 6.6 R & C in series.

It is clear that a similar impedance triangle may be formed, as shown in Fig. 6.7.

$$\therefore Z = \sqrt{R^2 + X_C^2}$$

and

$$\cos\theta = \frac{R}{Z} = \text{PF}$$

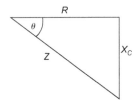

Figure 6.7 Impedance triangle.

Now we begin to combine these separate phasor diagrams.

R, *L* and *C* in series (Fig. 6.8)

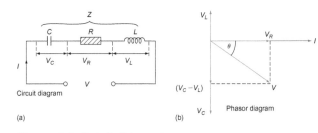

Circuit diagram
(a)

Phasor diagram
(b)

Figure 6.8 R, L & C in series.

The impedance triangle will be shown in Fig. 6.9.

$$\therefore Z = \sqrt{R^2 + (X_C - X_L)^2}$$

or, if V_L is greater than V_C,

$$Z = \sqrt{R^2 + (X_L - X_C)^2}$$

and

$$\cos\theta = \frac{R}{Z} = \text{PF}$$

Figure 6.9 Impedance triangle.

The following problems will be solved using different methods.

Example

From the circuit shown in Fig. 6.10, determine the value of the supply voltage and the PF.

Figure 6.10

Method 1, by phasors

$$V_C = I \times X_C = 2 \times 80 = 160\,\text{V}$$
$$V_L = I \times X_L = 2 \times 36 = 72\,\text{V}$$
$$V_R = I \times R = 2 \times 25 = 50\,\text{V}$$

From Fig. 6.11:

$$V = 101\,\text{V}$$

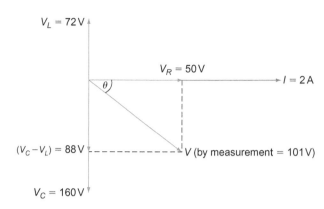

Figure 6.11

By measurement:

$\theta = 60.3°$

From cosine tables:

$\cos\theta = 0.495$

PF = 0.495 leading

Method 2, using the theorem of Pythagoras (impedance)

$$Z = \frac{V}{I}$$

But

$$Z = \sqrt{R^2 + (X_C - X_L)^2}$$
$$= \sqrt{25^2 + (80 - 36)^2}$$
$$= \sqrt{25^2 + 44^2}$$
$$= \sqrt{2561}$$
$$= 50.6\,\Omega$$

$$\therefore 50.6 = \frac{V}{2}$$
$$\therefore V = 2 \times 50.6$$
$$= 101.2\,V$$

$$PF = \cos\theta = \frac{R}{Z} = \frac{25}{50.6}$$
$$= 0.49 \text{ leading}$$

Method 3, using the theorem of Pythagoras (voltage)

From Fig. 6.12:

$$V = \sqrt{V_R^2 + (V_C - V_L)^2}$$
$$= \sqrt{50^2 + 88^2}$$
$$= \sqrt{10\,244}$$
$$V = 101.2\,V$$

$$PF = \cos\theta = \frac{\text{base}}{\text{hypotenuse}}$$
$$= \frac{50}{101.2}$$
$$= 0.49 \text{ leading}$$

Figure 6.12

Figure 6.13

Method 4, using trigonometry

From Fig. 6.13:

$$\tan\theta = \frac{\text{perpendicular}}{\text{base}}$$
$$= \frac{88}{50}$$
$$= 1.76$$

From tangent tables:

$\theta = 60.4°$

$$\cos\theta = \frac{\text{base}}{\text{hypotenuse}} = \frac{50}{V}$$
$$\therefore V = \frac{50}{\cos\theta}$$

From cosine tables:

$\cos 60.4° = 0.49 \text{ (PF)}$

$$\therefore V = \frac{50}{0.49}$$
$$= 101.2\,V$$

PF = $\cos\theta = 0.49$ leading

PF improvement

The magnetic effect of an inductor has many uses. It is, however, in equipment such as motors and fluorescent lighting that its effect on the PF is substantial enough to cause concern, and makes it necessary to improve the PF.

Consider the diagram of a resistive and inductive circuit (Fig. 6.14).

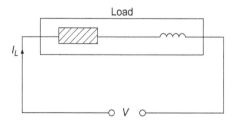

Figure 6.14 Resistive and inductive circuit.

The phasor diagram for this circuit is shown in Fig. 6.15.

We can show the supply current and voltage as in Fig. 6.16.

65

Figure 6.15 Phasor diagram.

Figure 6.16 Phasor diagram.

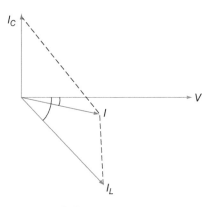

Figure 6.20 Full phasor diagram.

If we redraw this phasor diagram so that the voltage is drawn horizontally, it becomes as shown in Fig. 6.17.

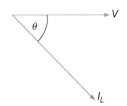

Figure 6.17 Phasor diagram.

If we now connect a variable capacitor across the supply terminals of the original load, we have the result shown in Fig. 6.18.

Figure 6.18 PF improvement.

It is a parallel circuit and the voltage is common to both the branches. We can therefore draw a current phasor diagram (Fig. 6.19). As V is common, we can combine both diagrams (Fig. 6.20).

Figure 6.19 Phasor diagram for (a) load; (b) capacitor.

The phasor resultant of I_L and I_C (Fig. 6.20) is clearly I, which is of course the supply current. Note that it is smaller than the load current I_L and that the angle between I and V is smaller than that between I_L and V. The closer an angle to zero the nearer its cosine to unity. Therefore, the addition of the capacitor has improved the PF of the system.

The actual current taken by the load does not change; it is the total supply current that decreases. This means that smaller supply cables may be used. With industrial loads, the supply authority's transformer and switchgear as well as their cable may be reduced in size.

In order to encourage PF improvement, the supply authorities make a higher charge to consumers who do not correct or improve their PF to a suitable level (usually about 0.95 lagging). It is not usual to improve the PF much beyond this point as the cost of providing extra capacitance required to gain a small decrease in current is uneconomic.

Capacitors are the most popular method of improving the PF although synchronous motors are used occasionally (this is discussed in greater detail in Chapter 8).

PF improvement capacitors may be fitted to individual plant or in banks connected to the supply intake terminals. The first method is more popular as the banked type needs automatic variation as plant is switched on and off.

Example

A 230 V single-phase motor takes a current of 8 A and has a PF of 0.7 lagging. A capacitor is connected in parallel with the motor and takes a current of 3.1 A.

Draw a scaled phasor diagram of the currents in the circuit and find the value of the supply current and the new PF.

From Fig. 6.21:
$$I_L = 8\,\text{A}$$
$$\text{PF} = 0.7$$
$$\text{As PF} = \cos\theta$$
$$\cos\theta = 0.7$$
$$\theta = 45.5°$$

Figure 6.21

Therefore the phasor diagram for the motor is as shown in Fig. 6.22.

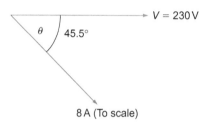

Figure 6.22

From the question:

$I_C = 3.1\,A$

I_C leads V by 90°

Therefore the phasor diagram for the capacitor is Fig. 6.23a. Combining both phasor diagrams, we have Fig. 6.23b.

(a)

(b)

Figure 6.23

By measurement:

$I = 6.2\,A$

$\alpha = 24.5°$

$\therefore PF = \cos\alpha$

$PF = \cos 24.5°$

$= 0.91$ lagging

Example

A 240 V, 50 Hz single-phase motor takes a current of 8 A at a PF of 0.65 lagging. Determine the value of

capacitor required to improve the PF to 0.92 lagging. What is the value of the new supply current?

Method 1, by phasors (Fig. 6.24)

Old PF $= \cos\theta = 0.65$

From cosine tables:

$\theta = 49.5$

New PF $= \cos\alpha = 0.92$

From cosine tables:

$\alpha = 23°$

The phasor diagram is now drawn to scale (Fig. 6.25). By measurement:

$I_C = 3.9\,A$

and

$I = 5.6\,A$

To find the capacitance required:

$$I_C = \frac{V}{X_C}$$

$$\therefore X_C = \frac{V}{I_C}$$

$$= \frac{230}{3.9}$$

$$= 59\,\Omega$$

Figure 6.24

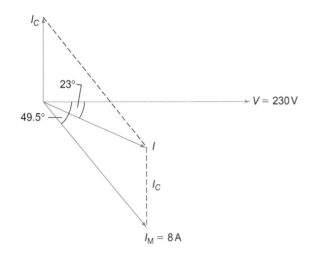

Figure 6.25

But

$$X_C = \frac{1}{2\pi f C}$$

$$\therefore C = \frac{1}{2\pi f X_C}$$

$$= \frac{1}{2\pi \times 50 \times 59}$$

$$= 54\,\mu\text{F}$$

Method 2, by trigonometry

In this method, I_c is found by calculating lengths PR and PQ in Fig. 6.26 and subtracting ($\text{PR} - \text{PQ} = \text{QR} = I_c$).

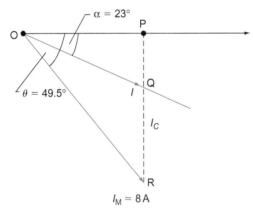

Figure 6.26

Triangle OPR:

$$\cos\theta = \frac{\text{base}}{\text{hypotenuse}} = \frac{\text{OP}}{\text{OR}}$$

$$\therefore \text{OP} = \text{OR} \times \cos\theta$$

$$= 8 \times 0.65$$

$$\text{OP} = 5.2\,\text{A}$$

This is called the active or horizontal component of I_M.

$$\sin\theta = \frac{\text{perpendicular}}{\text{hypotenuse}} = \frac{\text{PR}}{\text{OR}}$$

$$\text{PR} = \text{OR} \times \sin\theta$$

$$= 8 \times 0.76$$

$$\text{PR} = 6.08\,\text{A}$$

This is called the reactive or vertical component of I_M.

Triangle OPQ:

$$\tan\alpha = \frac{\text{perpendicular}}{\text{base}} = \frac{\text{PQ}}{\text{OP}}$$

$$\therefore \text{PQ} = \text{OP} \times \tan\alpha$$

$$= 5.2 \times 0.424$$

$$\text{PQ} = 2.2\,\text{A}$$

$$\therefore \quad \text{QR} = I_c = (\text{PR} - \text{PQ})$$

$$= 6.08 - 2.2$$

$$I_c = 3.88\,\text{A}$$

Calculation of C is as in method 1. To find current I:

$$\cos\alpha = \frac{\text{base}}{\text{hypotenuse}} = \frac{\text{OP}}{\text{OQ}}$$

$$\therefore \text{OQ} = I = \frac{\text{OP}}{\cos\alpha}$$

$$= \frac{5.2}{0.92}$$

$$I = 5.65\,\text{A}$$

Further examples of PF correction appear in Chapters 8 and 10.

The use of the previous methods of drawing and calculation may also be applied to power.

Power in a.c. circuits

The phasor diagram of voltages in an a.c. series circuit can be used to give a power triangle (Fig. 6.27).

$$\cos\theta = \frac{\text{W}}{\text{VA}} = \text{PF}$$

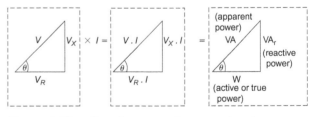

Figure 6.27   Development of power triangle.

Power may be added by phasor diagram or calculated by trigonometry.

Example

The following loads are connected to a factory supply: 5 kVA at 0.75 PF lagging; 8 kW at a PF of unity; 6.8 kVA at 0.6 PF lagging. Determine the total load taken from the supply and the overall PF.

Method 1, by phasor diagram (Fig. 6.28)

$$\cos\alpha = 0.75$$
$$\therefore \alpha = 41.4°$$
$$\cos\beta = 0.6$$
$$\therefore \beta = 53.1°$$

Figure 6.28

From Fig. 6.29:

$$\text{Total kVA} = 18.1\,\text{kVA}$$
$$\theta = 29°$$
$$\therefore \cos\theta = \text{PF} = 0.874$$

Figure 6.29

Method 2, by trigonometry (Fig. 6.30)

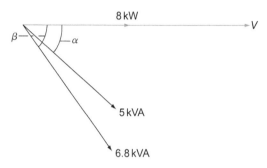

Figure 6.30

Active component of 8 kW = 8 kW
Active component of 5 kVA = 5 × cos 41.4°
$$= 5 \times 0.75$$
$$= 3.75\,\text{kW}$$
Active component of 6.8 kVA = 6.8 × cos 53.1°
$$= 6.8 \times 0.6$$
$$= 4.08\,\text{kW}$$
Total of active components = 8 + 3.75 + 4.08
$$= 15.83\,\text{kW}$$
Reactive component of 8 kW = 0
Reactive component of 5 kVA = 5 × sin 41.4°
$$= 5 \times 0.66$$
$$= 3.3\,\text{kVA}_r$$
Reactive component of 6.8 kVA = 6.8 × sin 53.1°
$$= 6.8 \times 0.8$$
$$= 5.44\,\text{kVA}_r$$

As both 5 and 6.8 kVA have lagging PFs, their reactive components are added (Fig. 6.31).

$$\therefore \text{Total reactive component} = 0 + 3.3 + 5.44$$
$$= 8.74\,\text{kVA}_r$$
$$\tan\theta = \frac{8.74}{15.83} = 0.55$$
$$\therefore \theta = 28.9°$$
$$\therefore \text{PF} = \cos\theta = 0.875$$

Figure 6.31

Also

$$\cos\theta = \frac{15.83}{\text{kVA}}$$
$$\therefore \text{kVA} = \frac{15.83}{\cos\theta}$$
$$= \frac{15.83}{0.875}$$
$$= 18\,\text{kVA at 0.875 PF lagging}$$

Self-Assessment Questions

1. A 230 V single-phase motor takes a current of 10 A and has a working PF of 0.5 lagging. Draw a scaled phasor diagram and from it determine the value of capacitor current required to improve the PF to 0.9 lagging. Calculate the value of the capacitor.

2. Determine the value of the voltage and the PF in the circuit shown in the following diagram.

3. A 230 V, 50 Hz single-phase motor takes 6 A at 0.56 PF lagging. Determine the value of capacitor required to improve the PF to unity.

4. A 230 V, 50 Hz fluorescent lamp unit takes a current of 0.6 A at a PF of 0.45 lagging. Calculate the capacitance required to correct the PF to 0.92 lagging.

5. Two 230 V, 50 Hz single-phase motors A and B are connected in parallel. Motor A takes a current of 8.6 A at 0.75 PF lagging and the total current taken from the supply is 16 A at 0.6 lagging. Calculate the current and the PF of motor B.

6. A consumer has the following loads connected to his or her supply: 3 kVA at 0.8 lagging; 4 kW at a PF of unity; and 5 kVA at 0.5 lagging. Calculate the total load in kVA and the overall PF.Resistance, Inductance and Capacitance in Installation Work

CHAPTER 7

Three-Phase Circuits

As we have seen in Chapter 4, a three-phase supply comprises three waveforms each separated by 120° and the resultant waveform is zero. Let us now consider how we can utilize this supply and how we can connect to it.

Star and delta connections

Figure 7.1a and b show the two main ways of connecting three-phase equipment.

The neutral conductor

Figure 7.2 shows the simple system of a star-connected load fed from a star-connected supply. The addition of the conductor between the star points converts the system into what is known as a 'three-phase four-wire system'. We can see that the currents supplied by the generator flow along the lines, through the load and return via the neutral conductor.

However, we have already seen that all line currents in a **balanced** three-phase system are equal, and add up to the neutral current, which is zero:

$$\therefore I_{Br} + I_{Bk} + I_{Gr} = I_N = 0$$

Hence the current flowing in the neutral is zero. Also, since no current flows between the star points, they must both be at the same potential, which is also zero. The star point of a transformer is earthed, as earth is also at zero volts.

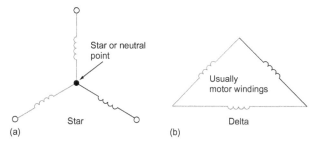

Figure 7.1 The two main ways of connecting three-phase equipment: (a) star; (b) delta.

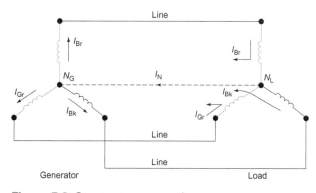

Figure 7.2 Star to star connections.

One reason for the connection of the neutral conductor is to provide a path for currents if the system became unbalanced. Another is that it enables single-phase loads to be connected to a three-phase system. The windings of most three-phase motors are connected in delta as the phase windings are perfectly balanced and no neutral is needed.

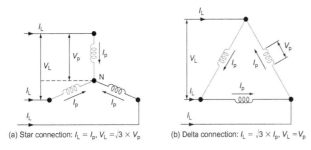

(a) Star connection: $I_L = I_p$, $V_L = \sqrt{3} \times V_p$ (b) Delta connection: $I_L = \sqrt{3} \times I_p$, $V_L = V_p$

Figure 7.3 Currents and voltages in star and delta systems.

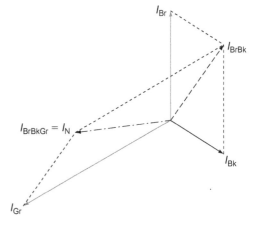

Figure 7.5 Phasor diagram for an un-balanced load.

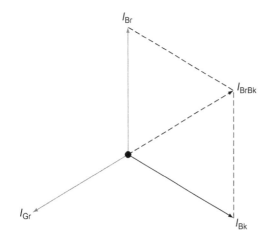

Figure 7.4 Phasor diagram for a balanced load.

Current and voltage distribution

Balanced three-phase a.c. systems (Fig. 7.3)

Use of phasors

Currents in three-phase systems may be added by the use of phasors. For balanced systems, the resultant current will, of course, be zero (Fig. 7.4).

The resultant of I_{Br} and I_{Bk} is I_{BrBk} which is clearly equal and opposite to I_{Gr} and hence the overall resultant will be zero.

Unbalanced three-phase a.c. systems

In an unbalanced system the resultant current will be the neutral current (Fig. 7.5).

The resultant of I_{Br} and I_{Bk} is I_{BrBk}, and the resultant of this (I_{BrBk}) and I_{Gr} is I_{BrBkGr} which is of course the neutral current I_N.

Example

The currents measured in a factory's three-phase supply were as follows: Brown phase, 70 A; Black phase, 50 A; Grey phase, 40 A. Determine, using a phasor diagram, the magnitude of the neutral current.

By measurement (Fig. 7.6):

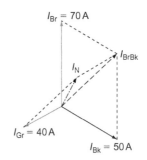

Figure 7.6

$$I_N = 28\,\text{A}$$

Power in three-phase systems (balanced only)

Power in a star-connected system

Since we are considering balanced systems, the total power is three times the power in one phase, and as

$$PF = \frac{W}{VA}$$

then

$$W = VA \times PF$$
$$\therefore W = V_p \times I_p \times PF$$

But

$$V_p = \frac{V_L}{\sqrt{3}} \quad \text{and} \quad I_p = I_L$$

$$\therefore W = \frac{V_L}{\sqrt{3}} \times I_L \times PF$$

$$\therefore \text{Total power } P = 3 \times \frac{V_L}{\sqrt{3}} \times I_L \times PF$$

$$= \sqrt{3}V_L \times I_L \times PF$$

Delta-connected system

Again W = VA × PF

$$\therefore W = V_p \times I_p \times PF$$

But

$$V_p = V_L \quad \text{and} \quad I_p = \frac{I_L}{\sqrt{3}}$$

$$\therefore W = V_L \times \frac{I_L}{\sqrt{3}} \times PF$$

$$\therefore \text{Total power } P = 3 \times V_L \times \frac{I_L}{\sqrt{3}} \times PF$$

$$P = \sqrt{3}V_L \times I_L \times PF$$

which is the same as for a star connection.

Hence for either star or delta connections the total power in watts is given by

$$P \text{ (watts)} = \sqrt{3}V_L \times I_L \times PF$$

Since

$$PF = \frac{W}{VA}$$

then

$$VA = \frac{W}{PF}$$

$$\therefore \text{Three-phase VA} = \sqrt{3}V_L \times I_L$$

$$\therefore \text{Line current} = \frac{VA}{\sqrt{3}V_L}$$

Example

A 15 kW, 400 V balanced three-phase delta-connected load has a power factor (PF) of 0.8 lagging. Calculate the line and phase currents.

From Fig. 7.7:

Figure 7.7

$$P = \sqrt{3}V_L \times I_L \times PF$$

$$\therefore I_L = \frac{P}{\sqrt{3}V_L \times PF}$$

$$= \frac{15\,000}{\sqrt{3} \times 400 \times 0.8}$$

$$I_L = 27 \text{ A}$$

For delta connections:

$$I_L = \sqrt{3}I_p$$

$$\therefore I_p = \frac{I_L}{\sqrt{3}}$$

$$= \frac{27}{\sqrt{3}}$$

$$I_p = 15.6 \text{ A}$$

Example

Three identical loads each having a resistance of 10 Ω and an inductive reactance of 20 Ω are connected first in star and then in delta across a 400 V, 50 Hz three-phase supply. Calculate the line and phase currents in each case.

$$Z \text{ for each load} = \sqrt{R^2 + X_L^2}$$

$$= \sqrt{10^2 + 20^2}$$

$$= \sqrt{500}$$

$$= 22.36 \, \Omega$$

For star connection (Fig. 7.8):

Figure 7.8

$$V_p = \frac{V_L}{\sqrt{3}} = 230\,V$$

$$I_L = I_p = \frac{V_L}{Z}$$

$$= \frac{230}{22.36}$$

$$I_L = I_p = 10.28\,A$$

For delta connection (Fig. 7.9):

Figure 7.9

$$V_L = V_p = 400\,V$$

$$\therefore I_p = \frac{V_p}{22.36}$$

$$= \frac{400}{22.36}$$

$$I_p = 17.9\,A$$

and

$$I_L = \sqrt{3}I_p$$

$$= \sqrt{3} \times 17.9$$

$$I_L = 31\,A$$

Measurement of power in three-phase systems

Single-phase circuit

Figure 7.10 shows the instruments required to determine the powers in watts and volt amperes, and the PF of a single-phase circuit.

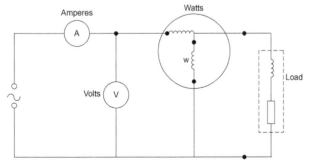

Figure 7.10 Ammeter, voltmeter and wattmeter connections.

The following values were recorded from a circuit similar to that in Fig. 7.10:

> Ammeter – 8 A
> Voltmeter – 230 V
> Wattmeter – 1.152 kW

Calculate the kVA and the PF of the load.

$$kVA = \frac{VA}{1000}$$

$$= \frac{8 \times 230}{1000}$$

$$= 1.84\,kVA$$

$$PF = \frac{kW}{kVA}$$

$$= \frac{1.152}{1.84}$$

$$= 0.626$$

Three-phase (balanced) four-wire circuit

In this case it is necessary to measure the power only in one phase. The total power will be three times this value. Also, the PF for one phase is the overall PF.

It is required to measure the PF of a three-phase star-connected balanced inductive load. Show how the necessary instruments would be arranged, a single voltmeter being used to measure the voltage across each phase.

If the readings obtained were 20 A, 230 V and 3680 W, calculate the total power in kilowatts and the PF (Fig. 7.11).

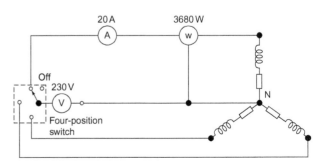

Figure 7.11

$$PF = \frac{W}{VA}$$

$$= \frac{3680}{230 \times 20} = 0.8 \text{ lagging}$$

Total power $= 3 \times 3680$

$$= 11\,040$$

$$= 11.04 \text{ kW}$$

Check:

$$V_L = V_p \sqrt{3}$$

$$V_L = 230 \times \sqrt{3}$$

$$P = \sqrt{3} V_L \times I_L \times PF$$

$$= \sqrt{3} \times \sqrt{3} \times 230 \times 20 \times 0.8$$

$$= 11.04 \text{ kW}$$

Self-Assessment Questions

1. What is meant by: three-phase generation and a four-wire system?

2. Explain with the aid of sketches the reason for the use of a neutral conductor.

3. The line current of a 400 V star-connected load is 10 A. Calculate the values of phase current and phase voltage.

4. A voltmeter, ammeter and wattmeter are arranged to measure the power in a single-phase circuit. Show how these instruments would be connected and calculate the circuit PF if the readings were 16 A, 230 V and 3600 W.

5. A three-phase star-connected load is supplied from the delta-connected secondary of a transformer. If the transformer line voltage is 190.5 V and the load phase current is 10 A, calculate the transformer phase current and the load phase voltage.

6. A 20 kW, 400 V three-phase star-connected load takes a line current of 34.8 A. Calculate the PF of the load.

7. A wattmeter, voltmeter and ammeter are arranged in a three-phase four-wire balanced system. Show how they would be connected, ready to determine the total circuit power in watts and the PF. Readings obtained were 2000 W, 230 V and 10 A. Calculate the circuit power in kW and kVA.

8. A small industrial unit has the following loadings measured at the intake position:
   Brown – 88 A
   Black – 72 A
   Grey – 98 A
   By means of a scaled phasor diagram determine the value of the neutral current.

7

CHAPTER 8

Motors and Generators

Motors play an important part in the modern domestic and industrial environment, and should therefore be of interest to the electrician.

Motors may be divided into two distinct kinds: those using direct current (d.c.) and those using alternating current (a.c.). Each of these categories is further divided into different types.

Direct-current motors

Simple single-loop motor

The effect of the force on a conductor in a magnetic field may be used to cause the rotation of a motor armature. Figure 8.1 illustrates a simple single-loop motor.

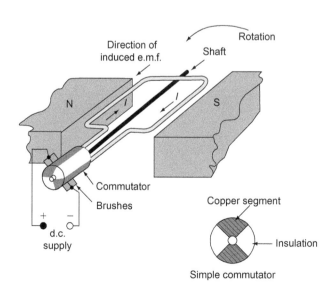

Figure 8.1 Single loop motor.

A single loop of conductor arranged as shown in this figure has its ends connected to a simple commutator, which comprises two copper segments insulated from each other. The commutator and loop are fixed to a central shaft which enables the whole assembly to be freely rotated. Two fixed carbon brushes bear on the surface of the commutator, enabling a supply to be connected to the loop.

Figure 8.2 shows a cross-section through the loop. The direction of movement may be determined using Fleming's left-hand rule.

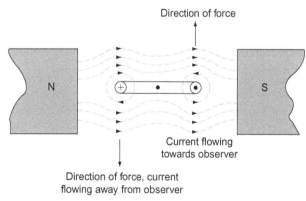

Figure 8.2 Cross section through single loop motor.

In practice, the d.c. motor comprises an armature of many loops revolving between electromagnetic poles. Both the armature and the field are supplied from the same source. The commutator has, of course, many segments to which the ends of the armature coils are connected, and the armature core is laminated to reduce eddy currents.

Electrical Installation Work. 978-1-138-84927-3 © Brian Scaddan. Published by Taylor & Francis.

Back e.m.f.: symbol, E – unit, volt (V)

It is interesting to note that as the armature revolves, its coils cut across the field flux, but we know that if a conductor cuts across lines of force, an electromotive force (e.m.f.) is induced in that conductor. This is the principle of the generator, as dealt with in this book. So, applying Fleming's right-hand rule to Fig. 8.1, we see that the induced e.m.f. is opposing the supply. This induced e.m.f. is called the 'back e.m.f.'. If the back e.m.f. were of the same magnitude as the supply voltage, no current would flow and the motor would not work. As current must flow in the armature to produce rotation, and as the armature circuit has resistance, then there must be a voltage drop in the armature circuit. This voltage drop is the product of the armature current (I_a) and the armature circuit resistance (R_a):

Armature voltage drop = $I_a \times R_a$

It is this voltage drop that is the difference between the supply voltage and the back e.m.f. Hence

$$E = V - (I_a \times R_a)$$

We also know (from Chapter 4) that induced e.m.f. is dependent on the flux density (B), the speed of cutting the flux (n) and the length of the conductor (l):

$$E = B \times l \times v$$

But

$$B = \frac{\Phi}{a} \left(\frac{\text{flux}}{\text{area}} \right)$$

$$\therefore E = \frac{\Phi}{a} \times l \times v$$

Both l and a for a given conductor will be constant and n is replaced by n (revs/s) as this represents angular or rotational speed.

$$\therefore E \propto n\Phi$$

So, if the speed is changed from n_1 to n_2 and the flux from Φ_1 to Φ_2, then the e.m.f. will change from E_1 to E_2.

$$\therefore \frac{E_1}{E_2} = \frac{n_1 \Phi_1}{n_2 \Phi_2}$$

Torque: symbol, T – unit, newton metre (Nm)

Work = force × distance

$\therefore$ Turning work or torque = force × radius

$$(F) \qquad (r)$$

We also know that

$$\text{Force} = B \times l \times I$$

$$\therefore \text{Torque } T = B \times l \times I_a \times r$$

$$\therefore T = \frac{\Phi}{a} \times l \times I_a \times r$$

Once again, for a given machine, a, l and r will all be constant.

$$\therefore T \propto \Phi \times I_a$$

Also, mechanical output power in watts is given by

$$P = 2\pi nT$$

where P is the output power in watts, n is the speed in revs/s and T is the torque in newton metres.

If we multiply $E = (V - I_a R_a)$ by I_a, we get

$$EI_a = VI_a - I_a^2 R_a$$

VI_a is the power supplied to the armature and $I_a^2 R_a$ is the power loss in the armature; therefore EI_a must be the armature power output. Hence

$$EI_a = P$$

$$\therefore EI_a = 2\pi nT$$

$$\therefore T = \frac{EI_a}{2\pi n} \, Nm$$

It is clear then that torque is directly proportional to armature current and inversely proportional to speed, that is if the mechanical load is lessened, the torque required is less, the armature current decreases and the motor speeds up.

Example

A 300 V d.c. motor runs at 15 revs/s and takes an armature current of 30 A. If the armature resistance is 0.5 Ω, calculate firstly the back e.m.f. and secondly the torque.

$$E = V - I_a R_a$$
$$= 300 - (30 \times 0.5)$$
$$= 300 - 15$$
$$E = 285 \, V$$

$$T = \frac{EI_a}{2\pi n}$$
$$= \frac{285 \times 30}{2\pi \times 15}$$
$$T = 90.72 \, Nm$$

Series motor

The series type of d.c. motor has its field windings and armature connected in series across the supply. It will be seen from this figure that the armature current I_a also supplies the field (Fig. 8.3). Therefore, when I_a is large (on starting, for example), the magnetic field will be strong, and the torque will be high. As the machine accelerates, the torque, armature current and field strength will all decrease. This type of motor should never be coupled to its load by means of a belt, because if the belt breaks the required torque from the armature will be removed, the armature and

Figure 8.3 Series motor.

field current will fall, reducing the magnetic field, and the motor will increase in speed until it disintegrates. Figure 8.4 shows the graphs of speed and torque to a base of load current.

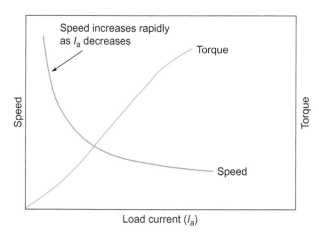

Figure 8.4 Load characteristics of a series motor.

Speed control

The most effective way of controlling the speed of a d.c. motor is to vary the strength of the magnetic field. On a series machine this may be achieved by diverting some of the current through a variable resistor (Fig. 8.5).

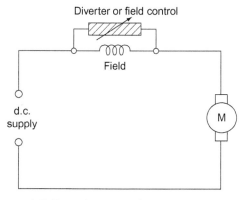

Figure 8.5 Speed control of a series motor.

Starting

A series motor is started by placing a variable resistor in series with the armature circuit (Fig. 8.6).

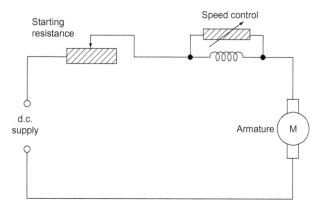

Figure 8.6 Starting a series motor.

Applications

The series motor is best used where heavy masses need to be accelerated from rest, such as in cranes and lifts.

Example

A 200 V series motor has a field winding resistance of $0.1\,\Omega$ and an armature resistance of $0.3\,\Omega$. If the current taken at 5 revs/s is 30 A, calculate the torque on the armature.

The total armature circuit resistance is $R_a + R_f$ (Fig. 8.7) as the two resistances are in series.

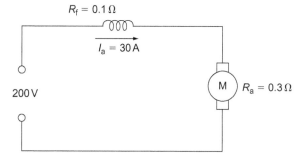

Figure 8.7

$$E = V - I_a R_a$$
$$= 200 - 30(0.1 + 0.3)$$
$$= 200 - 30 \times 0.4$$
$$= 200 - 12$$
$$= 188\,V$$

$$\text{Torque } T = \frac{EI_a}{2\pi n}$$
$$= \frac{188 \times 30}{2\pi \times 5}$$
$$= 180\,\text{Nm}$$

Shunt motor

In the case of a shunt motor, the motor winding is in parallel with the armature (Fig. 8.8). It will be seen here that the supply current $I = I_a + I_f$. Unlike the series motor, if the load is removed from the motor, only the armature current will decrease, the field remaining at the same strength. The motor will therefore not continue to speed up to destruction. Figure 8.9 shows the graphs of speed and torque to a base of load current.

Figure 8.8 Shunt-wound motor.

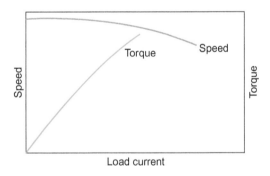

Figure 8.9 Load characteristics of a shunt motor.

Speed control

As in a series motor, speed control is best achieved by controlling the field strength, and in shunt motors a variable resistance is placed in series with the shunt winding (Fig. 8.10).

Starting

Modern methods use electronics; however, starting large d.c. shunt-wound motors can be carried out using a d.c. face-plate starter (Fig. 8.11).

The face-plate starter comprises the following items:

Figure 8.10 Speed control of a shunt motor.

Figure 8.11 The d.c. face-plate starter.

(a) a series of resistances connected to brass studs;
(b) a spring-loaded handle which makes contact with two brass strips and also the brass studs;
(c) a no-volt release;
(d) an overload release.

When the handle is located on the first stud, the field is supplied via the overload release, the top brass strip via the handle and the no-volt release (note that the field is continuously supplied in this way). The armature is supplied via the resistances.

As the handle is moved round, the resistance in the armature circuit is gradually cut out. On the final stud the handle is held in place by the no-volt release electromagnet. Should a failure in supply occur, the no-volt release will de-energize and the handle will spring back to the 'off' position. If a serious overload occurs, the overload release will energize sufficiently to attract its soft iron armature which will short out the no-volt release coil, and the handle will return to the 'off' position.

Note

The handle should be moved slowly from stud to stud.

Applications

As the speed of a shunt motor is almost constant over a wide range of loads, it is most suitable for small machine tools.

Example

A 400 V shunt-wound motor has a field winding resistance of 200 Ω and an armature resistance of 0.5 Ω. If the current taken from the supply is 22 A, calculate the back e.m.f.

From Fig. 8.12:

Figure 8.12

$$I = I_a + I_f$$
$$\therefore I_a = I - I_f$$
$$I_f = \frac{V}{R_f} = \frac{400}{200} = 2\,A$$
$$\therefore I_a = 22 - 2 = 20\,A$$
$$E = V - I_a R_a$$
$$= 400 - (20 \times 0.5)$$
$$= 400 - 10$$
$$= 390\,V$$

Compound motor

A compound motor is a combination of a series and a shunt-wound motor (Fig. 8.13a and b).

Figure 8.14 shows the speed–torque characteristics for the compound motor.

(a)   (b)

Figure 8.13 Compound-wound motor: (a) long shunt; (b) short shunt.

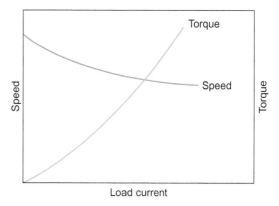

Figure 8.14 Speed and torque characteristics of a compound motor.

Speed control

Speed is usually controlled by variable resistors in the shunt field and armature circuit (Fig. 8.15).

Figure 8.15 Speed control of a typical compound motor (long shunt).

The series winding may be arranged such that it aids the shunt field (**cumulative** compound) or opposes it (**differential** compound).

Cumulatively compounded motors are similar in characteristics to series motors, while differentially compounded motors are similar to shunt motors (Fig. 8.16).

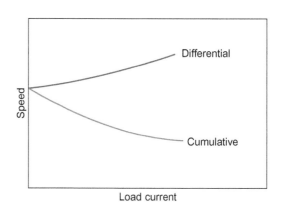

Figure 8.16 Load speed curves for compound motor.

Starting

The d.c. face-plate starter is suitable for the compound motor.

Applications

These motors can be used for applications where a wide speed range is required. However, the differentially compounded type is rarely used as it tends to be unstable. Cumulatively compounded types are suitable for heavy machine tools.

Reversing d.c. motors

The d.c. motors may be reversed in direction by altering the polarity of either the field or the armature. This is done by reversing the connection to the armature **or** the field winding.

The d.c. generators

If by some means a d.c. motor is supplied with motive power it will act as a generator. Connection of the field windings is the same as for motors; that is, series, shunt or compound.

E.m.f. generated: symbol E – unit volt (V)

When the armature is rotated in the field, an e.m.f. is induced or generated in the armature windings. When an external load is connected, current (I_a) will flow from the armature; this will cause a voltage drop of I_aR_a where R_a is the resistance of the armature circuit. Hence the voltage available at the load is less than the generated e.m.f.

$$\therefore E = V + I_aR_a$$

Example

A 240 V shunt-wound generator has a field resistance of 120 Ω and an armature resistance of 0.4 Ω. Calculate the generated e.m.f. when it is delivering 20 A to the load.

As the generator supplies its own field, then from Fig. 8.17:

$$I_a = I + I_f$$
$$I_f = \frac{V}{R_f} = \frac{240}{120} = 2\,A$$
$$\therefore I_a = 20 + 2 = 22\,A$$
$$E = V + I_aR_a$$
$$= 240 + (22 \times 0.4)$$
$$= 240 + 8.8$$
$$= 248.8\,V$$

Figure 8.17 Pole arrangement for a three phase motor.

Separately excited generator

The separately excited type of generator has its field supplied from a separate source (Fig. 8.18).

Figure 8.18 Separately excited generator.

The a.c. motors

There are many different types of a.c. motor operating from either three-phase or single-phase a.c. supplies. To understand the starting problems of the single-phase types, it is best to consider three-phase motors first.

Three-phase motors

A three-phase motor depends on the rotation of a magnetic field for its movement. Figure 8.19 shows how this rotation is achieved.

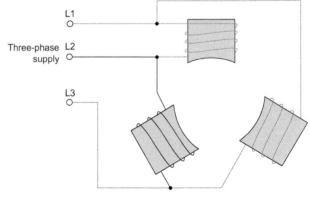

Figure 8.19 Magnetic field rotation.

If three iron-cored coils or poles are arranged at 120° apart and connected as shown to an alternating three-phase supply, then each pole will become fully energized at a different time in relation to the others. If the poles were replaced with light bulbs, it would appear as if the light was travelling around in a circular fashion from one bulb to another.

The iron core of each coil becomes magnetized as the coil is energized, and the arrangement gives the effect of a magnetic field rotating around the coils.

The speed of rotation of the magnetic field is called the **synchronous speed** and is dependent on the frequency of the supply and the number of **pairs** of poles. Hence

$$f = n_s p$$

where f is the supply frequency in hertz, n_s is the synchronous speed in revs/s and p is the number of pairs of poles.

Example

Calculate the synchronous speed of a four-pole machine if the supply frequency is 50 Hz.

$$f = n_s p$$
$$\therefore n_s = \frac{f}{p}$$
$$= \frac{50}{2}$$
$$= 25 \text{ revs/s} \quad \text{or} \quad 1500 \text{ revs/min}$$

Synchronous motor

If we take a simple magnetic compass and place it in the centre of the arrangements shown in Fig. 8.19, the compass needle will rotate in the same direction as the magnetic field, because the magnetized compass needle is attracted to the field and therefore follows it. A three-phase synchronous motor is arranged in the same way as a three-phase generator (Fig. 8.20).

This type of motor comprises the following:

(a) a stator, which supports the magnetic field poles
(b) the rotor, which is basically an electromagnet supplied from a d.c. source via slip rings.

The rotor will follow the rotating magnetic field at synchronous speed.

This type of motor is not self-starting and has to be brought up to or near to synchronous speed by some means, after which it will continue to rotate of its own accord. This bringing up to speed is usually achieved by providing the rotor with some of the characteristics of an induction motor rotor.

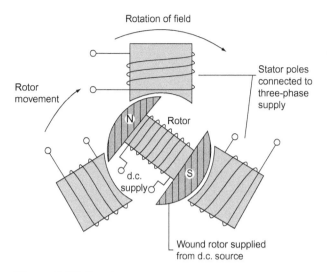

Figure 8.20 Synchronous motor.

Synchronous-induction motor

The synchronous-induction type of motor is essentially an induction motor with a wound rotor. It starts as an induction motor and when its speed has almost reached synchronous speed the d.c. supply is switched on and the motor will then continue to function as a synchronous motor.

This type of motor has various applications. For example, if the d.c. supply to the rotor is increased (when it is said to be 'overexcited'), the motor can be made to run at a leading power factor. This effect may be used to correct the overall power factor of an installation.

As it is a constant-speed machine, it is often used in motor–generator sets, large industrial fans and pumps. A great advantage of the synchronous-induction type is its ability to sustain heavy mechanical overloads. Such an overload pulls the motor out of synchronism, but it continues to run as an induction motor until the overload is removed, at which time it pulls back into synchronism again.

Three-phase induction motor

Squirrel-cage type

The squirrel-cage type of induction motor comprises a wound stator and a laminated iron rotor with copper or aluminium bars embedded in it, in the form of a cage (Fig. 8.21).

As the rotating magnetic field sweeps across the rotor, an e.m.f. is induced in the cage bars and hence a current flows. This current produces a magnetic field around the conductor and the magnetic reaction between this field and the main field causes the rotor

Copper or aluminium bars

Copper or aluminium end rings

Figure 8.21 Cage assembly for cage rotor.

to move. Since this movement depends on the cage bars being cut by the main field flux, it is clear that the rotor cannot rotate at synchronous speed. The speed of this type of motor is constant.

Wound-rotor type (slip-ring motor)

In the wound-rotor type of motor the cage is replaced by a three-phase winding which is connected via slip rings to a starter. The starter enables the rotor currents to be controlled which in turn controls, to a degree, the speed and torque. When the machine has reached the speed required the rotor windings are short-circuited and the brush gear, which is no longer required, is lifted clear of the slip rings. This type of motor is capable of taking extremely high rotor currents on starting and cables must be capable of carrying such currents.

Slip

As was previously mentioned, the rotor of an induction motor cannot travel at synchronous speed, as there would be no flux cutting and the machine would not work.

The rotor is, then, said to 'slip' in speed behind the synchronous speed. Slip (S) is usually expressed as a percentage and is given by

$$\text{Slip (\%)} = \frac{(n_s - n_r)}{n_s} \times 100$$

where n_s is the synchronous speed and n_r is the rotor speed.

Example

A six-pole cage induction motor runs at 4% slip. Calculate the motor speed if the supply frequency is 50 Hz.

$$S \text{ (\%)} = \frac{(n_s - n_r)}{n_s} \times 100$$

Synchronous speed $n_s = \dfrac{f}{p}$

$$= \frac{50}{3}$$

$$= 16.666 \text{ revs/s}$$

$$\therefore 4 = \frac{(16.66 - n_r)}{16.66} \times 100$$

$$\therefore \frac{4 \times 16.66}{100} = (16.66 - n_r)$$

$$\therefore n_r = 16.66 - \frac{(4 \times 16.66)}{100}$$

$$= 16.66 - 0.66$$

$$= 16 \text{ revs/s}$$

Example

An eight-pole induction motor runs at 12 revs/s and is supplied from a 50 Hz supply. Calculate the percentage slip.

$$n_s = \frac{f}{p}$$

$$= \frac{50}{4} = 12.5 \text{ revs/s}$$

$$S \text{ (\%)} = \frac{(n_s - n_r)}{n_s} \times 100$$

$$= \frac{(12.5 - 12)}{12.5} \times 100$$

$$= \frac{0.5 \times 100}{12.5}$$

$$= \frac{50}{12.5}$$

$$= 4\%$$

Frequency of rotor currents

As the rotating field is an alternating one, the currents induced in the rotor cage bars are also alternating. These are, however, not the same frequency as the supply. The frequency of the rotor currents f_s is given by

$$f_s = \text{slip} \times \text{supply frequency}$$

$$\therefore f_s = S \times f$$

Note

S here is expressed as a per unit value; that is for 4% slip,

$$S = \frac{4}{100} = 0.04$$

Example

An eight-pole squirrel-cage induction motor has a synchronous speed of 12.5 revs/s and a slip of 2%. Calculate the frequency of the rotor currents.

$$f = n_s \times p$$
$$= 12.5 \times 4$$
$$= 50\,Hz$$
$$f_s = S \times f$$
$$= \frac{2}{100} \times 50$$
$$= 1\,Hz$$

Note

Three-phase motors may be reversed by changing over any **two** phases.

Single-phase induction motors

With a three-phase motor the field is displaced by 120°. In the case of a single-phase supply there is no phase displacement and hence the rotor has equal and opposing forces acting on it and there will be no movement. The motor is therefore not self-starting. However, if the rotor is initially spun mechanically it will continue to rotate in the direction in which it was turned. Of course this method of starting is out of the question with all but the very smallest motors and is therefore confined to such items as electric clocks.

The creation of an artificial phase displacement is another and more popular method of starting.

Shaded-pole induction motor

The shaded-pole type of motor has a stator with salient (projecting) poles and in each pole a short-circuited turn of copper face is inserted (Fig. 8.22).

Figure 8.22 Shaded-pole arrangement.

The alternating flux in the pole face induces a current in the shading coil which in turn produces an opposing flux. This opposition causes a slight phase

displacement of the fluxes in the two parts of each pole which is enough to start the rotor turning.

As the phase displacement is very small the motor has a very small starting torque, thus limiting its use to very light loads.

Capacitor-start induction motor

With the capacitor-start induction motor the stator has a secondary winding, in series with which is a capacitor. This gives the effect of 90° phase difference and the motor will start. A second or two after starting, a centrifugal switch cuts out the secondary winding (Fig. 8.23).

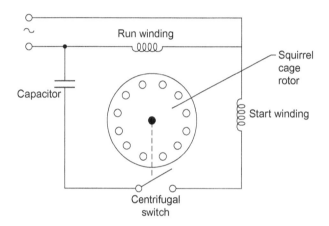

Figure 8.23 Capacitor-start motor.

This type of motor may be reversed in direction by reversing the connections to the start winding.

Reactance or induction-start induction motor

A phase displacement can be achieved by connecting an inductor in series with the start winding (Fig. 8.24). The centrifugal switch is as for the capacitor-start type.

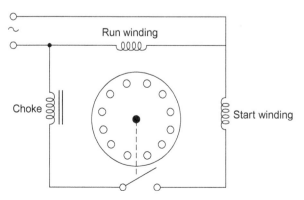

Figure 8.24 Reactance-start motor.

Larger motors of this type take heavy starting currents, and series resistances are used to limit this. Reversal of rotation is as for the capacitor-start type.

Resistance-start induction motor

In the case of a resistance-start induction motor a resistance replaces the choke or capacitor in the start winding to give a phase displacement (Fig. 8.25).

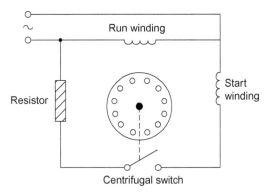

Figure 8.25 Resistance-start motor.

Capacitor-start capacitor-run induction motor

The most efficient of the range of single-phase induction motors is the capacitor-start capacitor-run type. The main feature is that the starting winding is not switched out but is continuously energized, the only change between starting and running being the value of capacitance. This change is achieved by using two capacitors and switching one out with the centrifugal switch (Fig. 8.26).

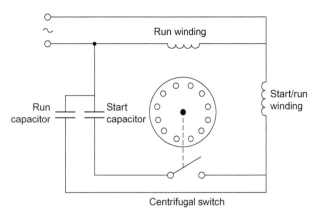

Figure 8.26 Capacitor-start capacitor-run motor.

Repulsion-start motor

Repulsion-start motors are of the wound-rotor type, the windings being terminated at a commutator, the brush gear of which is shorted out and arranged about 20° off centre. A transformer action takes place between the stator and rotor windings (mutual

inductance) and as both windings will have the same polarity they repel or repulse each other. Speed control is affected by slight movement of the brushes around the commutator (Fig. 8.27).

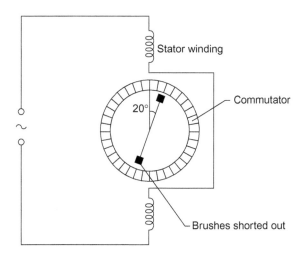

Figure 8.27 Repulsion motor.

A variant of this type starts as a repulsion motor; centrifugal gear then shorts out the commutator and lifts the brush gear clear, the motor then continuing to run as an induction motor.

The universal or series motor

The **universal** or **series** motor is simply a d.c.-type armature with commutator and an a.c. field. It is connected as for a d.c. series motor (Fig. 8.28).

Figure 8.28 Single-phase series motor.

This motor will operate on an alternating current because the polarity of the a.c. supply changes on **both** field and armature; the motor will therefore rotate in one direction. Reversing is achieved by reversal of either field or armature connections.

Starters

Direct-on-line starter

Figure 8.29 illustrates a typical three-phase direct-on-line (DOL) starter. When the start button is

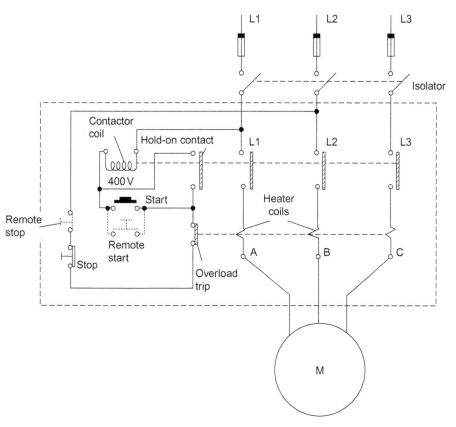

Figure 8.29 Three-phase direct-on-line starter.

pressed, the 400 V contactor coil is energized and the main and auxiliary contacts close and the motor will start. The auxiliary contact in parallel with the start button holds the coil on.

Both single-phase and three-phase types use the same control circuit, as illustrated in Figs 8.30 and 8.31. For Fig. 8.31 it is important to note that if a 240 V coil were to be used instead of a 400 V type, the coil connections would require a neutral conductor in the starter.

Overload or over current protection is provided by either thermal or magnetic trips.

Thermal overload protection relies on the heating effect of the load current to heat the thermal coils which in turn cause movement of a bimetallic strip. This trips out a spring-loaded contact in the control circuit. The speed at which the tripping takes place is adjusted to allow for normal starting currents, which may be four or five times as large as running currents.

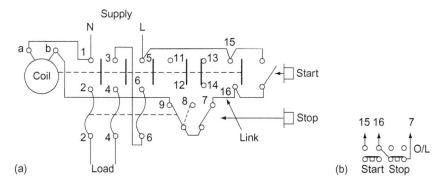

Figure 8.30 (a) Direct-on-line starter, single phase. (b) Connections for remote push-button (start/stop) control; omit link and connect as shown.

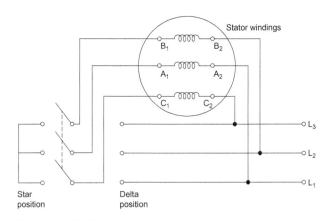

Figure 8.31 (a) Direct-on-line starter, three phase. (b) Schematic diagram. Control circuit supply: for line to line, connect as shown; for line to neutral, omit connection D and connect neutral to terminal a; for separate supply, omit D and E, and connect separate coil supply to terminals a and 15.

Magnetic protection uses the principle of the solenoid to operate the tripping mechanism. The time lag in this case is achieved by the use of an oil or air dashpot which slows down the action of the solenoid plunger (Fig. 8.32).

Figure 8.32 Oil dashpot damping.

Another form of thermal protection is given by the use of a **thermistor**, which is a temperature-sensitive semiconductor. It is embedded in the stator winding and activates a control circuit if the winding temperature becomes excessive.

Star–delta starter

If a motor's windings are connected in the star configuration, any two phases will be in series across the supply and hence the line current will be smaller (by 57.7%) than if the windings were connected in the delta arrangement. Hence larger type motors with heavy starting currents are first connected in star, and then, when the starting currents fall, in delta. This of course means that all six of the ends of the windings must be brought to terminations outside the casing (Fig. 8.33).

Figure 8.33 Basic star–delta starter.

The automatic version of this starter incorporates a timing relay which automatically changes the connections from star to delta.

Figures 8.34 and 8.35 show the wiring and schematic diagrams for a star–delta starter. This is clearly a more complicated system than the DOL type.

Nevertheless, reference to the schematic Fig. 8.35 will indicate how the system functions:

1. When the start is pushed, supply is given to the star contactor Ⓐ and electronic timer ET, from L1

(a)

(b) Start Stop

(c) Warning signal Neutral

(d)

Figure 8.34 (a) Automatic star–delta starter. (b) Connections for remote push-button (start/stop) control; omit link and connect as shown. (c) Connections for trip warning. (d) Motor windings: connect to appropriate terminals on starter.

8

Figure 8.35 Schematic diagram. Control circuit supply; for line to line, connect as shown; for line to neutral, omit connection D and connect neutral to terminal b; for separate supply, omit D and E, and connect separate coil supply to terminals b and 9. Connections for remote pilot switch control: remove connection 14 to 15 on delta contactor; connect between 14 and 16 on M contactor to terminal 14 on delta contactor; connect pilot switch in place of connection E.

to L3, and hence all contacts marked λ will operate. Supply to Ⓐ and ET is maintained, after the start is released, via λ 11 and 12, time contacts OR-Y, and λ 15 and 16. The main contactor Ⓜ is also energized via λ 11 and 12 and M15, and thereafter maintained via its own contacts M15 and 16.

2. The motor has of course started. After a predetermined time delay ET operates and its

contacts OR-Y change to OR-L. This cuts off supply to Ⓐ and ET. All λ contacts return to normal, and ET resets OR-L to OR-Y.

3. Supply is now given to the delta contactor Ⓐ I via M15 and 16, OR-Y, and λ 13 and 14. Delta contacts Δ 13 and 14 open and prevent further energization of the star contactor.

The reader will notice that the line and load terminal markings in Fig. 8.34 show letters in brackets; these are the continental equivalents.

Auto-transformer starting

A star-connected auto-transformer with tappings gives lower starting currents than the star–delta type (Fig. 8.36).

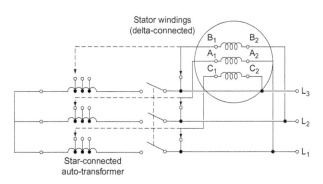

Figure 8.36 Basic auto-transformer starter.

Rotor-resistance starting

Figure 8.37 shows the rotor-resistance-type starter for use with wound-rotor induction motors.

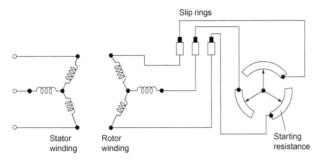

Figure 8.37 Rotor resistance starter.

Installing a motor

The correct handling, positioning, fixing and aligning of a motor are very important. Although it is a robust piece of machinery, great care should be taken, when transporting or positioning it, not to crack the casing or damage the feet.

Once the machine is in position it has to be fixed and aligned and this procedure will depend on the type of coupling used and the size of machine.

Very large machines are usually fixed to a concrete base or plinth. The concrete is a mixture of one part cement, two parts quartz sand and four parts gravel. Fixing bolts should be grouted in the base with a mixture of 'one to one' washed sand and cement.

Most motors are in fact mounted on iron-slide rails, the rails being fixed to the floor, wall or ceiling depending on the motor's use. In this way, the motor can be adjusted accurately for alignment, belt tensioning and direct coupling. Figure 8.38a–c indicates the correct methods for alignment.

> **Note**
>
> Always check that the insulation resistance of the motor is satisfactory before switching on the supply; dampness may have been picked up during storage.

Figure 8.38 (a) Incorrectly; (b) correctly aligned belt-driven machine; and (c) adjustment in the case of direct coupling.

Motor replacement

If a motor has been diagnosed as being faulty and is either repaired or replaced, always ensure that the reconnections are correct. A reversal of the connections to the windings of a three-phase star–delta motor could have serious implications for the motor as well as for the operation of the circuit protection.

Figures 8.39 and 8.40 show the terminal arrangements for a six-terminal motor and the correct winding connections.

Figure 8.39 Motor terminal block.

Figure 8.40 Motor windings.

If the connections to any winding are reversed, the magnetic fields will work against each other and a serious overload will occur, especially when the starter changes the motor to the delta configuration. For example, **never** allow A2 to be connected to B2 and A1 to C1.

This must not be confused with the reversal of any two **phases** of the supply when motor rotation needs to be reversed.

Motor enclosures

Once the most appropriate type of motor for a particular task has been selected, it is necessary to ensure that the motor enclosure is also suitable for its working environment. Various kinds of enclosure are listed in Table 8.1 and their applications are summarized in Tables 8.2 and 8.3.

Table 8.1 Applications of various types of motor enclosure

Type	Applications
Screen-protected	General purposes, engineering worktops, etc.
Drip-proof	Laundries, pump rooms, etc.
Pipe-ventilated	Flour mills, cement works, paper mills, etc.
Totally enclosed	Boiler houses, steelworks, outdoor winches, etc.
Flameproof	Gasworks, oil plants, chemical works, etc.

Table 8.2 Summary of the characteristics and applications of single-phase a.c. motors

Type	Main Characteristics	Applications
Universal (series)	Good starting torque, high PF	Small tools, drills, sanders, etc., vacuum cleaners
Repulsion	Good starting torque, low PF, and efficiency, speed control by brush shifting	Lifts, cranes, hoists, etc.
Repulsion-start	Good starting torque, low PF, and efficiency	Non-reversing load with heavy starting demands
Split-phase induction	Poor starting torque and PF	Used only with light starting conditions
Capacitor-start induction	Quite good starting torque and PF, quiet running	Refrigerators
Synchronous	Constant speed. Poor starting torque and PF	Clocks and timing devices

Table 8.3 Summary of the characteristics and applications of three-phase motors

Type	Main Characteristics	Applications
Small synchronous, no d.c. excitation	Constant speed, self-starting, light loads only, low PF	All drives needing synchronization
Synchronous with d.c. excitation	Constant speed, controllable PF, high efficiency with large outputs	Compressors, PF correction, ships' propulsion
Induction, squirrel-cage or wound-rotor	Starting performance good for small squirrel-cage but poor for large squirrel-cage motors good starting with all slip-ring motors	General service in engineering, pumps, machine tools, etc.

Screen-protected type

The most common enclosures in use are of the screen-protected type. The end covers are slotted and an internal fan draws cool air through the motor. Screen-protected enclosures can be used only in dust-free atmospheres.

Drip-proof type

The end plates on this type of motor are solid except for narrow slots on the underside. It can be used in damp and dust-free situations but is not waterproof.

Pipe-ventilated type

Cooling air is brought in via pipes from outside the building and circulated by an internal fan. This type is very suitable for extremely dusty environments.

Totally enclosed type

This type has no ventilation slots, its casing instead has ribs or fins to help cooling. The totally enclosed type is excellent for moist or dusty situations.

Totally enclosed flameproof type

This type is similar to the totally enclosed type but is more robustly built. It can withstand an internal explosion and prevent flames or sparks from reaching the outside of the casing.

Fault location and repairs to a.c. machines

Tables 8.4–8.8 indicate characteristics and fault diagnoses for motors.

Power factor of a.c. motors

Motors, being highly inductive pieces of equipment, have lagging power factors, some more so than others. In situations where a large number of

Table 8.4 Possible faults on polyphase induction motors

Symptoms	Possible Causes	Test and/or Rectification
Fuses or overcurrent trips operate at start	Premature operation of protective gear Overload Reversed phase of stator winding Short circuit or earth fault on stator circuit Short circuit or earth fault on rotor circuit	
Motor will not start	Faulty supply or control gear Overload or low starting torque Open circuit in one stator phase Reversed phase of stator winding Open circuit in rotor circuit	
Overheated bearing or noisy operation	Bearing or mechanical defects	
Periodical growl	Reversed stator coil or coils	
Humming of squirrel-cage motor	Loose joints on rotor conductors	
Fluctuating stator current	Open circuit in rotor circuit	
General overheating of case	Faulty ventilation, mechanical or electrical overload Rotor core not fully in stator tunnel Open circuit in one of two parallel stator circuits	Reassemble motor correctly
Overheating and overlabouring, two phases of star-connected stator or one phase of delta winding hotter than the rest Reduced speed	Single phasing owing to open-circuited supply line Open circuit in one phase of stator circuit Mechanical overload, low volts or low frequency Open circuit in rotor circuit	
Reduced speed of slip-ring motor	Rotor starter not fully operated Slip rings not short-circuited Voltage drop on cables to rotor starter	Overhaul protective gear to ensure correct operation Use slip-ring short-circuiting gear Fit rotor starter nearer motor, or use larger rotor circuit cables

Table 8.5 Possible faults on single-phase induction and capacitor types of motor

Symptoms	Possible Causes	Test and/or Rectification
Fuses or overcurrent trips operate at start	Premature operation of protective gear Overload Section of stator windings reversed Short circuit or earth fault on stator circuit Stator windings in parallel instead of series Short circuit or earth fault on slip-ring rotor circuit	

Table 8.5 (Continued)

Symptoms	Possible Causes	Test and/or Rectification
Motor will not start	Faulty supply or control circuit	
	Overload or low starting torque	
	Open circuit or reversed coils on stator winding	
	Open circuit on slip-ring rotor circuit	
	Centrifugal switch or relay sticking open	
Overheated bearing or noisy operation	Bearing or mechanical defects	
General overheating of case	Faulty ventilation, mechanical or electrical overload	
	Rotor core not fully in stator tunnel	Reassemble motor correctly
	Open circuit in one of two parallel stator circuits	
	Short circuit on auxiliary stator winding	Overhaul switch or relay and check operation
	Short circuit on centrifugal switch or relay	
	Centrifugal switch or relay sticking closed	
	Reversed section of stator windings	Overhaul switch or relay and check operation
	Prolonged or too frequent starting	
Reduced speed of all motors	Mechanical or electrical overload	
	Low volts or frequency	
	Open circuit in rotor	
Reduced speed of slip-ring motor	Rotor starter not fully operated	Overhaul protective gear to ensure correct operation
	Slip rings not short-circuited	Use slip-ring short-circuiting gear
	Voltage drop on cables to rotor starter	Fit rotor starter nearer motor, or use larger rotor circuit cables

Table 8.6 Possible faults on series a.c. and universal motors

Symptoms	Possible Causes	Test and/or Rectification
Fuses or overcurrent trips operate at start	Premature operation of protective gear	
	Overload	
	One field coil reversed	
	Short circuit or earth fault on field winding	
	Short circuit or earth fault on armature	
	Wrong brush position	
Motor will not start	Faulty supply or control circuit	
	Brushes not making good contact	
	Open circuit in field windings	
	Wrong brush position	
	Short circuit or earth fault on armature	
	Short circuit or earth fault on field windings	
	Reversed field coil	
Overheated bearing or noisy operation	Bearing or mechanical defects	
General overheating of the case	Faulty ventilation, mechanical or electrical overload	
	Short circuit, open circuit, or earth fault on armature	
	Short circuit or earth fault on field windings	
Reduced speed	Mechanical or electrical overload	
	Low voltage	
	Wrong brush position	
Increased speed	High voltage	
	Motor unloaded	It is inadvisable to run unloaded

(Continued)

Table 8.6 (Continued)

Symptoms	Possible Causes	Test and/or Rectification
Sparking at brushes	Faulty brushes or commutator	
	Mechanical or electrical overload	
	Wrong brush position	
	Incorrect brush spacing	
	Open circuit, short circuit or earth fault in armature	
	Reversed armature coil	

Table 8.7 Possible faults on repulsion-type motors

Symptoms	Possible Causes	Test and/or Rectification
Fuses or overcurrent trips operate at start	Premature operation of protective gear	
	Overload	
	Selection of stator windings reversed	
	Short circuit or earth fault on stator circuit	
	Short circuit or earth fault on stator circuit	
	Stator windings in parallel instead of series	
	Short circuit or earth fault on armature	
	Wrong brush position	
	Commutator short-circuiting gear sticking in running position	Overhaul centrifugal gear and check operation
Motor will not start	Faulty supply or control circuit	
	Overload or low starting torque	
	Short circuit or earth fault on stator circuit	
	Open circuit or reversed coils on stator windings	
	Brushes not making good contact	
	Short circuit or earth fault on armature	
	Commutator short circuiting gear sticking in running position	Overhaul centrifugal gear and check operation
	Wrong brush position	
Overheated bearing or noisy operation	Bearing or mechanical defects	
General overheating of case	Faulty ventilation, mechanical or electrical overload	
	Rotor core not fully in stator tunnel	Reassemble motor correctly
	Open circuit in one of two parallel stator circuits	
	Wrong brush position	Overhaul short-circuiting contacts
	Burnt contacts on commutator short-circuiting gear	
Overheating of repulsion-induction motor	No load	Motor usually runs hotter unloaded than on full load
Reduced speed	Low volts or frequency	
Wrong brush position		
Faulty commutator short-circuiting gear	Overhaul short-circuiting gear, check operation	
	Overload	
	Short circuit or earth fault on armature of plain repulsion motor	
	Open circuit in squirrel-cage of repulsion-induction motor	
Hunting of repulsion-start induction motor	Faulty commutator short-circuiting gear or brushes	Overhaul short-circuiting gear and brushes
Sparking at brushes	Faulty brushes or commutator	
	Mechanical or electrical overload	
	Wrong brush position	
	Incorrect brush spacing	
	Reversed armature coil	
	Short circuit, open circuit or earth fault on armature	

Table 8.8 Possible faults on synchronous types of motor

Symptoms	Possible Causes	Test and/or Rectification
Fuses or overcurrent trips operate at start	Premature operation of protective gear	
	Overload	
	Short circuit or earth fault on armature	
Motor will not start	Faulty supply or control gear	
	Low starting voltage	Adjust tappings on transformer
	Overload	
	Open circuit in one armature phase	
Motor fails to synchronize	External field resistance too high	Adjust field-regulating resistor
	Open circuit in field circuit	
	No excitation	Faulty exciter
Overheated bearing or noisy operation	Bearing or mechanical defects	
Vibration	Faulty supply	
	Open circuit in one armature phase	
General overheating	Faulty ventilation	
	Overload	
	High voltage	
	Short circuit, open circuit or earth fault on armature	
	Incorrect field strength	Adjust field-regulating resistor
	Unequal pole strength	Test field coils
	Unequal air gap	
Motor runs fast	High frequency	
Motor runs slow	Low frequency	
Motor pulls out of synchronism	Overload	
	External field resistance too high	Adjust field-regulating resistor
	Open circuit in field circuit	
	No excitation	Faulty exciter

machines are used, as in industrial premises, it is clear that some action should be taken to correct this lagging power factor. Where motors are used intermittently it is perhaps best to correct the power factor of each motor rather than the overall power factor of the installation, and capacitors connected across the terminals of each machine are used.

Figure 8.41

Example

A 230 V, 50 Hz single-phase induction motor takes a current of 13 A at a PF of 0.35 lagging. Calculate the value of capacitor required to correct the PF to 0.85 lagging.

The phasor diagram of the top branch (Fig. 8.41) is shown in Fig. 8.42, while the phasor diagram of the bottom branch is as shown in Fig. 8.43. Combining both phasor diagrams so that the resultant current is at 0.85 lagging (31.8°), we have Fig. 8.44.

Hence,

$$\cos\alpha = PF = 0.85$$
$$\therefore \alpha = 31.8°$$

Figure 8.42

Figure 8.43

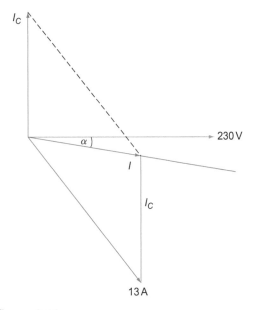

Figure 8.44

By measurement,

$I_c = 9.4\ \text{A}$

$X_c = \dfrac{V}{I_c}$

$= \dfrac{230}{9.4}$

$= 24.46\ \Omega$

$X_c = \dfrac{1}{2\pi f C}$

$\therefore C = \dfrac{1}{2\pi f X_c}$

$= \dfrac{1}{314.16 \times 24.46}$

$= 130\ \mu\text{F}$

Motor ratings

As motors are a.c. plant, their electrical input is rated in kVA. The mechanical output from the machine is rated in horsepower (hp) or kilowatts (kW) where 1 hp = 746 W, and this is the rating usually displayed on the motor.

> **Note**
>
> Horsepower is no longer used but will still be found on older machines.

As machines have moving parts there are mechanical as well as electrical losses and they will have an efficiency given by

$$\text{Efficiency (\%)} = \frac{\text{output} \times 100}{\text{input}}$$

> **Example**
>
> A 5 kW, 230 V, 50 Hz induction motor has a running PF of 0.7 lagging and an efficiency of 80%. Calculate the current drawn by the motor.
>
> $\text{Efficiency (\%)} = \dfrac{\text{output} \times 100}{\text{input}}$
>
> $\therefore \text{Input} = \dfrac{\text{output} \times 100}{\text{efficiency (\%)}}$
>
> $\quad = \dfrac{5 \times 100}{80}$
>
> $\quad = 6.25\ \text{kW}$
>
> $\text{PF} = \dfrac{\text{kW}}{\text{kVA}}$
>
> $\therefore \text{kVA} = \dfrac{\text{kW}}{\text{PF}}$
>
> $\quad = \dfrac{6.25}{0.7}$
>
> $\quad = 8.93\ \text{kVA}$
>
> $I = \dfrac{\text{VA}}{V}$
>
> $\quad = \dfrac{8.93 \times 10^3}{230}$
>
> $\quad = 38.8\ \text{A}$

> **Example**
>
> A 25 kW, 400 V, 50 Hz three-phase squirrel-cage induction motor is 87% efficient and has a PF of 0.92 lagging. Calculate the line current of the motor.
>
> $\text{Efficiency (\%)} = \dfrac{\text{output} \times 100}{\text{input}}$
>
> $\therefore \text{Input} = \dfrac{\text{output} \times 100}{\text{efficiency (\%)}}$
>
> $\quad = \dfrac{25 \times 100}{87}$
>
> $\quad = 28.73\ \text{kW}$

$$\text{Power (watts)} = \sqrt{3}V_L \times I_L \times \text{PF}$$

$$\therefore I_L = \frac{P}{\sqrt{3}V_L \times \text{PF}}$$

$$= \frac{28\,730}{\sqrt{3} \times 400 \times 0.92}$$

$$= 45\,\text{A}$$

Torque and output

Example

A four-pole cage induction motor is run from a 50 Hz supply and has a slip of 3%. The rotor shaft drives a pulley wheel 300 mm in diameter, which has a tangential force of 200 N exerted upon it. Calculate the power output from the rotor in watts.

From Fig. 8.45:

$$\text{Torque} = \text{force} \times \text{radius}$$

$$= 200 \times 150 \times 10^{-3}$$

$$= 30\,\text{Nm}$$

$$\text{Slip (\%)} = \frac{(n_s - n_r)100}{n_s}$$

and

$$n_s = \frac{f}{p}$$

$$= \frac{50}{2}$$

$$= 25\,\text{revs/s}$$

$$\therefore 3 = \frac{(25 - n_r)100}{25}$$

$$\therefore \frac{3 \times 25}{100} = 25 - n_r$$

$$\therefore n_r = 25 - \frac{(3 \times 25)}{100}$$

$$= 25 - 0.75$$

$$= 24.25\,\text{revs/s}$$

$$P = 2\pi n T$$

$$= 2\pi \times 24.25 \times 30$$

$$= 4.57\,\text{kW}$$

Figure 8.45

Points to Note

1. The voltage drop between the supply intake position and the motor must not exceed 5% of the supply voltage.
2. The motor enclosure must be suitable for the environment in which it is to work; for example, a flameproof enclosure is required for explosive situations.
3. Every motor must have a means of being started and stopped, the means of stopping being situated within easy reach of the person operating the motor.
4. Every motor should have a control such that the motor cannot restart after it has stopped because of mains voltage drop or failure (i.e. undervoltage protection). This regulation may be relaxed if a dangerous situation will arise should the motor fail to restart.
5. **Every** stopping device must have to be reset before a motor can be restarted.
6. A means of isolation must be provided for every motor and its associated control gear. If this means is remote from the motor, isolation adjacent to the motor must be provided, or the remote isolator must be capable of being 'locked off'.
7. If a single motor and/or its control gear in a group of motors is/are to be maintained or inspected, a single means of isolation for the whole group may be installed provided the loss of supply to the whole group is acceptable.
8. Excess current protection must be provided in control gear serving motors rated above 370 W and/or in the cables between the protection and the motor.
9. Cables carrying the starting and load currents of motors must be at least equal in rating to the full-load current rating of the motor. This includes rotor circuits of slip-ring or commutator motors.
10. The final circuit supplying a motor shall be protected by fuses or circuit breakers of rating not greater than that of the cable, unless a starter is provided which protects the cable between itself and the motor, in which case the fuses or circuit breakers may be rated up to twice the rating of the cable between the fuse and the starter (Fig. 8.46).

8

Final circuits

Fuse not greater than
cable rating

DB

M

Starter

Fuse up to twice
rating of cable

DB

M

Figure 8.46 DB = distribution board.

Self-Assessment Questions

1. (a) Explain what is meant by the term 'back e.m.f.' in a motor.
   (b) Outline the basic differences between series, shunt and compound-wound d.c. motors.

2. (a) With the aid of diagrams explain how the speed of series, shunt and compound-wound motors may be controlled.
   (b) A 440 V d.c. shunt-wound motor has a field resistance of 200 Ω and an armature resistance of 0.6 Ω. When its speed is 20 revs/s the current drawn from the supply is 12.2 A. Calculate its back e.m.f. at this speed. If the speed were decreased to 19 revs/s, the field flux remaining unchanged, calculate the new back e.m.f. and the new armature current.

3. A load of 6.4 kW at 240 V is supplied from the terminals of a shunt-wound d.c. generator. The field resistance is 180 Ω. Calculate the armature current.

4. (a) Explain with the aid of a diagram how a rotating magnetic field may be obtained.
   (b) What is meant by 'synchronous speed'? Calculate the synchronous speed of a 12-pole motor if the supply frequency is 50 Hz.

5. (a) Explain the action of a synchronous motor.
   (b) What methods are available to start a synchronous motor? Explain with diagrams.

6. (a) Explain the action of a cage induction motor.
   (b) What is meant by the term 'slip'? Calculate the percentage slip of a six-pole induction motor running at 16.2 revs/s from a 50 Hz supply.

7. Describe with sketches three different ways of starting a single-phase induction motor.

8. Explain with the aid of a sketch the action of a three-phase DOL starter. What are dashpots used for? How are remote start and stop buttons connected?

9. A 10 kW, 230 V, 50 Hz single-phase cage rotor induction motor is 85% efficient and has a PF of 0.68 lagging. Calculate the motor current and the value of capacitor required to raise the PF to 0.93 lagging.

10. Calculate the torque developed by an 18 kW four-pole induction motor run at 3.5% slip from a 50 Hz supply.

CHAPTER 9

Cells and Batteries

General background

In 1789, an Italian professor of botany, Luigi Galvani (1737–1798), noticed by chance that freshly skinned frogs' legs twitched when touched by two dissimilar metals. He called this effect as **animal electricity**.

It was, however, another Italian, Alessandro Volta (1745–1827), a professor of physics, who showed that the electric current which produced the muscular spasm was not due to the animal limb itself, but due to the moisture in it. In 1799 he developed a simple battery comprising copper and zinc discs separated by a brine-soaked cloth. This type of assembly is known as a **voltaic pile**. From this primitive beginning have come the cells and batteries we use today. The materials used may be more refined, but the basic concept has remained unchanged.

The primary cell

If two dissimilar metals are immersed in an acid or salt solution, known as an **electrolyte**, an electromotive force (e.m.f.) will be produced; this assembly is known as a **cell**. The e.m.f. may be used to supply a load, but will only do so for a limited time, as the chemical qualities of the electrolyte deteriorate with use. The chemicals have to be renewed to render the cell useful again.

The most common forms of primary cells in use are Leclanché's wet cell and the dry cell.

Figure 9.1 illustrates the component parts of these two types of cell.

The depolarizing agent is used to remove hydrogen bubbles from around the carbon rod. These bubbles,

(a)

(b)

Figure 9.1 (a) Wet cell; (b) dry cell.

Electrical Installation Work. 978-1-138-84927-3 © Brian Scaddan.
Published by Taylor & Francis. All rights reserved.

which are formed during the chemical action, impair the performance of the cell.

Applications

The dry cell has an obvious advantage over the wet cell because it is portable and so is commonly used for appliances such as torches, door bells, etc.

The wet cell, although almost obsolete, is used in larger bell and indicator circuits and for railway signalling.

The secondary cell

Unlike the primary cell, the secondary cell can be used again after it has discharged all its electrical energy. It can be recharged by **supplying** it with electrical energy. This reverses the chemical process which took place during discharge.

There are two types of secondary cell, the lead–acid and the alkaline cell.

The lead–acid cell

This cell consists of positive and negative lead electrodes, and an electrolyte of **dilute sulphuric acid**; all placed in an acid-resistant container.

The electrodes are made of several plates, the positive and negative being insulated from one another by separators of insulating material such as wood or ebonite.

The construction of the plates is of considerable importance and is discussed below.

Formed plates

Repeated charging and discharging of a cell under manufacture causes the plates to be covered in lead compounds, the negative plate with **spongy lead** and the positive plate with **lead peroxide**, these being important to the chemical action of the cell. This process is, however, both expensive and time consuming and for smaller types of cell, pasted plates are more popular.

Pasted plates

These plates are manufactured in the form of a grid, into which a compound of **sulphuric acid and red lead** is pressed; only a short initial charge is needed for the cell to be ready for use. These plates, however, disintegrate more easily than the formed type.

Figure 9.2 Lead–acid cell.

A combination of formed and pasted plates is used in large-capacity cells; the positive plate is formed and the negative plate is pasted.

When current is drawn from the cell the active chemicals on the positive plate expand and the plate tends to distort, especially under heavy loads. Some measure of protection against this distortion or buckling is achieved by arranging for each positive plate to be adjacent to two negative plates (Fig. 9.2).

Action of lead–acid cell

When an external load is connected to the terminals of the cell, electrical energy is delivered to the load. During this **discharge** period, a chemical reaction between the plates and the electrolyte takes place and a layer of **lead sulphate** is deposited on the plates. However, this process successively weakens the electrolyte until the cell is unable to deliver any more electrical energy.

If a d.c. supply is then connected to the cell terminals, and a current is passed through it, the **lead sulphate** is converted back into **sulphuric acid** and restores the cell to its original condition. This process is known as **charging**.

Care and maintenance of lead–acid cells

Provided that a lead–acid cell is maintained regularly and is cared for, it should last for an indefinite period of time. A **weekly** check on its condition is to be recommended.

Electrolyte level

The level of the electrolyte should never be allowed to fall below the tops of the plates. Any loss of electrolyte due to evaporation may be made up by the addition of distilled water.

Specific gravity of electrolyte

As a cell discharges, the electrolyte becomes weaker and its specific gravity (SG) falls, until the cell can no longer deliver energy. The state of charge of a cell can therefore be measured by the SG of the electrolyte. A hydrometer is used for this purpose. It consists of a glass syringe containing a weighted, graduated float. The syringe has a rubber nozzle for insertion into the electrolyte, and a rubber bulb at the upper end for sucking the liquid into the syringe.

The nozzle is inserted in the electrolyte and a sample is drawn up into the syringe by squeezing the rubber bulb. The level of the liquid in relation to the position of the float gives a direct reading of the SG of the electrolyte. The higher the float, the higher the SG (see Fig. 9.3); the lower the float, the lower the SG.

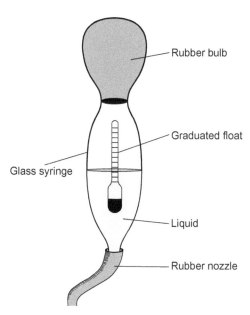

Figure 9.3 Hydrometer.

The following table indicates typical values of SG in relation to charge:

SG	Percentage of Charge
1.28	100
1.25	75
1.22	50
1.19	25
1.16	Fully discharged

A record of results obtained should be kept for each cell.

Note

Cells should not be allowed to fall below an SG of 1.18.

Terminal voltage

A check on the no-load terminal voltage should be made with a high-resistance voltmeter at the end of a discharge period. This reading should not be below 1.85 V. A fully charged cell should indicate about 2.2 V.

Plate colour

An indication of the state of charge is the colour of the plates. In a healthy cell the positive plate is chocolate brown and the negative plate is slate grey.

General maintenance procedures

1. When preparing an electrolyte, **always** add acid to water, **never** water to acid.
2. Ensure that any maintenance is carried out in a well-ventilated area.
3. Do not permit the use of any naked flame near the cells.
4. Cells which are to be taken out of commission for any time should be fully charged, the electrolyte left in, and a periodic charge given to keep the cell healthy until it is needed again.
5. Never leave a cell in an uncharged state, as a layer of whitish **sulphate** will form on the plates, which will increase the internal resistance and reduce the capacity of the cell. This process is known as **sulphation of the plates**.
6. Terminals should be coated with petroleum jelly to prevent corrosion.

Application of lead–acid cells

The most common use for this type of cell is the car battery (a battery is a group of cells). Other applications include standby supplies, alarm and control circuits. The electrolyte of such a battery has a negative temperature coefficient (Chapter 2) and hence a drop in temperature causes a rise in its resistance, so less current will be delivered. We have all experienced starting problems with vehicles in very cold weather.

The nickel–alkaline cell

There are two types of alkaline cell, the nickel–iron and the nickel–cadmium.

The nickel–iron cell

Here the positive plate is made of **nickel hydroxide**, the negative plate of **iron oxide** and the electrolyte is **potassium hydroxide**.

The nickel–cadmium cell

In this cell both the positive plate and the electrolyte are the same as for the nickel–iron cell; however, the negative plate is **cadmium** mixed with a small amount of iron.

The active chemicals in the plates of alkaline cells are enclosed in thin nickel–steel grids insulated from one another by ebonite rods. The whole assembly is housed in a welded steel container.

Care and maintenance of the nickel–alkaline cell

Unlike the lead–acid cell the nickel–alkaline needs minimal attention.

The open type only needs periodic topping up to compensate for the electrolyte lost by evaporation. The totally enclosed type needs no maintenance.

Applications

It has limited use owing to its cost and is mainly used in situations where a robust construction is needed, that is marine work.

Capacity of a cell

If a cell delivers, say, 10 A for a period of 10 h, it is said to have a capacity of 100 ampere hours (Ah) at the 10 h rate; taking any more than 10 A will discharge the cell in less than 10 h.

Efficiency of a cell

The efficiency of any system is the ratio of the output to the input. The efficiency of cells is given in two forms:

The ampere hour efficiency % =

$$\frac{\text{discharge amperes} \times \text{time} \times 100}{\text{charging amperes} \times \text{time}}$$

The watt hour efficiency % =

$$\frac{\text{discharge VA} \times \text{time} \times 100}{\text{charging VA} \times \text{time}}$$

Comparison of cell characteristics

	Lead–Acid	Alkaline
SG charged	1.28	1.2
SG discharge	1.18	1.2
P.d. charged	2.1 V	1.3 V
P.d. discharge	1.85 V	1.0 V

Ah capacity

The Ah obtainable from the alkaline cell at the higher discharge rates, that is 2 and 4 h, is much greater than that of the lead–acid cell (Fig. 9.4). This is because the SG of the electrolyte does not change during discharge.

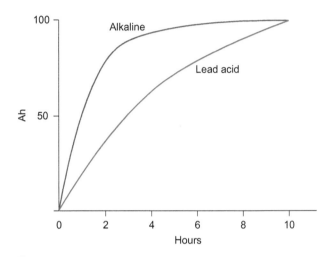

Figure 9.4 Ampere-hour capacity.

Charge and discharge

The dips in the charging curves (Fig. 9.5) are due to the reduction of the charging current, to prevent overheating of the cell.

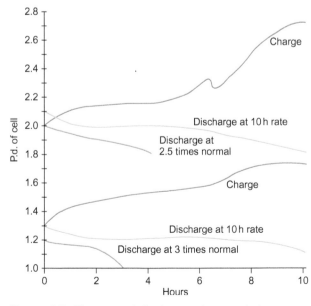

Figure 9.5 Charge and discharge characteristics.

	Advantages	Disadvantages
Lead–acid	Inexpensive	Fragile
	High discharge voltage	Self-discharges when not in use
	Uses plentiful materials	Requires regular maintenance
Alkaline	Very robust	Very expensive
	Retains its charge when not in use	Low discharge voltage
	Needs little or no maintenance	

Cell and battery circuits

E.m.f. of cell (E, volts)

This is the maximum force available (measured in volts) in a cell to produce current flow.

Internal resistance

When current flows through the cell there is some resistance to its flow (less than $1\,\Omega$ in a good cell) and hence a voltage drop across it.

Potential difference of a cell or terminal (V, volts)

This is the voltage measured at the terminals of a cell, and is less than the cell e.m.f. owing to the voltage drop across the internal resistance of the cell (Fig. 9.6). E = e.m.f. of cell; r = internal resistance of cell; R = resistance of load; I = circuit current and V = terminal voltage or potential difference (p.d.) across load.

The terminal voltage (V) available across the load is clearly the e.m.f. (E) less the voltage drop across the internal resistance (r):

$$V = E - (I \times r)$$

This is the same principle as the voltage available across the terminals of a load which is supplied by a long cable.

Example

If a cell with an e.m.f. of 2 V and an internal resistance of $0.2\,\Omega$ is connected across a $0.8\,\Omega$ load resistor, calculate the current that will flow (see Fig. 9.7).

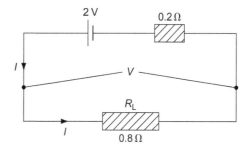

Figure 9.7

$$V = E - (I \times r) \quad \text{and} \quad V = I \times R_L$$
$$\therefore I \times R_L = E - (I \times r)$$
$$\therefore (I \times R_L) + (I \times r) = E$$
$$\therefore I \times (R_L + r) = E$$
$$\therefore I = \frac{E}{R_L + r}$$
$$= \frac{2}{(0.8 + 0.2)}$$
$$= \frac{2}{1}$$
$$= 2A$$

Cells in series

If a high p.d. is required, then cells are connected in series and internal resistances are added (Fig. 9.8).

$\circ\!\!-\!|\!\!-\!|\!\!-\!|\!\!-\!|\!\!-\!|\!\!-\!\circ$ Shown $\circ\!\!-\!|\!|\!|\!|\!|\!|\!-\!\circ$ Or $\circ\!\!-\!|\!\text{-}\text{-}\!|\!-\!\circ$

Figure 9.8 Cells on series.

Cells in parallel

For cells in parallel the p.d. is the same as that for one cell, but as the internal resistances are added in parallel their resultant internal resistance is less than for one cell and heavier currents can be drawn.

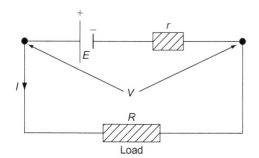

Figure 9.6 Internal resistance.

Battery charging

Cells and batteries are charged by connecting them to a controlled d.c. source. This source may be obtained in several ways: (1) rectified a.c.; (2) motor generator set; (3) rotary converter; and (4) d.c. mains supply.

The most commonly used method is rectified a.c. and there are two ways in which this system is used: (1) the constant-voltage method; and (2) the constant-current method.

Constant-voltage charging

In this method the d.c. charging voltage is kept constant at a value just above that of the final value of the battery e.m.f. The charging current is initially high, decreasing as the e.m.f. of the battery approaches that of the supply (Fig. 9.9).

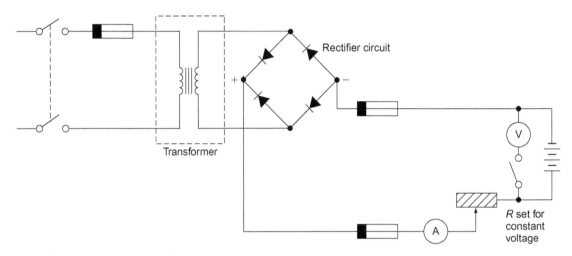

Figure 9.9 Constant-voltage charging.

Constant-current charging

In constant-current charging the current is kept constant by varying the d.c. input voltage as the battery e.m.f. increases (Fig. 9.10).

The more popular method, for everyday use, is the constant-voltage method.

Cells, batteries and their associated charging equipment are frequently used in installation work such as indicator and call systems in hospitals and hotels, fire alarm and burglar alarm systems, and emergency lighting installations.

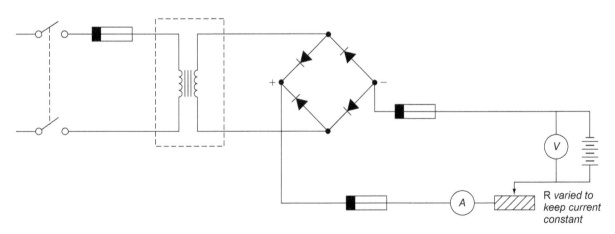

Figure 9.10 Constant-current charging.

Self-Assessment Questions

1. Explain the difference between a primary and a secondary cell.
2. Describe two methods of testing the state of charge of a secondary cell. What figures would you expect?
3. How does an alkaline cell differ from a lead–acid cell? What effect does discharging an alkaline cell have on the SG of its electrolyte?
4. (a) What is the difference between the e.m.f. of a cell and its terminal voltage?
   (b) A cell of e.m.f. 2 V and internal resistance 0.15 Ω delivers a current of 3 A to an external load. Calculate the resistance of the load and the terminal voltage of the cell.
5. State the two common methods of battery charging and explain the difference between them.
6. Why do batteries perform less well at low temperatures?

9

CHAPTER 10

Illumination and ELV Lighting

Light sources

The range of modern lighting fittings and lamps is so large that only the basic types will be considered here. Before we continue, however, it is perhaps wise to list the various units and quantities associated with this subject.

Luminous intensity: symbol, *I*; unit, candela (cd)

This is a measure of the power of a light source and is sometimes referred to as brightness.

Luminous flux: symbol, *F*; unit, lumen (lm)

This is a measure of the flow or amount of light emitted from a source.

Illuminance: symbol, *E*; unit, lux (lx) or lm/m²

This is a measure of the amount of light falling on a surface. It is also referred to as **illumination.**

Luminous efficacy: symbol, *K*; unit, lumen per watt (lm/W)

This is the ratio of luminous flux to electrical power input. It could be thought of as the 'efficiency' of the light source.

Maintenance factor (MF): no unit

In order to allow for the collection of dirt on a lamp and also ageing, both of which cause loss of light, a maintenance factor is used.

As an example, consider a new 80 W fluorescent lamp with a lumen output of 5700 lm. After about 3 or 4 months this output would have fallen and settled at around 5200 lm. Hence the light output has decreased by

$$\frac{5200}{5700} = 0.9$$

This value, 0.9, is the maintenance factor and should not fall below 0.8. This is ensured by regular cleaning of the lamps.

Coefficient of utilization (CU): no unit

The amount of useful light reaching a working plane will depend on the lamp output, the reflectors and/or diffusers used, position of lamp, colour of walls and ceilings, etc. The lighting designer will combine all of these considerations and determine a figure to use in his or her lighting calculations.

Lamp types

Tungsten-filament lamp

These are known as incandescent lamps and, because of their high energy consumption, are gradually being phased out and will ultimately become obsolete.

Figure 10.1 shows the basic components of a tungsten-filament lamp.

The tungsten filament is either single or double coiled (coiled coil). Figure 10.2 illustrates these types.

The efficacy of gas-filled lamps is increased by using a coiled-coil filament, as this type has in effect a

Electrical Installation Work. 978-1-138-84927-3 © Brian Scaddan.
Published by Taylor & Francis. All rights reserved.

Bulb

Filament

Glass stem

Glass bulb cemented into cap

Brass cap (bayonet type)

Cap pin

Insulation

Contact

Edison screw (ES) type

Figure 10.1 Tungsten-filament lamp.

(a)   (b)

Figure 10.2 Filament arrangements: (a) single coil; (b) coiled coil.

thicker filament which reduces the heat loss due to convection currents in the gas.

Filament lamps are of two main types: vacuum and gas filled.

Vacuum type

The filament operates in a vacuum in the glass bulb. It has a poor efficacy as it can operate only up to around 2000°C.

Gas-filled type

In this case the bulb is filled with an inert gas such as nitrogen or argon. This enables the operating temperature to reach 2500°C. The efficacy increases and the bulb is usually so bright that it is given an opaque coating internally. This type of lamp is usually called a 'pearl' lamp.

The following code refers to lamp caps:

BC – Bayonet

SBC – Small bayonet

SCC – Small centre contact

ES – Edison screw

SES – Small Edison screw

MES – Miniature Edison screw

GES – Goliath Edition screw

The efficacy of a tungsten lamp will depend on several factors, including the age of the lamp and its size, but tends to be around 12 lm/W for a 100 W lamp.

The colour of its light tends to be mostly red and yellow and in its basic form this type of lamp is used only in situations that do not require a high level of illumination.

Other lamps of the filament type include tubular strip lights, oven lamps, infrared heating lamps, spot- and floodlights, and tungsten–halogen lamps.

Tungsten-halogen lamp

These are also incandescent lamps with the glass envelope filled with an inert gas and a small amount of halogen. These lamps have a long life and emit an intense white light, hence, they are used for security lighting and anywhere when there is a need for high levels of illumination.

These lamps may be of the same shape as the common incandescent or the more familiar linear type, or the GU10 variety as shown.

Discharge lighting

This type of lighting relies on the ionization of a gas to produce light. As high voltages are present in such

Figure 10.3

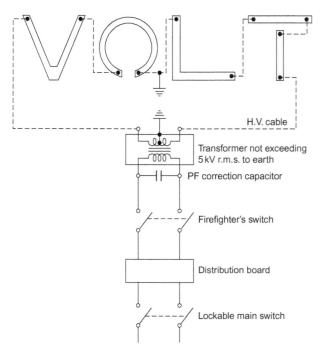

Figure 10.4 Single neon-sign circuit.

Figure 10.5 Low-pressure sodium-vapour lamp type SOX. (BC 5 bayonet cap)

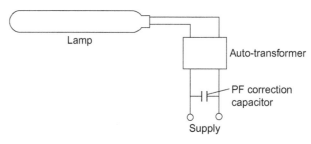

Figure 10.6 Control circuit.

lighting circuits, special precautions, outlined in the IET Regulations, must be taken. Typical discharge lamps include decorative neon signs, fluorescent lighting, and mercury and sodium-vapour lamps used for street lighting.

Neon tube

In the same way that the trade name 'Hoover' is colloquially used to indicate any make of vacuum cleaner, so 'neon' tends to be used to describe any sort of gas-filled tube. There are in fact several different gases used to give different colours, including helium, nitrogen and carbon dioxide.

Figure 10.4 shows the basic circuit for a cold-cathode neon-sign installation.

Sodium-vapour lamp

There are two types of sodium-vapour lamp available, working at high and low pressure, respectively.

The low-pressure type consists of a U-shaped double-thickness glass tube, the inner wall of which is of low-silica glass which can withstand attack by hot sodium. Inside the tube is a quantity of solid sodium and a small amount of neon gas (this helps to start the discharge process). An outer glass envelope stops too much heat loss from the inner tube. Figure 10.5 shows the components of a low-pressure sodium-vapour lamp, while Fig. 10.6 shows the control circuit for a sodium-vapour lamp.

The output from the auto-transformer is in the region of 480 V and the power factor (PF) correction capacitor is important, as the PF of the lamp and transformer can be as low as 0.3 lagging.

The recommended burning position of the lamp is horizontal, ±20°; this ensures that hot sodium does not collect at one end of the tube in sufficient quantities to attack and damage it.

The light output is almost pure yellow, which distorts surrounding colours, and as such is useful only for street lighting. The modern SOX type (superseding the SOH type) has a high efficacy, a 90 W lamp giving in the region of 140 lm/W. (The SOH type gives around 70 lm/W.)

The high-pressure type of sodium-vapour lamp differs from other discharge lamps in which the discharge tube is made of compressed aluminium oxide, which is capable of withstanding the intense chemical activity of the sodium vapour at high temperature and pressure. The efficacy is in the region of 100 lm/W, and the lamp may be mounted in any position. The colour is a golden white and as there is little surrounding colour distortion, it is suitable for many applications including shopping centres, car parks, sports grounds and dockyards.

High-pressure mercury-vapour lamp

This type consists of a quartz tube containing mercury at high pressure and a little argon gas to assist starting. There are three electrodes, two main and one auxiliary; the latter is used for starting the discharge (Fig. 10.7).

109

Figure 10.7 High-pressure mercury-vapour lamp. (ES: Edison screw type; GES: Goliath Edison screw type)

Figure 10.8 Control circuit.

Figure 10.8 shows the control circuit for a high-pressure mercury-vapour lamp. The initial discharge takes place in the argon gas between the auxiliary electrode and the main electrode close to it. This causes the main electrode to heat up and the main discharge between the two main electrodes takes place.

Several types of mercury-vapour lamp are available, including the following two popular types:

MB type – Standard mercury-vapour lamp; ES or GES cap; any mounting position; efficacy around 40 lm/W. Largely superseded by the MBF type.

MBF type – Standard, but with fluorescent phosphor coating on the inside of the hard glass bulb; ES or GES cap; efficacy around 50 lm/W. Used for industrial and street lighting, commercial and display lighting. Any mounting position.

The colour given by high-pressure mercury-vapour lamps tends to be blue-green.

Metal-halide lamp

This is a mercury-vapour lamp that has 'rare earth metals' added to the gas which increases its efficacy and produces high levels of illumination.

When the supply to these lamps is interrupted they take some time to re-start.

Together with the high-pressure mercury-vapour lamps these types are to be phased out in the EU by 2015.

Low-pressure mercury-vapour lamp

A low-pressure mercury-vapour lamp, more popularly known as a **fluorescent lamp**, consists of a glass tube, the interior of which is coated in fluorescent phosphor. The tube is filled with mercury vapour at low pressure and a little argon to assist starting. At each end of the tube is situated an oxide-coated filament. Discharge takes place when a high voltage is applied across the ends of the tube. Figure 10.9 shows the circuit diagram for a single fluorescent tube.

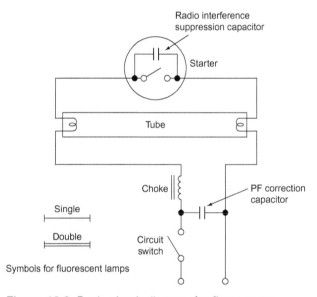

Figure 10.9 Basic circuit diagram for fluorescent lamp.

Practical operation

When the supply is switched on, the circuit is completed via: the choke, first lamp element, starter switch, second lamp element and the neutral. The elements, which are coated in oxide, become warm and the oxide coating emits some electrons and the gas ionizes at the ends of the tube (this helps the main ionization process). The starter contacts (usually of the bimetallic type) separate, owing to the current passing through them, and the choke is open-circuited.

As we have seen, breaking an inductive circuit causes high voltages to appear across the breaking contacts, and energy is released in the form of an arc. In this case, however, there is an easier way for the energy to dissipate, via the gas, and the high voltage appears across the ends of the tube.

When the gas is fully ionized, the choke limits the current to a predetermined value, and the light emitted, which is mostly ultraviolet, is made visible by the fluorescent powder coating.

The radio interference suppression capacitor is usually located in the starter. The PF correction capacitor is part of the control circuitry common to all fluorescent lighting installations.

Starters

Three methods are commonly available for starting the discharge in a fluorescent tube: the thermal start, the glow start and the quick start.

A **thermal starter** consists of two contacts (one of which is a bimetal) and a heater. Figure 10.10 shows how such a starter is connected.

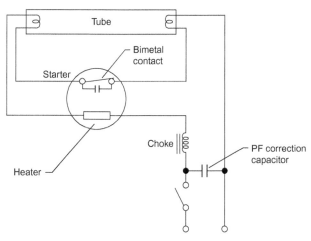

Figure 10.10 Thermal starter.

When the supply to the lamp is switched on, the heater is energized. Also, the lamp filaments are energized via the starter contact. The heater causes the contacts to part and the choke open circuits across the tube, so that discharge takes place.

The **glow starter** is the most popular of all the means of starting the discharge. It comprises a pair of open contacts (bimetallic) enclosed in a sealed glass bulb filled with helium gas. This assembly is housed in a metal or plastic canister. Figure 10.11 shows how this type of starter is connected.

When the supply is switched on, the helium gas ionizes and heats up, causing the contacts to close, and this energizes the tube filaments. As the contacts have closed, the discharge in the helium ceases, the contacts cool and part, open-circuiting the choke across the tube and discharge takes place.

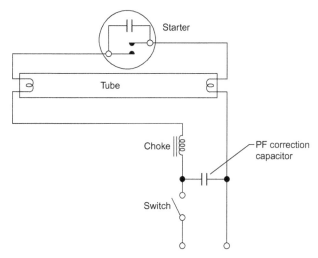

Figure 10.11 Glow starter.

In the case of the **quick start** or **instant starter**, starting is achieved by the use of an auto-transformer and an earthed metal strip in close proximity to the tube (Fig. 10.12).

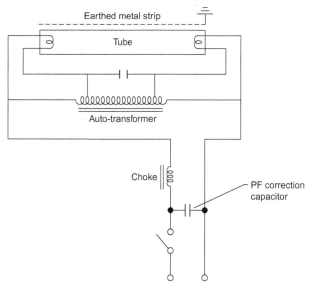

Figure 10.12 Quick start or instant starter.

When the supply is switched on, mains voltage appears across the ends of the tube, and the small part of the winding at each end of the transformer energizes the filaments that heat up. The difference in potential between the electrodes and the earthed strip causes ionization that spreads along the tube.

Fluorescent tube light output

There is a wide range of fluorescent tubes for different applications as Table 10.1 indicates.

111

Table 10.1 Tube colour and application.

Tube Colour	Application
White and warm white	General illumination requiring maximum efficacy, as in drawing offices
Daylight and natural	Any situation requiring artificial light to blend with natural daylight – jewellery, glassware, etc. (main shop areas)
Artificial daylight	Areas where accurate colour matching is carried out
Deluxe warm white	Offices and buildings requiring a warm effect (e.g. restaurants, furniture stores)
Northlight	Colour-matching areas such as in tailors and furriers
Deluxe natural	Florists, fishmongers, butchers, etc.
Green, gold, blue, red and pink	For special effects

The white tube has the highest efficacy, which for a 2400 mm, 125 W tube is around 70 lm/W.

Compact fluorescent lamp (CFL)

These work on the same principle as the fluorescent tube but the starting process is achieved electronically. These are the replacement for the incandescent lamp.

LED lamps

These comprise many LED's grouped together and produce a range of light intensities and hence are suitable for general lighting. They are extremely efficient, using little electrical energy and lasting some 60,000 h. They are however much more expensive to buy than CFL's.

Points to Note

1. If a switch, not designed to break an inductive load, is used to control discharge lighting, it should have a rating not less than twice the steady current it is required to carry, that is a 10 A switch for a 5 A load.
2. Although a discharge lamp is rated in watts, its associated control gear is highly inductive and therefore the whole unit should have a VA rating. It is on this rating that the current rating of the circuits is calculated. If no technical information is available, a figure of 1.8 is used to calculate the VA rating. That is,

   VA rating of 80 W fitting $= 80 \times 1.8$
   $$= 144\,VA$$

3. No discharge lighting circuit should use a voltage exceeding 5 kV r.m.s. to earth, measured on open circuit.
4. If a circuit exceeds low voltage and is supplied from a transformer whose rated input exceeds 500 W then the circuit must have protection such that the supply is cut off automatically if short-circuit or earth leakage currents exceed 20% of the normal circuit current.
5. All control equipment including chokes, capacitors, transformers, etc., must either be totally enclosed in an earthed metalwork container or be placed in a ventilated fireproof enclosure. Also, a notice must be placed and maintained on such a container or enclosure, reading 'DANGER – HIGH VOLTAGE'. The minimum size of letters and notice board is as shown in Fig. 10.13.

Figure 10.13 Danger notice (voltage).

6. Care must be taken to ensure that the only connection between discharge lamp circuits, operating at a voltage exceeding low voltage, and the mains supply is an earth conductor and/or the earthed neutral conductor of an auto-transformer having a maximum secondary voltage of 1.5 kV.
7. It is important that discharge lighting has a means of isolation from all poles of the supply. This may be achieved in one of the following ways:
   (a) An interlock device, on a self-contained discharge lighting unit, so that no live parts can be reached unless the supply is automatically disconnected (i.e. microswitch on the lid of the luminaire which will disconnect supply to a coil of a contactor when the lid is opened).
   (b) A plug and socket close to the luminaire or circuit which is additional to the normal circuit switch.
   (c) A lockable switch or one with a removable handle or a lockable

distribution board. If there is more than one such switch, handles and keys must not be interchangeable.

8. Every discharge lighting installation must be controlled by a firefighter's switch which will isolate all poles of the supply (it need not isolate the neutral of a three-phase four-wire supply).

9. The firefighter's switch should be coloured red and have fixed adjacent to it a notice as shown in Fig. 10.14. The notice should also display the name of the installer and/or maintainer of the installation.

Measurements are minimum sizes

Figure 10.14 Firefighters switch label.

10. The firefighter's switch shall have its ON and OFF positions clearly marked, the OFF position being at the top of the switch. The switch should be placed in a conspicuous and accessible position, not more than 2.75 m from ground level.

11. The firefighter's switch should be outside and adjacent to the installation for external installations and in the main entrance of a building for interior installations.

12. The Building Regulations require that new lighting installations in dwellings have at least 30% energy saving **fittings**. This may be achieved by having 1 per 25 m² floor area or one for every 4 fixed lighting points. This requirement is **not** satisfied by just installing low-energy lamps.

13. In general, cables used in discharge lighting circuits exceeding low voltage should be metal sheathed or armoured unless they are housed in a box sign or a self-contained luminaire or are not likely to suffer mechanical damage.

14. All cables should be supported and placed in accordance with the tables shown in Guidance notes to the Regulations.

15. If it is not clear that a cable is part of a circuit operating above low voltage, it should be labelled every 1.5 m as shown in Fig. 10.14.

Figure 10.15 Danger notice (voltage).

16. Extra-low voltage (ELV) lighting is now popular in many kitchens and bathrooms. Such systems comprise a 230 V/12 V transformer with 12 V dichroic 12 V lamps. This lighting is often incorrectly termed low voltage.

Calculation of lighting requirements

Inverse-square law

If we were to illuminate a surface by means of a lamp positioned vertically above it, measure the illumination at the surface, and then move the lamp twice as far away, the illumination now measured would be four times less. If it were moved away three times the original distance of the illumination would be nine times less. Hence it will be seen that the illuminance on a surface is governed by the square of the vertical distance of the source from the surface (Fig. 10.16).

$$\therefore \text{Illuminance } E \text{ (lux)} = \frac{\text{luminous intensity (cd)}}{d^2}$$

$$E = \frac{I}{d^2}$$

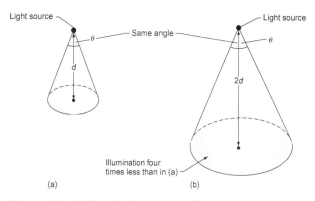

Figure 10.16 Illumination on a surface.

Example

A light source of 900 cd is situated 3 m above a working surface. (a) Calculate the illuminance directly below the

113

source. (b) What would be the illuminance if the lamp were moved to a position 4 m from the surface?

(a) $E = \dfrac{I}{d^2}$

$\quad = \dfrac{900}{9}$

$\quad = 100 \text{ lx}$

(b) $E = \dfrac{I}{d^2}$

$\quad = \dfrac{900}{16}$

$\quad = 56.25 \text{ lx}$

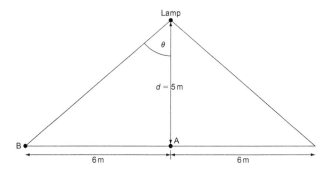

Figure 10.18

Cosine rule

From Fig. 10.17 it will be seen that point X is further from the source than is point Y. The illuminance at this point is therefore less. In fact the illuminance at X depends on the cosine of the angle θ. Hence,

$$E_X = \frac{I \times \cos^3 \theta}{d^2}$$

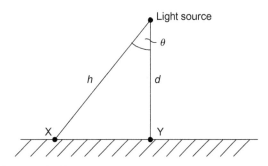

Figure 10.17

Example

A 250 W sodium-vapour street lamp emits a light of 22 500 cd and is situated 5 m above the road. Calculate the illuminance (a) directly below the lamp and (b) at a horizontal distance along the road of 6 m (Fig. 10.18).

From Fig. 10.18, it can be seen that the illuminance at A is given by

$$E_A = \frac{I}{d^2}$$

$$= \frac{22\,500}{25}$$

$$= 900 \text{ lx}$$

The illuminance at B is calculated as follows. Since the angle θ is not known, it can be found most simply by trigonometry:

$$\tan \theta = \frac{AB}{d}$$

$$= \frac{6}{5} = 1.2$$

From tangent tables:

$$\theta = 50.2°$$

and from cosine tables:

$$\cos 50.2° = 0.64$$

$$\therefore E_B = \frac{I \cos^3 \theta}{d^2}$$

$$= \frac{22\,500 \times 0.64^3}{25}$$

$$= 236 \text{ lx}$$

In order to estimate the number and type of light fittings required to suit a particular environment, it is necessary to know what level of illuminance is required, the area to be illuminated, the MF and the CU, and the efficacy of the lamps to be used.

Example

A work area at bench level is to be illuminated to a value of 300 lx, using 85 W single fluorescent fittings having an efficacy of 80 lm/W.

The work area is 10 m × 8 m, the MF is 0.8 and the CU is 0.6. Calculate the number of fittings required.

$$\text{Total lumens } (F) \text{ required} = \frac{E \text{ (lx)} \times \text{area}}{\text{MF} \times \text{CU}}$$

$$F = \frac{300 \times 10 \times 8}{0.8 \times 0.6}$$

$$= 50\,000 \text{ lm}$$

Since the efficacy is 80 lm/W:

Total power required $= \dfrac{50\,000}{80}$

$= 625\,\text{W}$

As each lamp is 85 W:

Number of lamps $= \dfrac{625}{85}$

$= 8$

PF correction

Example

The PF correction capacitor in a 230 V, 50 Hz fluorescent light unit has broken down and needs replacing. A test on the unit shows that, without the capacitor, the supply current is 0.86 A at a PF of 0.5 lagging. The values quoted on the original capacitor have faded and the only other information is that the working PF of the unit should be 0.95. Determine the value of the capacitor needed (Fig. 10.19).

Figure 10.19 (a) No capacitor; (b) with capacitor.

First the phasor diagram for the unit without a capacitor is drawn (Fig. 10.20). PF = 0.5 and cos $\theta = 0.5$. Therefore: $\theta = 60°$.

The phasor diagram of the unit showing the supply current at working PF is drawn (Fig. 10.21).

PF = 0.95

cos $\alpha = 0.95$

$\alpha = 18.2°$

Finally the combined phasor diagram can be drawn (Fig. 10.22).

Figure 10.20   **Figure 10.21**

Figure 10.22

The value of the capacitor current required to raise the PF to 0.95 lagging must be I_c, which is the same distance (ab). By measurement:

$I_c = 0.6\,\text{A}$

$\therefore X_c = \dfrac{V}{I_c}$

$= \dfrac{230}{0.6}$

$= 383.3\,\Omega$

But

$X_c = \dfrac{1}{2\pi fC}$

$\therefore C = \dfrac{1}{2\pi f X_c}$

$= \dfrac{1}{2\pi \times 50 \times 383.33}$

$= 8.34\,\mu\text{F}$

Note

By measurement, the supply current is 0.45 A.

Rating of fluorescent circuits

Fluorescent tubes are rated in watts, but as we have seen, the circuit of which the lamp part is inductive, and even after improvement has a lagging PF.

We know from Chapter 6 that plant is not rated in watts, but in volt amps (VA). It is recommended that, if no other information is available, **the lamp wattage may be multiplied by 1.8 in order to determine the VA rating**. For example, the VA rating of a fluorescent unit with an 80 W tube is 1.8 × 80 = 144 VA. Hence, when supplied at 230 V, the current taken would be

$$I = \frac{VA}{V}$$

$$= \frac{144}{230}$$

$$= 0.626\,A$$

When a fluorescent lamp is switched off, the choke is again open-circuited; this time the voltage appears across the switch contacts. This can damage the switch and unless it is specially designed to break an inductive circuit, **it should have a rating of not less than twice the total steady current it is required to carry**. For example, if a fluorescent light unit draws a current of 1 A, then the switch controlling it should have a rating of at least 2 A.

Example

A consumer has a work area that he or she wishes to illuminate with single 65 W, 230 V fluorescent fittings. The existing lighting points, which are to be removed, are controlled by a single-gang 5 A switch. This switch is to remain. How many fittings may be installed?

$$\text{VA rating of fitting} = 1.8 \times 65$$

$$= 117\,VA$$

$$\therefore \text{Current rating of fitting} = \frac{117}{230}$$

$$= 0.5\,A$$

$$\text{The 5 A switch must only carry } \frac{5}{2} = 2.5\,A$$

$$\therefore \text{Number of fittings permitted} = \frac{2.5}{0.5}$$

$$= 5$$

$$= 5 \text{ fittings}$$

Stroboscopic effect

While a fluorescent lamp is in operation the light may flicker. Under some circumstances this may make it appear that rotating machinery has slowed down or even stopped. This is called the **stroboscopic effect**. This is an undesirable state of affairs which is usually remedied by one of the two following methods.

Balancing the lighting load (three phase)

If a large lighting load is installed in a three-phase installation where there is some rotating machinery, the stroboscopic effect may be overcome by connecting alternate groups of lamps to a different

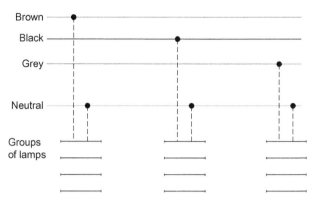

Figure 10.23 Lighting load.

phase. This also has the advantage of balancing the lighting load (Fig. 10.23).

The lead–lag circuit

In this method a capacitor is wired in series with every alternate lamp in a group. The value of the capacitor is such that the lamp unit is fitted to have an overall **leading PF**. This means that any pair of lamps have a lagging and a leading PF. This has the effect of cancelling out the resultant flicker, in the same way as two equal but **opposing** forces cancel each other out (see Fig. 10.24).

Figure 10.24 Lead–lag circuit.

Example

Two 230 V fluorescent lamp units A and B are arranged to overcome stroboscopic effects. Unit A has a series capacitor fitted and takes a current of 0.4 A at a PF of 0.985 leading. Unit B takes a current of 0.53 A at 0.5 PF lagging (Fig. 10.25). Draw a scaled phasor diagram showing these two currents and from it determine the total current and the overall PF. Ignore PF improvement.

Choose a suitable scale (Fig. 10.26).

$$\cos \alpha = 0.985$$
$$\therefore \alpha = 10°$$
$$\cos \theta = 0.5$$
$$\therefore \theta = 60°$$

Figure 10.25 IA 5 0.4 A at 0.985 lead; IB 5 0.53 A at 0.5 lag.

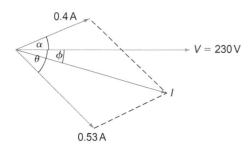

Figure 10.26

By measurement:

$$I = 0.77$$
$$\phi = 30°$$
$$\therefore PF = \cos \phi$$
$$= 0.866$$

Example

Two 230 V fluorescent lamps are arranged to overcome the stroboscopic effect. One unit takes 0.8 A at 0.45 PF leading, the other takes 0.7 A at 0.5 PF lagging (Fig. 10.27). Determine the total current drawn and the overall PF.

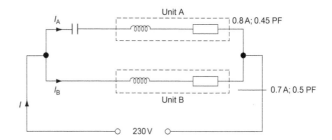

Figure 10.27

Method 1, by phasors

Unit A: $\cos \theta = 0.45$
$$\therefore \theta = 63.25°$$
Unit B: $\cos \alpha = 0.5$
$$\therefore \alpha = 60°$$

By measurement (Fig. 10.28):

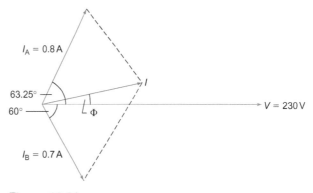

Figure 10.28

$$I = 0.71 A$$

and

$$\Phi = 8.5°$$
$$\therefore \cos \Phi = PF = 0.989 \text{ leading}$$

Method 2, by trigonometry

In this method, the active and reactive components of currents I_A and I_B are found (Fig. 10.29).

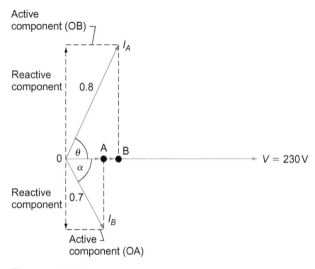

Figure 10.29

Active or horizontal component of $I_A = OB = I_A \times \cos \theta$.

$$\therefore OB = 0.8 \times \cos 63.25°$$
$$= 0.8 \times 0.45$$
$$= 0.36 \, A$$

Active component of $I_B = OA = I_B \times \cos \alpha$.
$$\therefore OA = 0.7 \times \cos 60°$$
$$= 0.7 \times 0.5$$
$$= 0.35 \, A$$

Since all the active components are in phase they may be added. Hence,

Total of active components $= 0.36 + 0.35 = 0.71 \, A$

Reactive or vertical components of $I_A = I_A \times \sin \theta$
$$= 0.8 \times \sin 63.25°$$
$$= 0.8 \times 0.893$$
$$= 0.714 \, A$$

Reactive component of $I_B = I_B \times \sin \alpha$
$$= 0.7 \times \sin 60°$$
$$= 0.7 \times 0.866$$
$$= 0.606 \, A$$

Since the two reactive components are opposite in phase they must be subtracted.

$$\therefore \text{Total of reactive components} = 0.714 - 0.606$$
$$= 0.108 \, A$$

The resultant of the total active and reactive components will be the total current taken (Fig. 10.30):

$$\tan \Phi = \frac{\text{perpendicular}}{\text{base}}$$
$$= \frac{0.108}{0.71}$$
$$\tan \Phi = 0.152$$
$$\therefore \Phi = 8.64°$$
$$\therefore PF = \cos \Phi = 0.988 \text{ leading}$$

and

$$\cos \Phi = \frac{\text{base}}{\text{hypotenuse}} = \frac{0.71}{I}$$
$$\therefore I = \frac{0.71}{\cos \Phi} = \frac{0.71}{0.988}$$
$$I = 0.718 \, A$$

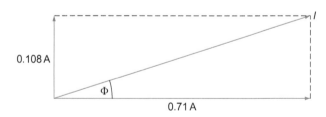

Figure 10.30

Self-Assessment Questions

1. With the aid of circuit diagrams explain the difference between a thermal and a glow-type starter for a fluorescent lamp.
2. Show with the aid of a sketch the construction of a high-pressure mercury-vapour lamp. Clearly label all parts.
3. Why should a low-pressure sodium-vapour lamp be mounted horizontally? What is the most common application for such a lamp? Why is this?
4. Compare the colour and efficacy of tungsten-filament, low-pressure sodium-vapour and low-pressure mercury-vapour lamps.
5. What colour of fluorescent lamp should be used for the following: (a) a butcher's shop, (b) a restaurant and (c) a jewellery shop?
6. Explain the purpose of the choke in a discharge lighting circuit.

7. Explain what is meant by (a) 'maintenance factor', (b) 'coefficient of utilization' and (c) 'the inverse-square law'.

8. (a) What is meant by 'illuminance'?
   (b) A light source of 850 cd is situated 2.5 m above a work surface. Calculate the illuminance directly below the light and 3 m horizontally away from it (at work surface level).

9. A small workshop 27 m × 17 m requires illuminance at bench level of 130 lx. Two types of lighting are available:
   (a) 150 W tungsten-filament lamps at 13 lm/W or
   (b) 80 W fluorescent lamps at 35 lm/W. Assuming that the MF in each case is 0.8 and that the CU is 0.6, calculate the number of lamps required in each case.

10. (a) A small supermarket 20 m long by 15 m wide is to be illuminated to a level of 600 lx by 2400 mm 125 W fluorescent lamps having an efficacy of 65 lm/W. The MF is 0.85 and the CU is 0.6. Calculate the number of fittings required and show their positions on a scale plan.
    (b) Calculate the total current taken by the lighting.

11. What are the recommendations with regard to (a) rating of fluorescent lamp units and (b) current rating of ordinary switches controlling fluorescent lighting circuits?

12. With the aid of a diagram explain the principles of operation of a fluorescent light unit.

13. What is meant by the **stroboscopic effect**, and how can it be minimized?

14. The following data relate to two 230 V fluorescent lighting units arranged to minimize stroboscopic effect:
    (a) Unit 1 – 0.75 A at 0.96 PF leading.
    (b) Unit 2 – 0.8 A at 0.6 PF lagging.
    By using a scaled phasor diagram determine the value of the total current taken by the two units. Check your answer by calculation.

10

CHAPTER 11

Electricity, the Environment and the Community

Environmental effects of the generation of electricity

In order to produce electricity by means of generators, we must provide a method of propulsion. This can be achieved in one of the following ways:

1. water power, or
2. steam power.

Water power (hydro-electric stations)

Where a natural and continuous flow of water is available (i.e. waterfalls or fast-flowing rivers), its potential energy can be used to turn a waterwheel coupled to a generator. This method is obviously inexpensive due to the fact that no cost is involved in providing the propulsion. However, in the United Kingdom the number of sites available for such generation on any useful scale is limited, and such sites are usually to be found in mountainous areas such as Scotland and Wales.

In some instances an artificial flow is produced by siting the generating station between two reservoirs, allowing the top one to discharge via the generator to the lower one, and then pumping the water back to the top reservoir (Fig. 11.1).

Figure 11.1 Hydro-electric system.

Finding such sites, however, where two natural reservoirs are available, is almost impossible, and valleys have to be flooded to provide the right conditions. This, of course, poses environmental problems, even if every effort is made to protect the local population from loss of agricultural land and hence livelihoods, and to avoid spoiling the landscape and adversely affecting tourist trade.

Siting of power stations

In view of the previous comments, it should be clear that the siting of a power station is extremely important in order for it to generate electricity at an economic level. Figure 11.2 shows the location of most power stations in England and Wales.

Steam power

Steam power involves the use of steam at very high pressure directed on to the blades of a wheel (turbine) coupled to a generator.

In order to produce this steam economically, readily available supplies of water and fuel are required.

Fuels

Fuels used in modern generating stations may be divided into two categories:

1. Fossil fuels (coal, oil, gas).
2. Nuclear fuel (uranium).

All fossil fuels have to be mined and transported to the power stations, and in view of this, it is clear that a

Electrical Installation Work. 978-1-138-84927-3 © Brian Scaddan. Published by Taylor & Francis.

Note: Operational power stations as at April 2002

Figure 11.2 Location and type of British power stations.

power station must be sited in an economic position. It would be pointless to locate a new station high in the Welsh mountains (conveniently out of sight) if it was impossible to transport the fuel to it. In this respect atomic or nuclear stations have an advantage in that the fuel is not bulky and transportation is relatively easy; hence it is possible to locate such stations in remote coastal areas.

Coal-fired stations

By far the greater part of electrical energy generated in Britain is provided by coal-fired stations.

The ideal situation for such a station would be adjacent to a coal mine and a large source of water (river or sea). If a mine is not at hand, the next best thing is good access for the delivery of the fuel, such as a railway or port.

Oil-fired stations

Many of the existing coal-fired stations have been converted to oil firing since the discovery of North Sea oil and the construction of overland pipelines. Another contributory factor to this changeover is the running down of the coal industry and the subsequent closure of many pits.

The oil is transported by rail, sea or by the pipelines previously mentioned.

However, it is an increasingly expensive fuel, and only a small percentage of stations are oil fired. Oil-fired stations should preferably be sited near ports.

Gas-fired stations

Gas-fired stations are rare. They are usually restricted to local generation and do not significantly contribute to the National Grid.

Nuclear stations

Although they are a subject of much controversy, new nuclear stations are being built, and modern technology is providing new techniques to make them more economic, relative to other types of power station.

The fuel is uranium, and relatively small quantities will produce vast amounts of heat. It is, however, a very expensive fuel, and the capital cost of building a power station to cope with all the dangers of using such fuel renders generation in this way, at present, uneconomic.

Nuclear power stations are usually sited in coastal areas.

Pollution

In any situation where a fuel is burnt, there is the attendant problem of pollution, and because of the large scale of the use of fuel in power stations, careful control of pollution is required.

Air pollution

When coal is burnt, the resulting smoke contains grit, dust and sulphur fumes, which have to be prevented from contaminating the environment. Much of the sulphur is washed out and the grit and dust remaining after treatment are prevented from reaching the ground, via low-level air currents, by ensuring that the chimney is very tall and that the discharge or emission of smoke into the air is at high velocities (15 to 25 m/s). In densely populated areas where high-rise buildings now exist, some older coal-fired stations have been converted to oil firing.

Water pollution

Pollution of sea- and especially river-water occurs when the water used in the station is recycled to its source at a higher temperature and with a considerable amount of impure solids suspended in it. The effect of this is to kill the river or marine life.

The discharge of polluted water into rivers is not illegal but the National Rivers Authority have the authority to pass by-laws controlling the amount discharged.

Control of pollution

Every month the National Grid Co. take readings from some 300–400 different gauges at 30 or more power stations in order to monitor the amounts of pollution directly due to the generation of electricity, and carry out research into grit removal and sulphur washing from smoke and the cleansing of waste water.

Waste disposal

Another major problem which has considerable environmental implications is the safe and effective disposal of waste.

Coal-fired stations

The waste from coal-fired stations is chiefly pulverized fuel ash, of which a controlled amount is allowed to be discharged into the sea or rivers. An increasingly large amount is used in the manufacture of building materials such as breeze block or Thermalite bricks, and the rest is deposited into old clay and gravel pits or built into artificial hills and landscaped.

Nuclear stations

Nuclear stations do not cause atmospheric pollution during normal operation, but there is always the attendant problem of the disposal of radioactive waste. A great deal of research is being carried out to investigate the possibility of using the waste, and recently some use has been found in a new type of nuclear station, the 'fast-breeder reactor'.

Disposal of radioactive waste is costly, as methods must be found to contain the material, with no risk of leakage, for the rest of its active life, which may be as long as 300 years. Lead-lined containers in steel shells and encased in concrete or glass many metres thick are effective in storing the contaminated waste; the whole is then deposited into the sea, or buried deep underground.

General waste and pollution

Clearly, power stations are not the only sources of waste and pollution, and there are several Acts and sets of Regulations that cover all aspects of protecting the environment.

1. **The Environmental Protection Act 1990:**
   This permits the 'Environment Agency' to place limits on the emission of polluting substances into the environment.
2. **The Pollution Prevention and Control Act 2000:**
   This has been superseded by the 'Environmental Permitting Regulations 2008'. Essentially, these regulations require an organization to have a permit to discharge waste.
3. **The Hazardous Waste Regulations 2005:**
   These regulations govern the production and disposal of hazardous waste.
4. **The Control of Pollution Act 1974:**
   This deals with the control of noise, smoke and collection and disposal of waste on land and into water.
5. **The Waste Electrical and Electronic Equipment Regulations 2009 (WEEE):**
   These regulations deal with the disposal of harmful toxic waste substances found in electrical and electronic equipment (e.g. TV's, batteries).
6. **The Packaging (Essential Requirements) Regulations 2003:**
   These are intended to reduce the impact that packaging has on the environment.
7. **The Environment Act 1995:**
   This Act created a number of new agencies and set new standards for the management of the environment.

Micro-renewable energy

There are various types of micro-renewable energy:
▶ Heat-producing
▶ Electricity-producing
▶ Co-generation
▶ Water conservation

Each of these types can be further sub-divided:

Heat producing

▶ Ground source heat pumps
▶ Air source heat pumps
▶ Biomass
▶ Solar thermal (hot water)

Electricity producing

▶ Solar photovoltaic
▶ Micro-wind
▶ Micro-hydro

Co-generation

▶ Micro-combined heat and power (heat-lead)

Water conservation

▶ Rainwater harvesting
▶ Greywater reuse

Heat-producing renewables
Ground source heat pumps (GSHP)

These work on the same principle as a refrigerator in that heat is extracted from the air inside the fridge by a heat exchanger and dissipated to the air outside.

In the case of a ground source heat pump, heat, which is caused by solar irradiation, is extracted from the ground via pipework filled with refrigerant (condenser/heat exchanger) and delivered to the property via an evaporator/compressor unit. This is known as the refrigeration cycle. The whole system is sealed.

The plastic pipework is commonly in the form of a coil, sometimes called a 'slinky' and is either buried horizontally in trenches at least 1m deep or, if land area is restricted, vertically in bore holes which may need to be up to 100m deep.

Figure 11.3 Slinky.

Planning permission: This may only be required if the land is in a conservation area or the property is a listed building.

Building Regulations: The Building Regulations Part 'L' for CO_2 emissions for energy efficiency would be fulfilled as the Coefficient of Performance (COP) claimed by manufacturers is around 4, i.e. 1kW of electrical energy (used to power the system) delivers 4kW of thermal heat, although in practice a COP of 2–3 is more likely. Any electrical work would need Part 'P' compliance

F- Gas Regulations 2006: These apply to the refrigerants used and are designed to limit the production and/or emission of hydro-carbons.

Pros and Cons

Pros:

▶ Reasonably efficient system especially if the electricity, needed for the system, is from a renewable source eg. Solar pv.
▶ Long life, around 40–50 years
▶ System can be reversed to provide cooling in summer
▶ Low maintenance as it is a sealed system
▶ Government grants are available to help with installation costs.

Cons:

▶ The building needs to be very well insulated having a low building heat loss
▶ No real use for those on mains gas
▶ Best for new build where building thermal insulation is highly efficient
▶ Low heat source so only useful for underfloor heating
▶ Needs electricity to power the system

▶ High installation costs
▶ Needs large area of ground unless bore holes are used

Air source heat pumps

These are similar to ground source heat pumps (GSHP) but in this case the heat is extracted from the outside air (they cease to function properly below −20°C) and either circulated via ducting (air to air) or used to heat the domestic water (air to water).

During cold spells the external pipework can freeze and to de-frost the system, hot water is taken from a 'buffer tank' rather than from the radiators or underfloor heating.

Planning permission: At present the installation of air source heat pumps are not permitted development and the advice of the Local Authority should be sought.

Building Regulations: As with GSHP's, compliance with Part 'L' will be achieved and where an air to air system is installed Part 'F' on ventilation will need to be considered. Any electrical work would need Part 'P' compliance

F- Gas Regulations 2006: These apply to the refrigerants used and are designed to limit the production and/or emission of hydro-carbons.

Pros and Cons

Pros:

▶ Less expensive to install than GHSP's
▶ Can provide heating and hot water
▶ Can provide air conditioning in the summer
▶ The COP is around 2–3

Cons:

▶ The building needs to be very well insulated having a low building heat loss

Figure 11.4 An air source heat pump.

125

▶ Lose efficiency in cold weather
▶ Noisy and blow cold air into the surrounding environment
▶ Needs electricity to power the system
▶ Not advisable for properties on mains gas
▶ Best for new build where building thermal insulation is highly efficient

Biomass

This refers to a fuel that is produced organically, either grown (wood) or as waste (agricultural etc.)

Energy from biomass is generated at different scales:

▶ Large/grid scale, from dedicated power stations
▶ Medium/district scale, power plants that provide energy to schools and small housing schemes
▶ Small/domestic scale, biomass boilers using pellets or logs

The small scale biomass boilers simply replace the energy sources derived from fossil fuels.

Clearly, there has to be storage facilities for fuel which needs to be kept dry. Pellets are usually kept in hoppers and fed to the boiler by an auger or in larger installations by hydraulics and/or conveyors.

Planning permission: This is permitted development unless:

▶ The flue exceeds 1m above the roof height, excluding the chimney or
▶ The flue is installed on the principle elevation and is visible from a road in buildings in Conservation Areas and World Heritage Sites.

Building Regulations: Part J of the Building Regulations 'Heat producing appliances' needs consideration and in particular the details of flue sizes which can prove difficult to comply with. Any electrical work would need Part 'P' compliance

DEFRA: Require smoke control exempted appliances

Pros and Cons

Pros:

▶ It is a sustainable fuel source
▶ It is almost carbon neutral
▶ Replacing conventional systems with biomass can reduce CO_2 emissions by 9.5 tonnes per annum

Cons:

▶ Biomass boilers are much larger than gas or oil boilers
▶ Additional heat storage may be required if the heat produced is greater than the demand
▶ There may be problems in 'smoke free' zones

Figure 11.5 Small, pellet fuel boiler, pellets made from 100% wood.

Figure 11.6 Log burning stove.

▶ Installation costs are higher than gas or oil
▶ Reliable source of fuel is essential

Solar thermal

Anyone who has a garden hose and left it un-rolled on a hot day will know how hot the water is when the hose is turned on. This effect is the basic principle of solar thermal heating.

Solar thermal panels absorb heat from sunlight either via evacuated glass tubes or flat collector plates and

Figure 11.7 Solar thermal heating system.

Figure 11.8 Evacuated tube panel (the manifold can be seen at the top).

transfers the heat to water which is distributed around the system by a circulation pump which is regulated by a differential temperature controller.

Evacuated tubes

Pipes containing water or an antifreeze, typically propylene glycol, are encased in an evacuated

Figure 11.9 Operation of evacuated tube.

(vacuum) glass tube which reduces heat loss. As the liquid in the pipes heats it rises to the top and evaporates dissipating the heat to a manifold which is connected to the hot water storage tank.

Flat plate collector

This system comprises a flat absorber plate or series of flat absorber strips to which flow tubes are fixed. The absorbers are usually copper with a black coating to increase absorption. The heat from the absorber/s transfer the heat to water flowing in the tubes. The assembly is similar to an automobile radiator.

There is not a vast difference between the performances of these two systems. The evacuated tube is more efficient but more expensive to install and repair where the flat plate type is virtually maintenance free.

Planning permission: For roof mounted systems this is permitted development in England unless the panels protrude more than 200mm when installed. There are additional restrictions in Scotland.

Stand alone systems are permitted development unless they are more than 4m high and/or are less than 5m away from any boundary and/or have an array area greater than 9m² or where they are visible from

Figure 11.10 Flat plate solar thermal panels.

Figure 11.11 Flat plate collector.

the highway in Conservation Areas and World Heritage Sites. There are additional restrictions in Scotland

Building Regulations: Part 'A' building structure needs to be considered as there may be issues regarding the strength of the roof and the weight of the panels. Also if pump and control systems are installed, compliance with Part 'P' is required.

As pipework will penetrate the building structure Part 'C' 'Resistance to moisture' will need consideration

Pros and cons

Pros:

▶ Environmentally friendly
▶ Efficient
▶ Government grants available towards installation costs
▶ Saving in energy costs
▶ Low maintenance

Cons:

▶ Only really suitable for south facing aspects
▶ Expensive to install
▶ Not so useful in winter
▶ Water storage needed when heat generation exceeds demand
▶ Electricity needed where pumps and controls are installed

Electricity-producing renewables

Solar photovoltaic (pv)

This is probably the most common and most widely advertised method of providing renewable energy.

Solar panels (modules) consist of a series of photoelectric cells that convert sunlight to d.c. electrical energy. The energy may be stored in a battery where no electricity supply is available or fed into an existing installation. An inverter is used to convert the d.c. voltage to 230V a.c. 50Hz.

The more sunlight that falls on the module (insolation) the greater the electrical output and hence the positioning and angle of the unit is of great importance. In the UK, for the best performance, modules should face due south and be at an angle of between 30^0 and 40^0. Modules are rated in kW_p which is the maximum output under Standard Test Conditions (STC) which in turn are an irradiance (insolation) of $1000W/m^2$ at 25^0C cell temperature.

Pv cells are constructed from silicone and are generally manufactured in three types,

Monocrystalline, Polycrystalline and Thin Film. The cells are arranged in groups to form modules.

Crystalline modules are manufactured to BS EN 61215 and Thin film to BS EN 61646.

Monocrystalline	Polycrystalline	Thin Film
15% – 21% efficient	12% 16% efficient	3 – 13% efficient

The PV cell, basically, comprises two layers of silicon which are 'doped' with impurities such that one layer is negatively charged and the other positively charged. Energy from the sun, in the form of Photons causes an agitation of the electrons between the layers which results in a d.c. voltage appearing between the front and back contacts.

There are two arrangements for pv systems, 'Grid tied' and 'Off-grid'

The 'Grid tied' system runs in parallel with the normal supply, the 'Off-grid' system is suitable where the is no normal supply, in this case the electricity generated is stored in batteries and converted to a.c. when required.

Figure 11.12 Monocrystalline Panel.

Figure 11.13 Polycrystalline Panel.

Figure 11.14 Monocrystalline PV array.

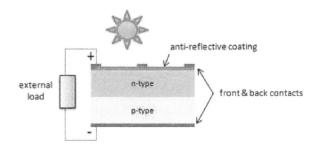

Figure 11.15 The PV cell.

Planning permission: For roof mounted systems this is permitted development in England unless the panels protrude more than 200mm when installed. There are additional restrictions in Scotland.

Stand alone systems are permitted development unless they are more than 4m high and/or are less than 5m away from any boundary and/or have an array area greater than 9m² or where they are visible from the highway in Conservation Areas and World Heritage Sites. There are additional restrictions in Scotland

Building Regulations: This is the same as for Solar thermal installations but with more emphasis on Part 'P'

Pros and Cons

Pros:

▶ Environmentally friendly
▶ No moving parts
▶ Saves on energy costs
▶ Government 'feed-in' tariffs (FiT) available

Cons:

▶ Expensive to install, more than solar thermal
▶ Only any use when facing south-west to south-east

▶ Shading from trees, chimneys and other building reduces efficiency
▶ Reduced output on cloudy days
▶ Large roof area needed. (10m2 for 1kW)

Micro-wind

Energy harnessed from the wind has been with us for many centuries in the form of windmills to grind corn etc. Wind turbines do the same thing by converting the power of the wind to d.c. electricity via a generator.

There are two types of domestic wind turbine:

▶ Mast mounted usually around 2.5kW – 6kW
▶ Roof mounted usually 1kW – 2kW

There are also two blade configurations

▶ Standard propeller type
▶ Vertical axis type

Planning permission: Yes this is required.

Building regulations: Part 'A' building structure needs to be considered as there may be issues regarding the strength of the roof and the weight and

Figure 11.16 Small wind turbine with two propellers and solar roof.

Figure 11.17 Mast mounted turbine.

Figure 11.18 Vertical axis turbine.

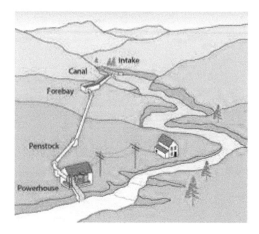

Figure 11.19 Micro-hydro system.

wind loading of the turbine. Also, compliance with Part 'P' is required as the electrical installation is similar to that for solar PV systems.

Pros and cons

Pros:

▶ No CO_2 emissions
▶ Reduction in energy costs
▶ Government 'feed-in' tariffs available

Cons:

▶ Only suitable in high areas with an average wind speed of 5m/s and low turbulence.
▶ Needs planning permission
▶ Expensive to install and maintain
▶ Can be noisy (manufacturers claim reduced noise from vertical axis types)
▶ Moving parts so not maintenance free
▶ Unsightly

Micro-hydro

Using the energy of flowing water to drive a water-wheel has been in use for thousands of year and we are all familiar with Hydro-electric power stations. Micro-hydro is just a smaller version of these.

Such systems require a continual supply of moving water eg. rivers, large steams, waterfalls etc. For rivers and small steams there needs to be an appropriate gradient from the intake of the system to the outlet/powerhouse, as seen in the figure below. The forebay is a small storage reservoir and the penstock is the pipe valve that controls the flow of water.

Mechanisms used to drive the turbine/generator include the Archimedes Screw, the traditional waterwheel and turbine blades in an enclosure.

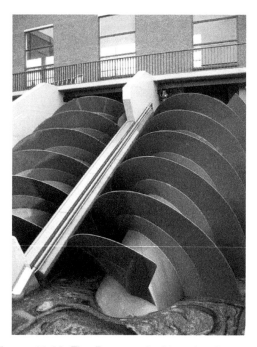

Figure 11.20 The Reverse Archimedes Screw. Picture by M.A. Wijngaarden. Source: Wikimedia Commons.

Figure 11.21 Traditional waterwheel.

Impulse hydro-turbines require a high head and high water flow. The Archimedes Screw type, usually called a 'Reverse Archimedes Screw' is used where there is a low head.

Planning permission: Yes this is required, as is contact with developers, the Environment Agency and other organisations such as Natural England.

Building Regulations: In this instance it is likely that only Part 'P' will be involved.

Pros and cons

Pros:

- Free energy source
- No CO_2 emissions
- Reasonably low maintenance costs
- Savings on energy costs (some systems can replace normal electricity supplies.
- Government feed-in tariffs available

Cons:

- Installations are limited to locations where rivers, small steams or waterfalls exist.
- Very expensive to install
- At dry times of the year the water flow may not be enough to supply all the electricity needs.

Co-generation

Micro-combined heat and power CHP (heat led)

Basically, Micro-CHP is an engine that drives a generator. The heat from the engine is used to heat water whist the generator produces electricity.

There are two types of engine used

- The internal combustion engine
- The Stirling engine

The internal combustion engine in this case is basically a diesel engine powered by mains gas or Low Pressure Gas (LPG).

The Stirling engine invented in 1816 by Robert Stirling, comprises two pistons in a closed system filled with gas. When heated, the gas expands, lifting one piston whilst compressing the other. The compressed gas then pushes that piston back up gives and the cycle repeats. The compression part of the cycle causes the cylinder to give off heat.

Whichever system is used, it is incorporated in a unit about the same size as a conventional boiler and the main output is heat with some electricity generation.

Figure 11.22 CHP Efficiency.

Figure 11.23

Figure 11.24

Planning permission: No these can be direct replacements for conventional boilers.

Building Regulations: Part 'J' Heat producing appliances as well as Part 'P' will need to be complied with.

Pros and cons

Pros:

▶ Produces electricity as a by product of heat
▶ Inexpensive to install (similar to a standard gas boiler)
▶ Very efficient
▶ Government 'Feed-in' tariffs available

Cons:

▶ Not completely carbon neutral as electricity or gas is required as a pre- heat source.
▶ Cost of the unit can be up to twice that of a conventional boiler

Water conservation

This is more 'reusable' rather than 'renewable' energy and is divided into two areas:
▶ Rainwater harvesting
▶ Greywater reuse

Rainwater harvesting

This is simply collection of rainwater, usually from roofs, in an above ground or underground storage tank. The water is filtered and pumped to a tank in the loft and distributed for use with toilets, sinks, washing machines, car washing, garden watering etc.

This system i.e. gravity fed does mean slower cistern and washing machine filling and less pressure for car washing and garden hose use and mains back-up is

Figure 11.26 pumped system.

usually required. Many modern systems in fact use a pump to create mains water pressure.

Planning permission: May only be required with regards to storm-water and mains drainage.

Building Regulations: Compliance with Part 'G' Sanitation, Hot water safety and Water efficiency and Part 'H' Drainage and Waste Disposal is required. Where electrical control systems, pumps, valves are installed, Part 'P' will apply.

Water Supply (Water fittings) Regulations 1999 Section 6: This requires the prevention of back-flow of 'unwholesome water' contaminating the mains water.

Pros and cons

Pros:

▶ Free resource
▶ Significantly reduces costs where water metering is installed
▶ Helps to reduce flooding during heavy storm run-off
▶ Little or no lime-scale build up
▶ Excellent for garden use.

Cons:

▶ Cannot be used for drinking unless specially treated
▶ For retro-fit, existing plumbing arrangement will have to be modified and all external drainage i.e. gutters and down pipes will need to be altered to divert all water to a central point.

Greywater reuse

Waste water from showers, baths, and hand basins is treated to a reasonable level of cleanliness and re-cycled to flush toilets and in some instances to fill washing machines. There are various systems for the treatment and use of greywater, these include:

Figure 11.25 Gravity fed system.

Figure 11.27 Greywater reuse system.

▶ **Direct reuse.** Here there is no storage, the greywater is used directly for use in gardens.
▶ **Short retention.** In this case the greywater is stored for a short period and surface solids such as soap particles and hair etc. are removed by a 'surface skimmer'. Useful for flushing toilets, car washing etc.
▶ **Biological.** Here the greywater passes through sand or soil containing plants such as reeds. The microbes in the soil/plants break down the organic matter into nutrients for the plants and the cleansed water is used for irrigation etc.
▶ **Bio-mechanical.** This is similar to sewage treatment where filtration is carried out mechanically and biologically and with disinfectants. Can then be used in washing machines.

Planning permission: Not required.

Building Regulations: Compliance with Part 'G' Sanitation, Hot water safety and Water efficiency is required. Where electrical control systems, pumps, valves are installed, Part 'P' will apply.

Water Supply (Water fittings) Regulations 1999 Section 6: This requires the prevention of back-flow of "unwholesome water" contaminating the mains water.

Pros and cons

Pros:

▶ Reduces the mains water requirements for toilet flushing
▶ Can be used for garden watering

Cons:

▶ Very expensive especially in existing properties
▶ Long time to recover installation costs
▶ Use usually limited to toilet flushing

The purpose and function of the national grid

Historical background

In the early days of electricity supply, each British town or city had its own power station which supplied the needs of its particular area.

Standardization was not evident and many different voltages and frequencies were used throughout the country. By the time of the First World War (1914–1918), there were some 600 independent power stations in use. However, the heavy demands made by the war industry showed the inadequacies of the system and several select committees were set up to investigate possible changes. Little was achieved until 1926 when it was suggested that 126 of the largest and most efficient power stations should be selected and connected by a grid of high-voltage transmission lines covering the whole country, and, at the same time, the frequency standardized at 50 Hz. The remaining power stations would be closed down, and local supply authorities would obtain their electricity in bulk from the grid, via suitable substations. By 1932 the system was in operation with the addition of several large, new power stations. The system voltage was 132 000 V (132 kV), and the supply frequency 50 Hz.

On 1 April 1948 the whole of the electricity supply industry was nationalized and in 1957 the 'Central Authority' responsible for the generation of electricity was renamed the 'Central Electricity Generating Board' (CEGB).

Since then, of course, the electricity industry has become privatized and the CEGB has been replaced by the National Grid Co. who buy, at the lowest price, generated electricity from such companies as National Power, PowerGen, Nuclear Electric, French Electric, Scottish Hydro Electric, etc.

The purpose of the grid system

The purpose of the grid system is to maintain a secure supply of electricity at a standard voltage and frequency to consumers throughout the country. Having stated its purpose, we can now list several advantages that have resulted from its introduction:

1. Security of supplies.
2. Standardization of frequency and voltages.
3. Economy.
4. The ability to transmit very large loads for considerable distance without loss.
5. The ability to transfer electricity to and from different parts of the country.

The function of the grid

In order to fulfil its purpose, the grid system must function in the following way.

The National Grid Control Centre in Wokingham, in association with the various grid control centres around the country, estimates the load required in different areas each day. This information is then used to arrange to purchase the country's power depending on the demand. In this way stations are used to their maximum efficiency, which in turn reduces the cost of generation. Due to the fact that the system is interconnected, bulk supply points can be fed from other areas, should a failure of the usual supply occur.

Generation, transmission and distribution systems

The very nature of the grid system is such that power has to be transmitted over large distances. This immediately creates a problem of voltage drop. To overcome this problem, a high voltage is used for transmission (400 or 132 kV), the 400 kV system being known as the **Super Grid**. We cannot, however, generate at such high voltages (the maximum in modern generators is 25 kV), and transformers are used to step up the generated voltage to the transmission voltage. At the end of a transmission line is a grid substation where the requirements of the grid system in that area can be controlled, and where the transmission voltage is stepped down via a transformer to 132 kV.

It is at this voltage that the different Distribution Network Operators (DNOs) transmit the power required by their consumers around that particular area. The system voltage is then further reduced at substations to 33 000, 11 000 and 415/240 V.

The declared voltage at consumer terminals is 400 V three-phase/230 V single-phase. However, the measured voltage is still likely to be 415/240 V for many years.

Overhead lines and underground cables

The only economic method of transmitting power at grid voltages is by means of overhead lines; cables are used only in very short lengths. However, the cost of producing cable rated at the lower voltages 33 kV, 11 kV and 415 V enables its use to be widespread. The topic of lines and cables is dealt with in greater detail later.

System layout

Figure 11.3 illustrates how electricity is made and conveyed to the consumer.

The aesthetic effects of the siting of generation and transmission plant

An unfortunate situation has arisen in modern times from the public's conflicting desires for:

1. An ever-increasing supply of electrical energy.
2. The beauty of the countryside to be preserved.

The task of completely satisfying both of these requirements will remain impossible until man can find a way of generating and transmitting the country's energy needs without the use of power stations and overhead lines.

Contrary to popular opinion, however, the National Grid Co. and DNOs **cannot** build power stations and erect lines wherever they please. A large body of organizations oppose or approve decisions made by the Boards until a satisfactory compromise can be reached.

Siting of power stations

The siting of power stations in relation to fuel and water resources was discussed on pp. 201–203, and, of course, these are primary considerations in deciding where a new station should be located. However, the aesthetic aspects must also be considered.

How can something as large as a modern power station be hidden?

Obviously it cannot, but could the design be made to blend with the surrounding countryside? Once again this would be almost impossible to achieve, as the efficient working of such a building depends on a very definite design that does not lend itself to the irregular contours of the landscape.

It only remains, therefore, to try to make the best of a difficult job. Within the design limits, the buildings

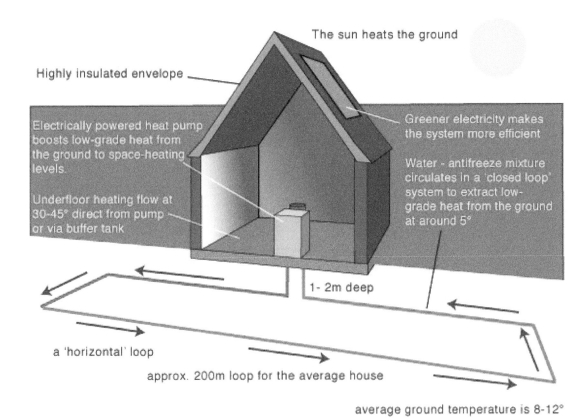

The sun heats the ground

Highly insulated envelope

Electrically powered heat pump boosts low-grade heat from the ground to space-heating levels.

Underfloor heating flow at 30-45° direct from pump or via buffer tank

Greener electricity makes the system more efficient

Water - antifreeze mixture circulates in a 'closed loop' system to extract low-grade heat from the ground at around 5°

1- 2m deep

a 'horizontal' loop

approx. 200m loop for the average house

average ground temperature is 8-12°

Figure 11.28 An example of a GSHP powered space heating system.

are made streamlined using modern materials and as much glass as possible, together with an off-white finish to the main structures. This, at least, is more acceptable to the majority of people than was the older type of brick-built station.

Transmission lines

Opinion is divided as to the beauty of transmission lines; some think pylons graceful, others think them ugly. Once again, within the design limits, an attempt has been made to make them as graceful as possible. Why, the public asks, cannot the power be transmitted by underground cables? The simple answer is, economy.

Enormous costs would be involved in manufacturing a cable whose insulation was capable of withstanding 400 or 132 kV between phases. The total cost of manufacturing and installation of 400 kV oil-cooled cable, together with all the necessary oil pressure equipment along the route, and compensation to farmers for destruction of agricultural land, is 12 times as much per mile as for transmission lines.

If this information is coupled with the fact that locating and repairing a fault on underground cable would involve many times the cost compared with that for a fault on an overhead line, it will be seen that overhead transmission lines are the only economic proposition.

CHAPTER 12

Health and Safety

Whilst enjoying leisure pursuits, especially those of a sporting nature, most of us wear the accepted clothing suited to the sport in question, such as cricket and football pads, knee and elbow protectors and crash helmets for skateboarding, and padded jackets and visors for fencing. However, in the hazardous and sometimes potentially lethal environment of our place of work, many of us choose to ignore the warnings and carry out our duties without the use of the recommended protective clothing. It is this careless attitude that accounts for thousands of lost man-hours in industry every year, not to mention the pain and suffering of the casualty.

By the same token many employers are seriously at fault in not providing safe working conditions, and it is up to us as potential casualties to bring the lack of safety to our employers' notice. Remember accidents don't just happen, they are caused either by our own carelessness or by that of others.

It is for these reasons that over the years a number of safety regulations have come into being, designed to protect us from injury at work. The Health and Safety at Work Act 1974 is the overriding legislation from which spring a number of other Regulations covering specific areas. The following list indicates some of these Regulations, the PPE, COSHH, EAWR, CDM and Building Regulations being most closely related to electrical installation work.

Safety regulations

In addition to the Health and Safety at Work Act, the following regulations also apply:

1. **Management of Health and Safety at Work Regulations 1999**:
   This requires employers to carry out risk assessments, appoint competent personnel and arrange for appropriate training and development.
2. **Workplace (Health, Safety and Welfare) Regulations 1992**:
   This deals with issues such as heating, ventilation, lighting, seating etc.
3. **Health and Safety (Display Screen Equipment) Regulations 1992**:
   This covers the requirements for VDU's.
4. **Personnel Protective Equipment at Work Regulations (PPE) 1992**:
   This requires employers to provide suitable PPE for their employees.
5. **Provision and Use of Work Equipment Regulations (PUWER) 1998**:
   This requires all work equipment to be safe.
6. **Manual Handling Operations Regulations 1992**:
   This covers the movement of objects by bodily force.
7. **Health and Safety (First Aid) Regulations 1981**:
   This covers first aid requirements.
8. **The Health and Safety Information for Employees Regulations 1989**:
   This deals with the display of posters informing employees about health and safety.
9. **Employers Liability (Compulsory Insurance) Act 1969**:
   This requires employers to insure their employees against accident or illness.

Electrical Installation Work. 978-1-138-84927-3 © Brian Scaddan. Published by Taylor & Francis. All rights reserved.

10. **Electricity at Work Regulations 1989 (EAWR)**:
This places a responsibility on any person who has control of an electrical system to ensure safety.

11. **Reporting of Injuries, Diseases, and Dangerous Occurrences Regulations 1995 (RIDDOR)**:
This requires employers to report injuries, diseases etc.

12. **Control of Substances Hazardous to Health Regulations 2002 (COSHH)**:
This requires employers to assess the risks of and take appropriate precautions regarding hazardous substances.

13. **Noise at Work Regulations 2005**:
This requires employers to protect employees from damage to hearing.

14. **Dangerous Substances and Explosive Atmosphere Regulations 2002**:
This requires employers and the self-employed to assess the risks of working with dangerous substances.

15. **Construction (Design and Management) Regulations 2007 (CDM)**:
This deals with safe systems of work on construction sites.

16. **Control of Major Accident Hazards 1999**:
This requires notification to the relevant authority by those who move, store, manufacture etc. hazardous substances.

17. **Chemicals (Hazardous Information and Packaging for Supply) Regulations (CHIP) 2009**:
This requires suppliers of dangerous chemicals to provide data sheets, labels etc for such items.

18. **Gas Safety (Installation and Use) Regulations 1998**:
This covers the safe installation and maintenance of domestic and commercial gas systems.

19. **Working at Heights Regulations 2005**:
This details the requirements relating to the prevention of falls when working at heights.

20. **Control of Asbestos Regulations 2006**:
This deals with the use, disturbance and removal, etc. of materials containing asbestos.

These are known as statutory regulations, which means they have the force of law, and non-compliance with them can result in a fine, imprisonment, or, in some cases, both.

The regulations issued by the Institution of Electrical Engineers (IET Wiring Regulations), which are not statutory, are designed to ensure minimum safety standards in the installation of electrical equipment in buildings, and are based on an internationally agreed plan.

There are other statutory Acts which cover situations such as railways, mines and quarries and, for the time being, all these separate Acts will remain in force, even though the Health and Safety at Work Act 1974 deals with the welfare of virtually every person at work.

This Act is extremely important and the next few sections give a general outline of its requirements.

The Health and Safety at Work Act 1974

The Act outlined

The purpose of the Health and Safety at Work Act 1974 is to provide the legislative framework to promote, stimulate and encourage high standards of health and safety at work.

The minister primarily concerned with the Act is the Secretary of State for Employ-ment, and he or she makes most of the regulations in consultation with other appropriate ministers.

The Health and Safety at Work Act makes one comprehensive and integrated system of law to deal with the health and safety of virtually all people at work, and the protection of the public where they may be affected by the activities of people at work. The Health and Safety Commission and the Health and Safety Executive administer the legislation and are a focus of initiative for all matters relating to health and safety at work.

All 'persons at work', whether employees or self-employed, are covered with the exception of domestic servants in a private household. About five million people, such as those employed in education, medicine, leisure industries and in some parts of the transport industry who have not previously been covered by safety legislation, are protected for the first time.

Duties of employers

Employers must safeguard, as far as is reasonably practicable, the health, safety and welfare of all the people who work for them. This applies in particular to the provision and maintenance of safe plant and systems of work, and covers all machinery, equipment and appliances used.

Some examples of the matters which many employers need to consider are as follows:

1. Is all plant up to the necessary standards with respect to safety and risk to health?
2. When new plant is installed, is latest good practice taken into account?

3. Are systems of work safe? Thorough checks of all operations, especially those operations carried out infrequently, will ensure that danger of injury or to health is minimized. This may require special safety systems, such as 'permits to work'.

4. Is the work environment regularly monitored to ensure that, where known toxic contaminants are present, protection conforms to current hygiene standards?

5. Is monitoring also carried out to check the adequacy of control measures?

6. Is safety equipment regularly inspected? All equipment and appliances for safety and health, such as personal protective equipment, dust and fume extraction, guards, safe access arrangements, monitoring and testing devices, need regular inspection (sections 2(1) and 2(2) of the Act).

No charge may be levied on any employee for anything done or provided to meet any specific requirement for health and safety at work (section 9).

Risks to health from the use, storage or transport of 'articles' and 'substances' must be minimized. The term **substance** is defined as 'any natural or artificial substance whether in solid or liquid form or in the form of gas or vapour' (section 53(1)).

To meet these aims, all reasonably practicable precautions must be taken in the hand-ling of any substance likely to cause a risk to health. Expert advice can be sought on the correct labelling of substances, and the suitability of containers and handling devices. All storage and transport arrangements should be kept under review.

Safety information and training

It is now the duty of employers to provide any necessary information and training in safe practices, including information on legal requirements.

Duties to others

Employers must also have regard for the health and safety of the self-employed or contractors' employees who may be working close to their own employees; and for the health and safety of the public who may be affected by their firm's activities.

Similar responsibilities apply to self-employed persons, manufacturers and suppliers.

Duties of employees

Employees have a duty under the Act to take reasonable care to avoid injury to themselves or to others by their work activities, and to co-operate with employers and others in meeting statutory requirements. The Act also requires employees not to interfere with or misuse anything provided to protect their health, safety or welfare in compliance with the Act.

Enforcement

If an inspector discovers a contravention of one of the provisions of the Act, he or she can:

1. issue a prohibition notice to stop the activity,
2. issue an improvement notice to remedy the fault within a specified time,
3. prosecute any person instead of or in addition to serving a notice.

Contravention of some of the regulations can lead to prosecution summarily in a Magistrates' Court. The maximum fine is £20 000. Imprisonment for up to 2 years can be imposed for certain offences. In addition to any other penalty, the court can make an order requiring the cause of the offence to be remedied.

The Health and Safety Commission

The Health and Safety Commission consists of representatives of both sides of industry and the local authorities. It takes over from government departments the responsibility for developing policies in the health and safety field.

The Health and Safety Executive

This is a separate statutory body appointed by the commission which works in accordance with directions and guidance given by the commission. The executive enforce legal requirements, as well as provide an advisory service to both sides of industry.

Electricity at Work Regulations 1989

Persons on whom duties are imposed by these regulations

1. Except where otherwise expressly provided in these regulations, it shall be the duty of every:

   (a) employer and self-employed person to comply with the provisions of these regulations in so far as they relate to matters which are within his or her control

   (b) manager of a mine or quarry (within either case the meaning of section 180 of the Mines and Quarries Act 1954) to ensure that all requirements or prohibitions imposed by or under these regulations are complied with in so

139

far as they relate to the mine, quarry or part of a quarry of which he or she is the manager and to matters which are within his or her control.

2. It shall be the duty of every employee while at work:
   (c) to co-operate with their employer so far as is necessary to enable any duty placed on that employer by the provisions of these regulations to be complied with
   (d) to comply with the provisions of these regulations in so far as they relate to matters which are within his or her control.

Employer

1. For the purposes of the regulations, an employer is any person or body who (a) employs one or more individuals under a contract of employment or apprenticeship or (b) provides training under the schemes to which the H & SW Act applies through the Health and Safety (Training for Employment) Regulations 1988 (Statutory Instrument No. 1988/1222).

Self-employed

2. A self-employed person is an individual who works for gain or reward otherwise than under a contract of employment whether or not he or she employs others.

Employee

3. Regulation 3(2)(a) reiterates the duty placed on employees by section 7(b) of the H & SW Act.
4. Regulation 3(2)(b) places duties on employees equivalent to those placed on employers and self-employed persons where these are matters within their control. This will include those trainees who will be considered as employees under the regulations described in paragraph 1.
5. This arrangement recognizes the level of responsibility which many employees in the electrical trades and professions are expected to take on as part of their job. The 'control' which they exercise over the electrical safety in any particular circumstances will determine to what extent they hold responsibilities under the regulations to ensure that the regulations are complied with.
6. A person may find him or herself responsible for causing danger to arise elsewhere in an electrical system, at a point beyond their own installation. This situation may arise, for example, due to unauthorized or unscheduled back feeding from

the installation on to the system, or to raising the fault power level on the system above rated and agreed maximum levels due to connecting extra generation capacity, etc. As such circumstances are 'within his or her control', the effect of Regulation 3 is to bring responsibilities for compliance with the rest of the regulations to that person, thus making them a duty holder.

Absolute/reasonably practicable

7. Duties in some of the regulations are subject to the qualifying term 'reasonably practicable'. Where qualifying terms are absent the requirement in the regulation is said to be absolute. The meaning of reasonably practicable has been well established in law. The interpretations below are given only as a guide to duty holders.

Absolute

8. If the requirement in a regulation is 'absolute', for example if the requirement is not qualified by the words 'so far as is reasonably practicable', the requirement must be met regardless of cost or any other consideration. Certain regulations making such absolute requirements are subject to the defence provision of Regulation 29.

Reasonably practicable

9. Someone who is required to do something 'so far as is reasonably practicable' must assess, on the one hand, the magnitude of the risks of a particular work activity or environment and, on the other hand, the costs in terms of the physical difficulty, time, trouble and expense which would be involved in taking steps to eliminate or minimize those risks. If, for example, the risks to health and safety of a particular work process are very low, and the cost or technical difficulties of taking certain steps to prevent those risks are very high, it might not be reasonably practicable to take those steps. The greater the degree of risk, the more the weight that must be given to the cost of measures needed to prevent that risk.
10. In the context of the regulations, where the risk is very often that of death (e.g. from electrocution) and where the nature of the precautions which can be taken are so often very simple and cheap (e.g. insulation), the level of duty to prevent that danger approaches that of an absolute duty.
11. The comparison does not include the financial standing of the duty holder. Furthermore, where

someone is prosecuted for failing to comply with a duty 'so far as is reasonably practicable', it would be for the accused to show the court that it was not reasonably practicable for him or her to do more than he or she had in fact done to comply with the duty (section 40 of the H & SW Act).

Whilst the Health and Safety at Work Act and the Electricity at Work Regulations are the major legislative documents applicable to electrical operatives, there are others that must seriously be considered by such operatives.

Personal Protective Equipment Regulations (PPE) 1992

These regulations detail the requirements for safety regarding protective clothing, tools, etc., for example hard hats, protective footwear, rubber mats and insulated tools. Signs are usually posted on site to indicate the need for the use of such equipment.

Construction (Design and Management) Regulations

Whilst these regulations are important to all operatives 'on-site', they are the result of risk assessment made by architects, designers, planners, etc., at the initial design stage of a project. They have to ensure that all safety matters are addressed and followed for the duration of any contract.

Control of Substances Hazardous to Health Regulations

In this case these regulations deal with the use, storage, transportation and disposal of any substance which may be dangerous to the health of persons and/or livestock. These may include bottled gases, flammable liquids, corrosive materials: for example propane or oxy-acetylene for plumbing or welding, petrol or paraffin for motors or burners, caustic soda for cleaning, paint strippers and the powder in fluorescent tubes.

Clearly such substances must be stored somewhere on site and provision must be made to ensure that such storage is safe. In the case of the disposal of used substances, once again such materials will need to be collected and stored ready for collection by an appropriate disposal company. In any event, different substances should not be stored in the same area (e.g. toxic and explosive materials). All dangerous materials should be housed and locked in a secure area away from normal work activities. Records should be kept of the movement and storage of materials and only authorized personnel should have access.

It should always be remembered that incorrect disposal of hazardous waste can have a devastating effect on the environment not only for humans but for all other life.

There are some everyday items that present a serious risk such as fluorescent tubes, compact fluorescent lamps (CFL's), sodium- and mercury-vapour lamps, etc.

There is enough mercury in the fluorescent powder of one fluorescent tube to contaminate 30 000 l of water beyond safe-drinking levels.

There are many other items of electrical equipment, laptops, televisions, batteries, etc. that contain hazardous materials such as cadmium, nickel and lead and many carry the sign of the 'wheely-bin crossed out' to indicate that they should not be disposed of with ordinary household waste.

Always arrange for proper re-cycling of hazardous waste in order to protect the environment.

Chemicals (Hazardous Information and Packaging for Supply) Regulations (CHIP) 2009

CHIP applies to those who supply dangerous chemical substances and places a duty on them to provide information on the packaging regarding potential hazards. This information is usually provided, simply, in the form of a label such as the following:

	Very Toxic A substance which if swallowed, inhaled or absorbed through the skin can cause extremely serious – acute or chronic health effects – including even death.		**Explosive** A substance which can be detonated by the action of fire, friction, shock, etc.

12

	Toxic		Oxidizing
	A substance which if swallowed, inhaled or absorbed through the skin can cause serious – acute or chronic health effects – including even death.		A substance which, though not flammable itself, can ignite combustibles such as paper, organic compounds, cotton, etc.
	Harmful		Extremely Flammable
	A substance which if swallowed, inhaled or absorbed through the skin can cause limited health effects.		Substances that have a flash point below 0°C and a boiling point (or initial boiling point) of lower than or equal to 0°C.
			(Flash point – temperature at which a flammable liquid gives off enough vapour to form a mixture with air, above the surface of the liquid, capable of ignition by an external source such as a flame.)
	Corrosive		Highly Flammable
	A substance which can destroy living tissue on contact.		A substance which has (1) a flash point below 21°C but is not classified as being extremely flammable or (2) is spontaneously flammable in air.
	Irritant		Flammable
	A non-corrosive substance which, following skin contact, can cause inflammation.		A substance which has a flash point below or equal to 55°C.
	Harmful to the Environment		
	Discharge of this material may cause adverse environmental effects.		

Working at Height Regulations 2005

These regulations require employers and employees to carry out risk assessments and work accordingly with regard to the dangers of working at any height, even 'hop-ups'.

The section in this chapter on access equipment gives more details with regard to ladders, scaffolds, etc.

Control of Asbestos Regulations 2002

These regulations prohibit the use new asbestos and place a responsibility on those whose work involves the disturbance of materials that contain old asbestos.

The ingestion of asbestos fibres, even in small quantities, can cause respiratory problems leading to premature death.

Building materials containing asbestos have been around from 1930 to around 1980 and disturbance of items such as insulation board, asbestos lagging, asbestos-cement, plastic floor tiles, textured finishes such as artex, etc. may cause the release of air-borne particles.

So, for instance, installing downlighters in an artexed ceiling could be a health hazard. If there is even the slightest doubt, **seek advice**.

The Building Regulations

Clearly these regulations encompass all aspects of building construction and are known as Approved documents (Ads) or Parts. There are some Ads that are relevant to electrical installations. These are as follows:

Ad Part A. This, overall, deals with the building structure and, to some extent, on how the fabric of the building may be violated to accommodate electrical systems via chases, notching, drilling, etc. (see pp. 263–283 regarding joists).

Ad Part B. This concerns fire safety. All electrical systems should be constructed and installed to prevent the start of and the spread of fire. Such considerations would include: heat resistant sleeving on conductors, hoods over downlighters, fire barriers in trunking and sealing of holes made in the building fabric during the installation process. Added to this there is a requirement to provide smoke and fire alarm detection systems.

Ad Part F. Ventilation is important, especially in kitchens and bathrooms, and in consequence particular attention should be paid to the positioning and effectiveness of extract fans and the number of appliances installed.

Ad Part L. This is about the efficient use of fuel and energy. Generally in domestic dwellings, lighting is probably most abused and in consequence the types and control of luminaries have to be carefully considered.

Ad Part M. Disabled persons need access to buildings and facilities and hence the position of sockets and switches is important. The Part M document indicates reasonable heights of such accessories where there is access/use by the disabled.

Ad Part P. There is now a requirement to inform the Local Authority Building Control (LABC) of any electrical work carried out in a domestic dwelling. This is generally confined to kitchens, bathrooms and gardens, where the risk of shock is considerably higher. This does not extend to changing old for new fittings or accessories. Those who carry out electrical work in such areas must be competent and registered with an authoritative body before certification can legitimately be issued.

General safety

None of us enjoys the discomfort of injury. It is, therefore, clearly sensible to avoid such distress by working in a safe manner and ensuring that our colleagues and employers do likewise.

Working on 'site', even if the site is a domestic premises where rewiring is taking place, may involve considerable risk to oneself and/or any occupants or other trades. Hence the type and size of the site is of no real consequence: danger in a working environment is always present in one form or another.

All employees and employers should be aware of the risks and should be alert to danger at all times. No situation, however seemingly innocent, should be ignored as it may develop into the cause of a serious accident.

If the situation or environment requires it, the PPE must be used at all times. On 'bona fide' construction sites, signs and notices are displayed and all personnel are required to comply. Failure to do so will result in exclusion from the site.

The CDM requirements for a site will include all the procedures for dealing with emergency situations such as fire, explosion and leakage of toxic materials, and should be available for all to acquaint themselves with. Everyone should know the basic commonsense actions to take together with the specific requirements for each particular site.

Basically, in the event of discovering a fire:
1. raise the alarm by operating relevant call points and/or shouting **fire**.
2. call emergency services.
3. make sure no one ignores the alert and evacuation is in process.
4. only if safe to do so and the fire is relatively small, attempt to control using an appropriate extinguisher.

When an explosion has occurred:
1. alert all personnel to evacuate the area.
2. call emergency services.
3. do **not** attempt to approach the area as other explosions may occur.

The same procedure should be adopted if there is a leakage of toxic material. Only trained personnel should deal with such situations. If in any doubt, **all work should be stopped and the area evacuated**. A good example of this is the cutting, drilling and disposal of any substance that looks like, but has not been cleared of containing, asbestos. In no circumstances should work continue until experts have investigated.

Listed below are some important DOs and DON'Ts associated with working conditions.

DO	Know the site procedures for evacuation in the event of an emergency.
DO	Know where any emergency exits are located.

DO   Know where telephones are located in order to contact emergency services.

DO   Wash hands after accidental contact with any corrosive or irritant substance.

DO   Always ensure that there is enough working space and adequate levels of lighting when working, especially where electrical equipment is present: the EAWR require this. Items of electrical equipment should be sited such that work can be carried out without difficulty and in such a location to afford easy access. Hence distribution boards should not be installed in areas used for storage of non-electrical equipment (consumer units in understairs cupboards!). Control switches for washing machines, dishwashers, etc., should be visible and accessible. Switch rooms housing electrical equipment should have good lighting, preferably of the maintained emergency variety with switches inside the room.

DO   Wear the correct protective clothing for the job, for example goggles when using a grindstone, safety helmet when working on a building site.

DO   Adopt the correct posture when lifting.

DO   Use the correct access equipment (e.g. ladders and scaffolding).

DO   Ensure that any machinery is mechanically and electrically safe to work on or with.

DO   Report any unsafe situation that you personally cannot remedy.

DO   Ensure that when a job of work is completed, everything is put back to normal, for example fixing floorboards and tacking down carpets, and replacing trunking covers, etc.

DO   Know where first aid is available.

DON'T   Leave pools of liquid on floors, or objects lying in access ways.

DON'T   Wear loose clothing that may catch or snag, especially when using rotating machinery.

DON'T   Work in an untidy fashion.

DON'T   Misuse tools and equipment or use damaged tools.

DON'T   Attempt to lift and carry objects that are too heavy, or that obstruct vision.

DON'T   Take short-cuts or take risks.

DON'T   Ignore warning signs or alarm bells.

DON'T   Play the fool in a working environment.

In order to carry out some of the DOs and DON'Ts, it is necessary to have a greater understanding of the subject we are dealing with. Others, of course, are common sense, like wearing goggles to protect the eyes.

Risk assessment and method statements

On many sites, an assessment of all potential risks has to be made and a method statement prepared indicating all the risks, how they should be minimized and controlled and the actions necessary in the event of an emergency.

Gases

Those operatives who work on construction sites may well encounter various gases that are used and which must be treated with the care that any explosive/volatile substance requires.

Typical gases and common uses are:

▶ Butane   used for gas torches/heaters and as a propellant in nail guns

▶ Propane   used for industrial heaters and as a fuel for some machinery

▶ Oxy-acetylene   generally used in welding

▶ Nitrogen   used in thermal spray coatings and in liquid form to cool concrete. (the gas itself is not harmful but may become so when mixed with other gases or substances)

Clearly all the safety precautions regarding usage and storage must be observed, and reference to manufacturers' instructions should always be made.

Excavations

Any hole in the ground can be dangerous, be it a pothole in the road or a pit or trench in a field.

Construction sites in particular are likely to have pits or trenches for drainage, and services such as gas, water, electricity etc.

Electrical operatives may need to work in or near excavations and the law requires that no work should take place until the excavation is safe.

Excavations can collapse unless the relevant supports are in place and trench sheets, props etc. are available for use.

Loose materials should be prevented from falling into the excavation by the use of toe boards and trench sheets.

When inside an excavation, head protection should be worn.

Where there is a chance of persons falling into excavations, substantial barriers should be erected.

All excavations should be regularly inspected by a person competent to do so.

Confined spaces

Working in a confined space carries the risk of death or serious injury where there is a lack of oxygen or where hazardous substances such as paint or chemicals are used.

Examples of confined spaces are:

▶ Tanks
▶ Silos
▶ Pits
▶ Trenches
▶ Sewers
▶ Ductwork
▶ Unventilated areas etc

A risk assessment should be carried out to determine what precautions need to be implemented in order to avoid danger and to have in place arrangements for emergency rescue.

Wherever possible:

1. Avoid entering a confined space,
2. Follow a safe system of work if you need to be in a confined space,
3. Make sure emergency arrangements are in place.

The mechanics of lifting and handling

Simple levers

Figure 12.1a–c illustrates three forms of the simple lever and how the principles of leverage are used in practice. We can see from the diagrams that the load is as close to the fulcrum as possible, and the effort is as far from it as possible. In this way heavy loads can be lifted with little effort. In Fig. 12.1a and c first-order levers are seen, and Fig. 12.1b is a second-order lever.

The inclined plane

The inclined-plane method of moving a load is best illustrated as in Fig. 12.2a and b. Most of us at sometime have had to push or help push a car up a gentle slope. Imagine, then, trying to get a car from the road onto the pavement. Would it be possible for us physically to lift it from one level to another? Doubtful. More than likely we would arrange two wooden planks from the pavement to the road, and then, without too much effort, push the car up.

Variations of the inclined plane

The wedge

From Fig. 12.2b it can be seen that the arrangement of an inclined plane has a wedge shape.

Figure 12.3a and b shows the use of wedges.

Simple screw jack

As a screw thread is at an angle, the screw jack is a variation of the inclined plane (Fig. 12.4a). Another application of this tool is a bearing extractor (Fig. 12.4b).

12

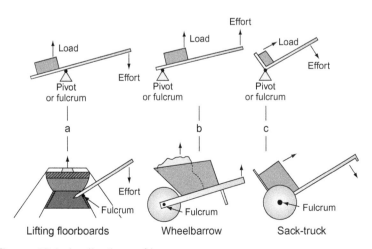

Figure 12.1 Applications of levers.

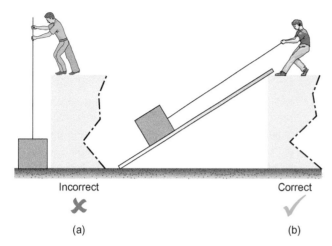

Figure 12.2 The inclined plane.

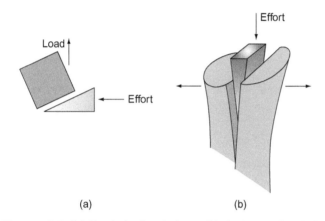

Figure 12.3 (a) Single inclined plane; (b) double inclined plane.

Figure 12.4 (a) Screw jack; (b) bearing extractor.

Pulleys

Figure 12.5 shows some simple pulley systems. The figure shows multiple sheave blocks, the most common form of rope and pulley system. In practice, all the wheels in a pulley system are of the same size.

Manual lifting and handling

Many injuries in industry and in the home are the direct result of incorrect lifting or handling. Figure 12.6 illustrates some of the DOs and DON'Ts of lifting and handling.

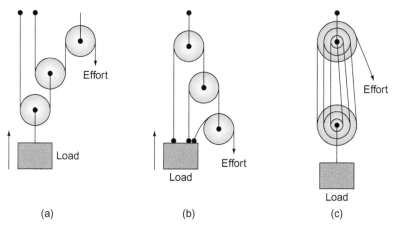

(a)   (b)   (c)

Figure 12.5 Pulleys.

Figure 12.6 Lifting techniques.

Apart from considering one's own welfare when lifting and handling, the safety of others must be taken into account.

Always ensure that others are in no danger whilst you are lifting, lowering or moving items of equipment especially full lengths of conduit, tray or trunking. Never throw items to fellow workers, always be **alert.**

Work, load and effort

Work: symbol, *W*; unit, joule (J)

If an object is moved from one place to another, work is done. The heavier the object and the greater distance it is moved, the more work is done.

147

Therefore work is a product of the amount of force or effort used to move the object, and the distance it is moved:

Work = force × distance
$$W = F \times l$$

If we ignore friction and any other losses, then the force we exert is equal to the force the load exerts against us.

Load force: symbol, F; unit, newton (N)

The mass of a load is a measure of the amount of material making up the load and the units we use are kilograms (kg). The weight or force that a mass exerts on the surface of the Earth is measured in newtons (N), and mass in kilograms can be converted into load force by multiplying by 9.81.

Load force (N) = load mass (kg) × 9.81

Effort

Effort is the force we require to move a load and as we have already seen, it is equal to the load force or weight (ignoring friction).

Example

A concrete block has a mass of 100 kg. Ignoring friction, calculate the work done in moving the block 8 m.

Work (J) = force (N) × distance (m)

First we must convert the load force into newtons.

Load force = mass × 9.81
$$= 100 \times 9.81$$
$$= 981 \text{N}$$
$$\therefore \text{Work} = 981 \times 8$$
$$= 7848 \text{ J}$$
or   7.848 kJ (kilojoules)

Calculations involving simple levers

The effort required to lift a load is calculated as shown in Fig. 12.7.

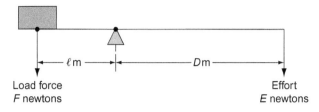

Figure 12.7 ($E \times 3D = F \times 3l$).

Example

Calculate the effort required, using a simple first-order lever, to lift a load of mass 500 kg if the effort is exerted 3.5 m from the fulcrum and the load is 0.5 m from the fulcrum (Fig. 12.8).

Figure 12.8 ($E \times 3D = F \times 3l$).

First we must convert the load mass into load force:

$$F = 500 \times 9.81$$
$$= 4905 \text{ N}$$

Now

$$E \times D = F \times l$$
$$\therefore E = \frac{F \times l}{D}$$
$$= \frac{4905 \times 0.5}{3.5}$$
$$= 700.7 \text{ N}$$

Access equipment

In order for us to carry out certain tasks in installation work, it is often necessary to work above ground level (installing conduit and trunking, etc.). It is obvious, then, that a safe means of access must be used. The following items are typical of access equipment available:

Ladders	No-bolt scaffolding
Swingback steps (step-ladders)	Bolted tower scaffold
Ladder and step-ladder scaffold	Standard independent scaffold
Trestle scaffold	Putlog scaffold

Ladders

There are various types of ladder: wooden or aluminium; single, double and triple extension; rope-operated extension and telescopic.

Safe use of ladders

1. Ladders must not be used with:
   (a) broken, missing or makeshift rungs.
   (b) broken, weakened or repaired stiles.
   (c) broken or defective ropes and fittings.
2. Rungs must be clean and free from grease.
3. Ladders must never be painted as this may hide defects.
4. Aluminium ladders must not be used near low- or high-voltage cables.
5. Ladders over 3 m long must be secured at the top or bottom, or a second person must 'foot' the ladder on the bottom rung.
6. Ladders should be erected at an angle of 75° to the ground.
7. Ladders used for access to working platforms must extend five rungs or 1 m beyond the working surface (Fig. 12.9).

Figure 12.9 Securing a ladder top.

Swingback steps (step-ladders)

The safe use of wooden or aluminium step-ladders is as follows:
1. As for items 1–4 for ladders.
2. Hinges must be secure.

Ladder or step-ladder scaffold

Two ladders may be used to support scaffold boards (Fig. 12.10):

Figure 12.10 Ladder scaffold.

1.5 m span – use 38 mm scaffold board
2.5 span – use 50 mm scaffold board
Above 2.4 m – use two scaffold boards
Maximum span is 3 m
Maximum height is 1.9 m
Maximum overhang of scaffold boards is four times the thickness of board
Minimum overhang of scaffold boards is 50 mm
Access to the platform should be by a third ladder

The same span distances, etc., apply to scaffold boards used with step-ladders.

Trestles (wooden or aluminium)

Trestles are used to support scaffold boards (Fig. 12.11) and should not be used in the closed position as a ladder.

Span lengths and overhangs as for ladders
Safety precautions as for ladders
Maximum platform height 4.57 m

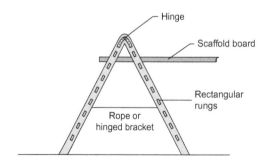

Figure 12.11 Trestle.

Scaffold boards

Scaffold boards (Fig. 12.12) are made to the British Standard BS 2482/70 and are the only boards that should be used. Their maximum length is usually no greater than 4 m; beyond this, special staging is used.

Scaffold boards should be:

1. clean and straight
2. free from decay or damage
3. free from paint, dirt or grease.

149

Galvanized iron strip

Figure 12.12 Typical scaffold board ends.

Tower scaffolding

Tower scaffolding can have either the modern no-bolt interlocking type of construction or the conventional bolted type.

Safety precautions are the same for both types:

1. Beyond a height of 6.4 m, the tower must be tied to the building or have an outrigger fitted.
2. Maximum working platform height should not exceed:

   3 times the width of the narrow side out of doors
   3½ times the width of the narrow side indoors.
3. Working platforms above height of 2 m must have a toe board and hand rail (Fig. 12.13).

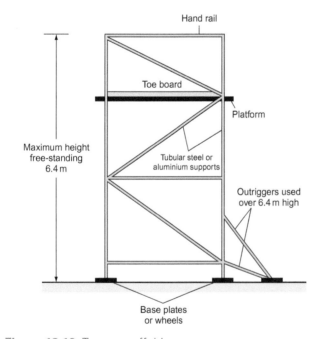

Hand rail

Toe board

Platform

Maximum height
free-standing
6.4 m

Tubular steel or
aluminium supports

Outriggers used
over 6.4 m high

Base plates
or wheels

Figure 12.13 Tower scaffold.

> **Note**
>
> For any kind of scaffolding:
>
> 1. Toe boards must be at least 150 mm high.
> 2. Hand rails must be between 920 and 1150 mm above the working platform.
> 3. The distance between boards must not exceed 25 mm.
> 4. Mobile towers must have locking devices fitted to the wheels.
>
> Overlaps and spans of scaffold boards are as for ladders.

Independent scaffolding

Independent scaffolding (Fig. 12.14) is one of the typical systems used on building sites.

Putlog scaffolding

The maximum safe height of putlog scaffolding (see Fig. 12.15) is 45.5 m.

The distance between supports for platforms on both putlog and independent scaffolding should be as follows:

990 mm for 32 mm board thickness
1520 mm for 38 mm board thickness
2600 mm for 50 mm board thickness

All scaffolding should be inspected once every 7 days and after bad weather.

The joining of materials

It is often required to bond two materials together, for example cables or metal sheets. Three typical methods used in the electrical industry are soldering, riveting and crimping.

Soldering

Care must be taken when soldering to avoid the risk of burns or starting a fire, or inhaling toxic fumes.

When using molten solder to pour over joints or into lugs, wear protective gloves and goggles.

When using soldering irons, always use a fireproof stand to hold the iron. Never flick surplus solder from the iron; wipe it off.

Figure 12.14 Independent scaffolding.

Figure 12.15 Putlog scaffolding.

12

Riveting

There are many different designs of riveting tool available, the most potentially dangerous being the powered types which rely on compressed air or an explosive charge. These must **never** be operated incorrectly. Follow the written operating procedure and if this is not available, seek instruction from a person competent in the use of the tool.

Crimping

Crimping is a method used to join cables together. The ends of the conductors to be connected are placed in a sleeve which is then crushed to provide a mechanical weld between sleeve and conductors.

Once again, the most hazardous of the tools available are the powered types and the same safety precautions as for riveting should be observed.

Fire safety

Fire prevention

It is obviously more sensible to prevent a fire than to be forced to put one out. Most establishments carry notices prohibiting the use of naked flames or smoking in certain areas, and provided that we comply with these instructions, and also make sure that all appliances are switched off after use and that all highly combustible materials are kept away from any source of heat, then the risk of fire will be minimized.

Fire detection and control

In order for fire to exist, three properties must be present (Fig. 12.16). These are fuel (oil, paper, wood, etc.), oxygen and heat (naked flame, spark, etc.).

Figure 12.16 Components of a fire.

Detection of the outbreak of fire is normally confined to three methods: human detection, smoke detection and heat detection.

Human detection is limited to an individual noticing the fire and raising the alarm manually at a break-glass call point. Both smoke and heat detection rely on automatic means of sensing the fire and operating the alarms.

The control of fire is limited to two methods, one manual (hand-held extinguishers) and the other automatic (sprinkler system). Both systems, however, use the same substances (water, foam, etc.), depending on the type of fire.

Classes of fire

Fires may be divided into six classes depending on the location and the fuel:

Class	Types of Fire
A	Solids – wood, paper, textiles, etc.
B	Flammable liquids – oil, petrol, paint, etc.
C	Flammable gases – acetylene, butane, methane, etc.
D	Metals – magnesium, sodium, etc.
E	Flammable liquids – fats, solvents, etc.
F	Kitchen products – cooking oils, etc.

Types of extinguisher

It is extremely important to understand that **not all fires can be extinguished by the same kind of extinguishing media.** It would be dangerous to use water on a liquid fire as the burning fluid would only spread on the surface of the water. It would also be hazardous to use water on an electrical fire as water is a conductor of electricity. Table 12.1 shows the correct kind of extinguisher, its class and colour for a particular type of fire.

Table 12.1 Fire extinguishers

Type	Most Suited for	Class	Colour
Water	Solids	A	Red
Carbon dioxide (CO_2)	Liquids, electrical	B, E	Black
Foam (AFFF)	Solids, liquids	A, B	Cream
Powder	Solids, liquids, gases	(A), B, C, D	Blue
Wet chemical	Kitchen oils etc.	A, (B), F	Yellow

Current legislation requires that **all** fire extinguishers are coloured red but with a block or band of colour denoting the type.

Remember, some kinds of materials such as plastics and certain chemicals give off highly toxic fumes when ignited. Under these circumstances, in order to minimize the danger to the fire fighter, special breathing apparatus would be needed.

Electrical safety

Electricity, with the exception of extra-low voltage, is potentially lethal. It must be treated with respect at all times and used only in approved ways. Many people die every year as a result of the misuse of electricity. They die as a direct result of electric shock, or as an indirect result of faulty or misused electrical equipment that causes fire.

Listed below are some important points to remember when working with electricity:

1. Never work on live equipment (unless a special live test is required, for which you will need to be an experienced and competent person).
2. Always ensure, by using approved test instruments, that equipment is dead.
3. Never accept another person's word that a circuit is safe to work on; always check.
4. Ensure that all supplies to equipment to be worked on are isolated at the appropriate places and locked 'OFF' if possible, and that all supply fuses are removed and retained in a safe place. Place 'CAUTION MEN AT WORK' notices at positions of isolation.
5. If work is to be carried out on dead equipment which is adjacent to live supplies, ensure that barriers are used to define safe areas, or 'DANGER LIVE APPARATUS' notices are placed on all adjacent live equipment.

Figure 12.17a shows the ways in which a test lamp is used to indicate the circuit condition, and Fig. 12.17b and c shows how the human body can become part

Figure 12.17 Use of a voltage indicator.

of a circuit. Either position in Fig. 12.17a will indicate whether the live terminals are live or dead. In effect the body replaces the test lamp of Fig. 12.17a. At the normal domestic voltage of 230 V, it needs only 0.05 A (50 mA) of current flowing through the body to cause death.

Safe isolation of supplies

The Electricity at Work Regulations require that **no live** working is undertaken unless it is impracticable to work **dead.**

All electrical circuits, apparatus, etc., must, therefore, be switched off, isolated (these can sometimes be achieved by one action, e.g. switching of a circuit breaker also isolates the circuit) and, whenever possible, locked in the 'off' position.

In order to prove the system dead and safe to work on, an approved voltage indicator should be used (this is usually a test lamp, see p. 401). Neon screwdrivers, 'volt sticks' and the like should **not** be used. The following steps should be observed:

1. Identify the circuit/equipment to be worked on.
2. Ensure that it is safe/convenient to switch off.
3. Isolate the supply.
4. Using an approved voltage indicator:
   – check the indicator works, on a known supply
   – then use it to check the system is dead
   – then re-check the indicator is still working, on the known supply.
5. Lock off the isolator or remove the fuse and keep in a safe place. If no locking facilities are available, post warning signs or notices.

Safe use of electrical equipment

When one is using electrical equipment such as drills, saws, sanders, etc., on site or in a workshop, great care must be taken to ensure that the tools are in good condition and that the cables supplying them are not damaged in any way and are adequate for the job they have to do.

Any connections of cables must be carried out by a competent person using approved tools and equipment.

For work on building sites, tools using a voltage lower than usual (110 V instead of 230 V) are recommended.

All current-using and current-carrying apparatus used on sites must be inspected and checked at regular intervals. A 3-month period is recommended, but the user should always check before use that all electrical apparatus is in good condition.

Ensure that all cables exposed to mechanical damage are well protected.

First aid

No matter how careful we are, there are times when we or our workmates have an accident. Prompt action can relieve unnecessary suffering, prevent permanent disability and, in extreme cases, save life. This action is **first aid**. Table 12.2 gives the first-aid action recommended in some common cases; electric shock is dealt with separately.

First aid facilities, no matter how small, should be always available in the workplace and should also be kept fully stocked. Accident books must be provided and **used**.

Also, any injury, dangerous occurrence, etc. needs to be reported as required by RIDDOR 1995, and facilities must be in place to enable such a report to be made.

12

Table 12.2 Recommended first-aid action

Symptoms	Action
Shock (suffered by most casualties)	Move casualty as little as possible. Loosen all tight clothing (belts and ties, etc.). If possible lay him/her down and raise legs over head level. Keep casualty warm with blankets or coats. Reassure casualty. Send for medical help.
Burns	Immerse burnt areas in cold water. Treat for shock. Do not remove burnt clothing. Do not apply oils or grease to burns. Cover burns with clean dry cloth. Arrange for medical help if burns are severe.
Cuts and grazes	Small cuts and grazes should be washed in running water, and a dressing applied. More severe bleeding of deep cuts should be stopped by applying pressure to the wound. Treat for shock. Call for medical assistance.
Bruising	Apply an ice bag or damp cloth until the pain stops.
Falling	Make casualty lie still. Check for obvious injuries. If necessary treat for shock. If in doubt call for medical help.
Contact with chemicals	Wash affected area with water, cover with dry dressing. Get casualty to hospital.
Contact with toxic fumes	Remove casualty from gas-laden or oxygen-deficient area. Loosen tight clothing. Apply resuscitation if breathing has stopped.

Electric shock

An electric shock is experienced when a current passes through the body. Not everyone would have the same level of shock from the same source; it would depend on variables such as the individual's body resistance, his or her health, etc. However, it is generally accepted that 50 mA (0.05 A) at 230 V can be lethal. Below this level, contact with a live source throws us away from the source. Above 50 mA the muscles contract or freeze and we are unable to break contact. Also, interruption of the heart's rhythm takes place and its beating may stop altogether (ventricular fibrillation). Burns to the parts of the body in contact can occur, together with burning of internal organs and loss of breathing.

It is clearly essential that prompt action be taken in the case of severe electric shock. The following procedure must be adopted:

1. **Do not touch the casualty with the hands.**
2. Switch off the supply.

3. If this is not possible, pull the casualty away from contact using **insulating material** (i.e. a scarf, piece of wood, newspaper, etc.).
4. If heart or breathing or both have stopped, apply resuscitation and cardiac massage until the casualty recovers.
5. Treat for burns.
6. Check for other injuries; treat as necessary.
7. Treat for shock.
8. Call for medical assistance.

Cardiopulmonary resuscitation (CPR)

Place victim flat on back on a hard surface and check the mouth for foreign bodies, false teeth, etc. and remove any loose objects.

A. Open the airway.

If unconscious and not breathing tilt head back and support jaw.

B. Breathing.

Look, listen and feel for breathing. If not breathing give mouth to mouth resuscitation.

Close nostrils with fingers and blow into mouth. Make sure there is no air leak and the chest expands. Give two normal breaths.

Feel for carotid pulse in neck. If no pulse:

C. Perform chest compressions.

Place both hands on the lowest third of breastbone. Depress sternum to one-third of the depth of the patient's chest at a rate of 100 compressions per minute. Alternate two breaths with 15 compressions. Continue cycle until the victim shows signs of recovery by coughing, vomiting or making signs to move.

Figure 12.18 Resuscitation techniques.

Note

If possible get a third person to call for medical help as soon as the casualty is discovered.

Method of resuscitation is shown in Fig. 12.18.

CHAPTER 13

The Electrical Contracting Industry

Let us imagine that we are about to start an electrical installation company employing half a dozen or so staff. We will assume that initial finance is not a problem and that suitable premises are available. What, then, are the organizations, rules and regulations, and standards that could affect the way in which we will be working?

Just as all working processes are governed by one or more sets of regulations, the use of electricity is no exception to this, the main body of regulations being as follows:

▶ Electricity at Work Regulations 1989
▶ Electricity Safety, Quality and Continuity Regulations 2002
▶ IET (Institution of Electrical Engineers) Wiring Regulations 17th edition
▶ British Standards Specification (BS)
▶ British Standards Codes of Practice (CP)

Electricity at Work Regulations 1989

These are statutory regulations (i.e. they may be enforced by law) and are the basis of the IET Regulations. They lay down, in broad terms, the measures to be taken to ensure the safe installation and use of electrical equipment.

Electricity Safety, Quality and Continuity Regulations 2002

Once again these are statutory regulations. In this case, they ensure the safety and welfare of the public

and that a proper and sufficient supply of electricity is provided.

IET Wiring Regulations 17th edition

The IET Regulations are not statutory, but are based on the statutory acts and internationally agreed codes of safety. They are also a British Standards, BS 7671:2008. They lay down recommendations for the safe installation and use of electrical equipment in buildings, and are therefore of great importance to members of the electrical contracting industry.

British Standards Specifications

British Standards (BS) ensure a national uniformity in the quality, performance, dimensions and listing of materials. Many items of equipments have a BS EN number which signifies European acceptance.

British Standards Codes of Practice

Codes of Practice (CPs) are issued by the British Standards Institution and recommend standards of good practice. In the case of electrical installations, they follow, in general, the recommendations of the IET Regulations. In some cases, when a choice of methods is available, they select one as preferred practice.

All sources of information mentioned are important to the installation electrician and he or she should be familiar with all of them.

Electrical Installation Work. 978-1-138-84927-3 © Brian Scaddan.
Published by Taylor & Francis. All rights reserved.

Also important is a knowledge of the many different boards, associations, unions and schemes which influence the electrician's work.

Joint Industrial Board

The Joint Industrial Board (JIB) is a national organization which basically acts as a means of liaison between unions and employers in the subject of grading and rates of pay for employees. Hence a JIB-graded electrician will have attained a certain academic and practical standard and will receive a set basic wage wherever he or she is working.

Construction Skills Certification Scheme

This organization was set up to help improve health and safety in the workplace. An operative may apply for and obtain a CSCS card which is an indication of occupational competence.

Local Authority Building Control (LABC) Competent Persons Scheme

One way of complying with the requirements of Part 'P' of the Building Regulations is to demonstrate competence by belonging to one of the following approval bodies, all of which operate the Competent Persons Scheme for Domestic Installers.

- ▶ **BSI** (Britsh Standards Institute)
- ▶ **ECA** (Electrical contractors Association)
- ▶ **ELESCA** (Part of the ECA group)
- ▶ **NAPIT** (National Association of Professional Inspectors and Testers)
- ▶ **NICEIC** (National Inspection Council for Electrical Installation Contractors)

Trade unions

There are several unions to which the electrician can belong. Whichever he or she chooses, if indeed there is a choice, he or she should remember that the unions' basic role is to ensure that employees enjoy satisfactory working conditions and rates of pay. They are the employees' voice to speak to management, and in the event of any serious dispute with the employer, it is best to let the union deal with the matter in the correct way, using the approved disputes procedure and conciliation machinery.

Electrical Contractors' Association

The Electrical Contractors' Association (ECA) was established in 1901 and requires its members to have a high standard of workmanship.

Anyone employing the services of an ECA-registered firm can expect top-quality work carried out in an efficient manner. Contractors can become members only after their work has been carefully examined by ECA experts.

A customer has a certain comfort in employing an ECA company, because if the company were to cease trading in the middle of a contract, the ECA would ensure that the work was completed by another ECA firm at no extra cost.

National Inspection Council for Electrical Installation Contractors (NICEIC)

The National Inspection Council for Electrical Installation Contractors (NICEIC) concerns itself solely with standards of workmanship, and contractors can apply to be placed on the NICEIC role as an approved contractor. Regular inspections of the contractors' work are undertaken to ensure that high standards are maintained. It is also an approval body for Part P.

National Association of Professional Inspectors and Testers (NAPIT)

NAPIT is an organization similar to the NICEIC.

Contracts and tenders

The majority of electricians are employed by a contracting firm and therefore are not likely to become involved with the administrative side of the business. However, it is important that the electrician is aware of the procedures involved in obtaining the work he or she is to carry out.

Usually, the first step in obtaining a contract to carry out an installation is to tender a price for the work.

Tenders

A tender is, by dictionary definition, an offer to supply goods and/or services at a fixed rate. In many cases, this is a simple procedure but on larger jobs the tender can become complicated, and considerable experience is necessary to complete such a tender correctly.

Contracts

The law relating to contracts is extremely complicated and involved, and hence only the most basic concepts will be considered.

In simple terms, for any job there is a **main contractor**, which can be an electrical installation firm, building firm or a decorating firm, etc., depending on the work to be done. This main contractor is responsible to the client (i.e. the person ordering the work to be done), either directly or via an agent such as an architect.

Should the main contractor employ the services of another firm, this firm is called the **subcontractor** and is responsible to the main contractor.

A typical sequence of events is as follows:

1. The client approaches an architect with a view to having, say, a hotel designed and built.
2. The architect designs the building and the design is approved by the client.
3. A specification and a bill of quantities are prepared.
4. The work is put out to tender, and eventually one is selected – not necessarily the lowest priced.
5. The architect may nominate the subcontractors, that is, painters, electricians, plumbers, etc., or leave it to the main contractor. In any event the subcontracts will go to tender. This is where the ability to read and interpret drawings, bills of quantities and specifications is so important.

Specifications

The specification indicates the quantity of sockets, lighting points and other fixed accessories required for each room, and sometimes the preferred manufacturer. It may also indicate the type and size of cable to be used (see sample specification).

The architect's drawings normally show elevations and plan views, with details of recommended positions for all accessories.

It is from these two sources of information, specification and drawing, that a design for the electrical installation can be prepared and a competitive tender submitted.

Bills of quantities

Usually prepared by a quantity surveyor, bills of quantities indicate, for each trade concerned with the work, the quantity and sometimes cost of the materials to be used.

Variation order

In the case of electrical installation work, there is every chance, on a big job, that some variation from the programme will occur. It is very important that the site electrician notifies his or her superiors **immediately** of any change. A **variation order** can then be made out which will enable the new work to be carried out without breaking any of the terms of the contract.

Daywork

It may be necessary to carry out work in addition to that referred to in the contract and this work will be the subject of daywork.

Daywork is normally charged at a higher rate than the work tendered for on the main contract, and the charges are usually quoted on the initial tender.

Typical additional charges are as follows:

120% – labour
 20% – materials   } added to normal rates
  5% – plant

It is important that the members of the installation team on site record on 'daywork forms' all extra time, plant and materials used.

It must always be remembered that employer and employees are both essential for a business to exist. For it to succeed, the two sides must work in harmony. From the point of view of the electrician on site, this involves – apart from doing a good wiring job – accurate recording of time on time sheets and daywork sheets, recording and checking deliveries of materials on site, ensuring that all materials stored on site are safe, and keeping in constant contact with the employers, the main contractor and other subcontractors.

Sample specification

Phone: Waterlooville 2583

J. A. Smith (Waterlooville) Ltd.
Building Contractor
Registered Office
88 Jubilee Road, Waterlooville, Hants. PO7 7RE
Company No 662680 Registered in England

Your Ref:

Our Ref: FB/HI

3 May 2007

M Jones & Co
34 Queensway
Anytown
England

Dear Sirs

Re: Electrical Installation at Hunters Meadow Estate

We should be pleased to receive by return your lowest price for materials/subcontract work as detailed below, delivered to/executed at the above project.

Workmanship and materials must comply with the requirements of BS 7671 and relevant Building Regulations.

As the main developers for this project we have the order to place for these items.

Yours faithfully
F A Smith (Waterlooville) Limited

Fred Bloggs

Fred Bloggs
Buyer
E/O for connecting central heating.

Wiring To be carried out in PVC cable; lighting wired in 1.0 mm²: ring main in 2.5 mm².

Boxes Metal boxes to be used for all fittings on plastered walls.

Sheathing PVC sheathing to be fixed on walls under plaster.

Fittings To be of cream flush Crabtree or equivalent; 13 mm socket outlets for ring main.

HUNTERS MEADOW ESTATE, ACACIA AVENUE, ANYTOWN

TYPE 'B' 3 BED LINKED

This spacious Link/Detached three-bedroomed house with attractive elevations has a larger than average lounge with separate dining area, good-sized bedrooms and downstairs cloakroom making the ideal family home.

Heating is by means of gas-fired central heating and an electric fire in hardwood surround to lounge. Thermoplastic floor tiles to choice from our standard range for the whole of the ground floor.

ACCOMMODATION

Lounge	6 m × 4 m
Diningroom	3 m × 3.5 m
Kitchen	3.5 m × 3.5 m
Hall	2 m × 2 m
Cloaks	2 m × 1 m
Bedroom	4 m × 4 m
Bedroom	3 m × 4 m
Bedroom	3 m × 2.5 m
Bathroom	2.5 m × 2.5 m
Airing cupboard	1 m × 0.6 m

EXTERNAL DOORS Are softwood glazed in two squares to match windows

GARAGE DOORS Metal up and over

INTERNAL DOORS Sapele hardwood flush

CEILINGS Artex stipple finish

DOOR AND WINDOW FURNITURE In UPVC

CURTAIN BATTENS To all windows

LOUNGE
2 Ceiling light points
3 – Twin 13 A socket outlets
1 T.V. aerial socket outlet with down lead from roof space
1 Point for electric fire
2 Radiators

DINING ROOM
1 Ceiling light point
2 Twin socket outlet
1 Radiator

KITCHEN
1 Fluorescent ceiling light
3 – Twin 13 A socket outlets
1 Cooker control with 1 additional socket incorporated
1 Immersion heater switch
1 – 13 A point for gas boiler
Wall-fitted gas boiler for central heating
3 Gas points
Double drainer, stainless steel sink unit, mixer taps and cupboard under
1 Large base unit
1 Broom cupboard
1 High level cupboard

HALL
1 Ceiling light point
1 – Twin 13 A socket outlet

CLOAKROOM
1 Ceiling light point
1 Low-level W.C. suite
1 Corner hand basin

GARAGE
Metal up and over door
1 Ceiling light point

TYPE 'B' 3 BED LINKED (Continued)
1 – Twin 13 A socket outlet
Electric and gas meters

LANDING
1 Ceiling light point
1 – Twin 13 A socket outlet
1 Full height airing cupboard with slatted shelves

BEDROOM 1
1 Ceiling light point
2 – Twin 13 A socket outlets
1 Radiator

BEDROOM 2
1 Ceiling light point
2 – Twin 13 A socket outlets
1 Radiator

BEDROOM 3
1 Ceiling light point
2 – Twin 13 A socket outlets
1 Radiator

BATHROOM
1 Ceiling light point
Coloured bathroom suite comprising low-level suite, pedestal basin and 2 m bath
1 Chromium-plated towel rail (electric)
Half-tiled walls to match suite colour
Bathroom cabinet with mirror front
1 Shaver point

EXTERNAL
Tarmac drive to garage
Paved concrete paths to front and back entrance 2' 6" wide
Turfed open-plan front garden
Rear and side boundaries are defined by various means to suit the development as a whole, for example, screen walls, chain link fences, etc.
Interwoven fencing panels erected adjacent to the property

EXTERNAL COLOUR SCHEME
Woodwork is generally white but details are featured in colours which are designed for the development as a whole.

INTERNAL COLOUR SCHEME
A limited choice of internal decoration is available from our standard range.

VARIATIONS
The house has been designed as an integrated unit and as a general rule it is not possible to accept variations.

TYPE 'B' 3 BED LINKED (Continued)
THIS INFORMATION IS GIVEN AS A GUIDE TO PROSPECTIVE PURCHASERS, THE DETAILS MAY VARY FROM ONE PLOT TO ANOTHER AND FULL INFORMATION CAN BE OBTAINED FROM OUR SALES PERSONNEL. THIS SPECIFICATION DOES NOT FORM PART OF ANY CONTRACT AND MAY BE REVISED WITHOUT PRIOR NOTICE.

The design team

This may consist of one or more people, depending on the size of the contracting firm and the complexity of the installation. The example in question (a domestic dwelling) would require only one designer. This designer will need to be competent and experienced, not only in electrical work, but also in other trades, and be able to interpret the architect's drawings. He or she will also, in the absence of any protective information, provisionally locate all electrical accessories using symbols to BS EN 60617. These symbols are extremely important, especially to the installation team.

The installation team

Once again, the team will comprise one or more people, depending on the contract. For the work in this example, two people would normally be employed: an approved electrician and perhaps an apprentice.

A good electrician will be competent and have a reasonable knowledge of the basic principles of the other trades involved in the building and servicing of the house, and in this respect, the electrician will often need to discuss, competently, problems which may occur due to the location of his or her materials in relation to those of different trades. The electrician must, of course, be familiar with the symbols in BS EN 60617, and be able to communicate intelligently with the client or their representative.

The client

The client, in this case usually the builder, has to rely on the expertise of the design and installation teams; it is therefore important that a good liaison be maintained throughout the duration of the contract.

It is the satisfied client who places further contracts with the installation firm.

Working relationships

Working with the building trade on site can present its own problems not found in, say, private rewiring.

Good working relationships with other trades go a long way to overcoming any problems. The phrase 'good working relationships' does not mean just having a pleasant attitude to other workers; it involves liaising with them either directly or via a site foreman, and endeavouring not to hinder their work progress. A schedule of intended work with proposed dates can help a great deal. There is nothing more damaging to working relationships than, for example, a plastering team arriving on site to find that the cable drops are not complete and the electrician is not on site.

Customer relationships

In a private dwelling requiring a rewire or simply additional lighting or power points, the electrician(s) must have the correct attitude to the customer and his or her property; after all, it is the customer who is paying for the work and who, justifiably, expects the best possible service. The following are some examples of conduct when working on other people's property:

DO	be polite under **all** circumstances.
DO	be presentable in dress and manner.
DO	be tidy, clear away all unnecessary debris, replace furniture to its original position.
DO	consult the customer if positions of accessories are not clearly defined.
DON'T	use bad language.
DON'T	place tools on furniture.
DON'T	use furniture as steps.
DON'T	leave without informing customer (unless he or she is out).
DON'T	leave floorboards and/or carpets unsecured when work is completed.

Requisitions and estimates

Before an estimate can be prepared a requisition or materials list must be compiled. This is where experience is invaluable in choosing the correct, and also the most economic, materials for the job. The dwelling we are considering could be wired in MICC (mineral-insulated copper-clad) cable, but although perfectly acceptable, this would involve the client in very high costs, so much so, that the tender for the work would stand little chance of being accepted.

The choice of cable and accessories is in fact quite easy in this case, but in more complex contracts the designer may be faced with extreme conditions, and will be involved in a considerable number of calculations for voltage drop and current rating and evaluation of external influences before a choice of materials can be made.

Cost of materials and systems

Materials

The type of dwelling in the specification would only require PVC twin with circuit protective conductor (cpc) cable, with drops to switches, etc. run in an oval PVC conduit and buried in the plaster, and run unenclosed under the floor. To run the wiring in single-core PVC cable enclosed in a conduit would be pointless and rather expensive.

Systems

The choice of a wiring system is just as important as the choice of material. The system dictates the material quantity. For example, the specification might call for 10 socket outlets (downstairs). Wired as a ring system requires only one 30/32 A way in the consumer unit. Wiring on a radial system would require a calculation of floor area and a probable increase in cable size to $4.0\,mm^2$.

In many instances switch-gear and accessories are denoted not by name but by symbols on a drawing. All such symbols should be to BS EN 60617. Figure 13.1 illustrates some of those symbols commonly used.

Materials list

Once a system has been decided on and all the calculations have been completed, a list or requisition of materials can be compiled. The example of such a list shown in Table 13.1 is typical for the dwelling we are concerned with.

To complete the quantity column, information must be obtained from the specification and plan, the cost being obtained from the wholesalers, or manufacturer's catalogue.

Equipment on site

Let us assume that the tender has now been accepted, and work has to commence on site. In the case of a large site, a contracting firm may well have a site hut erected for the storing of materials; for smaller contracts, equipment would be stored at base and transported as required.

In either event, materials have to be delivered and an accurate check on goods delivered must be made and records must be kept. Starting a job only to find that

Figure 13.1 Part 11 Architectural and topographical installation plans and diagrams.

items have been incorrectly sent is totally inefficient and does not make for good customer relationships.

The correct procedure is as follows:

1. Keep a copy of the original material order.
2. Delivered goods should be accompanied by an **advice note**. Ensure that there is one.
3. Then, while the delivery driver is with you, check that the goods correspond with those stated on the advice note.
4. Check the advice note against the original order.

Security

This presents a major problem on the larger site, where much equipment may be stored. There is not always a nightwatchman, and security patrols cannot maintain a constant vigil.

The only answer is to have good padlocks and an alarm system. Apart from the risk of bulk theft of materials, there is always a danger of smaller amounts being stolen, which have been left at the point of

Table 13.1 Bill of quantities

Description	Manuftr.	Code No.	Price Ea.	Qty.	Total Cost
Cable 1.0 mm² 6242Y 1.0 mm² 3 core 6243Y 2.5 mm² 6242Y 6.0 mm² 6242Y 16 mm² tails 6491X 10.0 mm² bonding 0.85 mm² bell wire 0.5 mm² lighting flex 1.5 mm² butyl 2.5 mm² butyl Low-loss TV co-axial					
Oval conduit 12 mm 20 mm 25 mm					
Metal boxes (KO) Plaster depth 25 mm deep 1 gang 25 mm deep 2 gang 35 mm deep 2 gang					
Cable clips 1.0 mm 2.5 mm 6.0 mm					
Power Switched socket outlet (twin) Switched fused connection unit Cooker control box					
Lighting Energy saving luminaires 1-gang 1-way plate switch 1-gang 2-way plate switch 2-gang 2-way plate switch 3-gang 2-way plate switch Pull cord switch Ceiling rose Lampholder Battenholder					
Others Shaver point Heated towel rail Immersion heater Thermostat 8-way consumer unit housing circuit breaker (cb's), RCBO's and/or RCD's Earth clamp Bell Bell transformer Bell push 4 ft fluorescent fitting TV outlet Surface boxes (S/O) Surface boxes (switch) Earth sleeving					

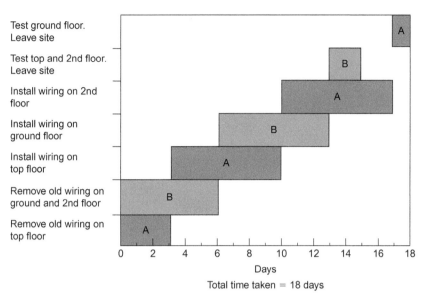

Figure 13.2 Bar chart.

work. The simple remedy for this is: do not leave any tools, materials or other equipment lying around after work. Lock everything away.

Protection of materials

All materials used in installation work should be in perfect condition. This cannot be achieved if the materials are carelessly stored or handled on site. All equipment should be kept away from damp or corrosive conditions and equipment involving delicate mechanisms, that is, thermostats, contactors, relays, etc., should be stored or handled so as to prevent mechanical damage.

If these recommendations are not observed, it is likely that it will be necessary to return to the installation, after completion, to carry out repairs or replacements.

Disposition of equipment on site

In order that a job runs smoothly and efficiently, the positioning of materials on site must be considered. The correct procedure may be summed up as: **Always ensure that all the relevant tools and materials are taken to the place of work**. In this way needless journeys to and from the site hut or base can be

eliminated and the correct amount of time can be spent on the installation.

Bar charts

In order that a job may be carried out in the most efficient manner, some job programmers use a bar chart. This is simply a method of showing graphically each stage of work to be completed on a job. This is best illustrated by an example, though it should be borne in mind that there are several ways of drawing a bar chart, this being only one of the possible methods.

Example

An electrical contracting firm has the job of rewiring an old three-storey dwelling, each floor of which is to be converted into a self-contained flat. Two pairs of men will do the work and the estimated time for each stage per pair is as follows:

Removal of old wiring per floor – 3 days
Installing new wiring per floor – 1 week
Testing and inspection per floor – 1 day

By using a bar chart (Fig. 13.2) estimate the least time in which the work may be completed.

CHAPTER 14

Installation Materials and Tools

Before any wiring system is installed, account must be taken of the environment in which the system is to operate. There are many influences that may contribute to the final design of the installation. The IET Regulations give details of such influences which include, for example, ambient temperature, water, dust, vibration and corrosion.

The IET Regulations employ an alphanumeric system with three categories: environment (A), utilization of the environment (B) and the building structure (C). Within these categories there are subdivisions of A, B and C denoting specific conditions, and then numbers to indicate the severity of the influence.

So, for example, in the environment category A, the second letter, say D, denotes a water influence, and the associated number the severity of the influence. For example, an AD4 code indicates an environment that is subject to splashes of water, whereas an AD8 environment is total submersion in water. In the same way, the second letter E denotes the presence of dust. Hence AE1 is negligible dust and AE6 is heavy dust.

The same system is used for utilization B and building structure C.

The most likely environmental conditions that may be encountered in installation work are as follows:

AA	Ambient temperature
AD	Water
AE	Foreign bodies and dustτ
AF	Corrosion
AG	Impact

AH	Vibration
AK and AL	Flora and fauna
AN	*Solar*

With regard to enclosures used to protect against, in particular, moisture and dust, the IP code provides the degree of protection needed. Hence an AD8 environment would require an IPX8-rated enclosure, and an AE6 environment would need an IP6X enclosure. There are also a set of codes for impact, these are called the IK codes (see IP and IK codes, p. 352 & 353).

Cables

A great number of types of cables are available, ranging from the very smallest single-core wire used in electronic circuits to the huge oil- and gas-filled cables used in high-voltage transmission systems. In this book we are concerned only with cables used in low-voltage systems (50–1000 V).

A cable comprises two parts: the conductor or conductors and the sheathing and insulation.

Conductors

A conductor may be defined as the conducting portion of a cable, which consists of a single wire or group of wires in contact with each other.

As we have seen in previous chapters, the ability of a material to be a good or bad conductor of electricity depends on the composition of that material (i.e. its

Electrical Installation Work. 978-1-138-84927-3 © Brian Scaddan.
Published by Taylor & Francis. All rights reserved.

resistivity). The following table indicates the resistivity of some common conducting materials:

Material	Resistivity at 20°C (µΩ cm)
Silver	1.58
Copper	1.72
Gold	2.36
Aluminium	2.6
Tungsten	5.6

Silver is clearly the best conductor in the list shown, but its cost prohibits its use as a conductor material on any large scale. Gold is also a material too expensive for use in the construction of conductors.

It can be seen that copper, quite a plentiful mineral, has a low enough resistivity to make it suitable as a conductor material, and, in fact, its use in the manufacture of cable is widespread.

Aluminium, although cheap and with a relatively low resistivity, is not as suitable as copper. It has to have a large cross-sectional area (c.s.a.) to pass the same current and is mechanically inferior to copper.

Tungsten, because of its high resistivity, is used mainly in heating elements and light-bulb filaments.

Conductor construction

Conductors may be divided into two groups:

1. Solid conductors.
2. Stranded conductors.

Solid conductors are either circular or rectangular in cross-section and are used for fixed wiring. Circular solid conductors are restricted mainly to cable cores up to 2.5 mm², although c.s.a.'s of up to 25 mm²

are sometimes used in trunking, from the ground floor to the top floor of a block of flats, to provide a supply point for each floor. These conductors are called **risers**. Rectangular conductors (usually called **bus-bars**) are used in distribution boards or specially constructed bus-bar chambers designed to allow many different circuits to be 'tapped' off (Fig. 14.1). Risers are a version of this type.

Stranded conductors are used in both fixed wiring cable and flexible cords, the latter being flexible cables not exceeding 4.0 mm² in c.s.a.

Conductors for fixed wiring up to 25 mm² have seven strands; for example, a 6 mm² conductor has seven strands each of 1.04 mm diameter (7/1.04). Conductors of c.s.a. above 25 mm² have more strands depending on their size. Flexible cords have conductors comprising a great many fine strands. This type of construction gives the conductor its flexible quality.

Sheathing and insulation

With the exception of bare conductors (i.e. bus-bars and bare risers) all conductors have some sort of insulation and/or sheathing.

Cables for fixing wiring

In many existing installations, old cables can still be found insulated with rubber with an outer sheathing of lead, cotton or rubber. This type of cable for fixing wiring is no longer manufactured. A modern insulating material for cables is polyvinyl chloride (PVC). Some cables also have a PVC sheath (Fig. 14.2).

PVC-insulated single-core cables (singles) are used when the installation is to be run in conduit or trunking.

Figure 14.1 Bus-bar chamber.

Figure 14.2 PVC-insulated PVC-sheathed flat twin with earth. *Colour code*: Brown for *line*, blue for *neutral*, bare copper for *earth*.

Another type of fixed wiring cable is mineral-insulated metal-sheathed (m.i.m.s.) cable. The construction of this type of cable is shown in Fig. 14.3.

Figure 14.3 Twin-core m.i.m.s. cable.

The most popular version of this cable has copper conductors and a copper sheath (m.i.c.s.). It is also available with an overall PVC covering. This is a very strong and long-lasting type of cable.

Armoured cable

Armoured cable is an extension of the type of cable used in fixed wiring. Here, the inner PVC sheath is in turn sheathed in strands of steel wire and an overall PVC sheath is fitted (Fig. 14.4).

Figure 14.4 Three-core armoured cable.

Cables for flexible cords

In this case rubber and cotton are still used as well as PVC for insulation and sheathing. Applications include high-temperature PVC for drops to lamp-holders; butyl rubber for supply to water heaters; cotton-covered rubbers for leads to irons.

> **Note**
>
> The colour code for flexible cords is as follows:
> Brown – line
> Blue – neutral
> Green and yellow – earth

Other cable types

There are many other types of cable, mostly for specialized applications, such as:

▶ **FP** range (e.g. FP100, 200, 400, etc.) it is a fire retardant cable used in fire detection and alarm systems. It is aluminium sheathed with inner and outer heat-resistant insulation and a tinned copper cpc which is in contact with the sheathing throughout its length.

▶ **SY** flexible cable used in instrumentation, control, data, etc., applications. It is multi-core ranging from 2 to 60 core and is easily recognized, as the outer sheath is transparent and the galvanized metal screen is visible.

▶ **Structured data** cable used extensively in the Information Technology world for connection of computers, data processing equipment, etc. to peripheral items.

▶ **Pre-fabricated** cable used in modular wiring systems (see cable management systems). Used in commercial and industrial installations.

▶ **Fibre optic** cable used extensively in the telecommunications industry. The cable comprises strands of glass fibre that transmit data at very fast speeds using light rather than electricity.

Reasons for sheathing and armouring

The insulation immediately surrounding a cable conductor is designed to withstand the cable's working voltage in order to prevent danger. The additional sheathing and/or armouring is added to protect the insulated conductors from mechanical damage. Some environments are more hazardous than others, and cables must be chosen carefully to suit those environments; for example, a PVC-sheathed steel wire armoured (s.w.a.) cable would be used for running underground for house services.

Jointing and terminations

Wherever conductors are to be joined together, or to accessories, or to bus-bars, a safe and effective termination or joint must be made.

167

Jointing

There are many different ways of joining two conductors together. Here we discuss some of the more popular methods.

The screw connector

The screw connector (Fig. 14.5) is probably the simplest method of joining conductors. Because of their simplicity these devices are often used incorrectly. Connectors of this type should always be fixed to a base, allowing the conductor no movement. They should **not** be used to connect two flexible cords together.

Figure 14.5 Simple connector.

> **Note**
>
> These methods would, in modern times, only be used as a last resort.

Junction box

A junction box (Fig. 14.6) is an extension of the connector block type of joint. It is enclosed with a lid, and screwed to a base (joint, wall, etc.) and is designed for fixed wiring systems.

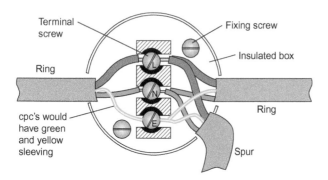

Figure 14.6 Tee joint or spur using a ring final circuit junction box.

Soldered joints

This type of joint may be achieved in several ways by:

1. interlacing (like the fingers of the hands) the cable conductors (stranded) and running solder over the joint. This is called a married joint (straight-through joint).
2. a tee joint, as shown in Fig. 14.7a and b.
3. a soldered straight-through joint using a split ferrule (Fig. 14.8).
4. a crimped joint (Fig. 14.9a–f).

(a)

(b)

Figure 14.7 Soldered tee joint.

Figure 14.8 Joint using split ferrule.

Terminations

When a cable conductor is finally connected to the apparatus it is supplying, a safe and effective termination of that conductor must be made.

PVC singles into screw terminals

The insulation should be removed **only** far enough to allow the conductor to enter the terminal. **Do not leave bare conductor showing outside the terminal** (Fig. 14.10a and b). Take care **not** to score

(a) Crimping sleeve

(b)

(c) Cable
Insulation
Aluminium tube

(d) Flattened with crimping tool

(e) Eye crimping

(f) Pin crimping

Figure 14.9 Crimped joints (a) before crimping; (b) after crimping.

Screw terminal

(a) Correct ✓

(b) Incorrect ✗

Figure 14.10 Screw terminal connections.

the conductor surface when removing the insulation as this may cause the conductor to break if moved.

PVC twin and three-core with earth

In this case, although the final connection into screw terminals is the same as for singles, care must be taken not to damage the conductor insulation when removing the outer sheath. Two methods used are as follows:

1. Using a sharp knife, slice the cable lengthwise, open the cable up and cut off the sheath (Fig. 14.11).

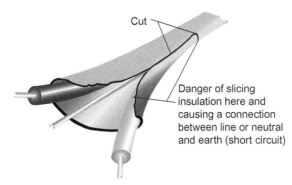

Cut

Danger of slicing insulation here and causing a connection between line or neutral and earth (short circuit)

Figure 14.11 Flat cable sheath removal.

2. With a pair of side cutters, split the end of the cable and hook out the end of the earth conductor.

This may then be held with a pair of pliers and pulled backwards along the cable, splitting open the sheath. Excess sheath may then be cut away (Fig. 14.12).

Pull to split cable sheath

Figure 14.12 Flat cable sheath removal.

Mineral-insulated metal-sheathed cable

As the magnesium oxide insulation is absorbent, the termination of m.i.m.s. cable has to be watertight. The main method of terminating this type of cable is by using a 'screw-on' seal. The preparation of the cable is as follows:

1. The sheath is stripped using a stripping tool (Fig. 14.13).
2. Loose powder is removed by tapping cable.
3. The gland nut is slipped over the cable.

Figure 14.13 Stripping tool.

169

Figure 14.14 Screw-seal termination.

After this the remainder of the termination is shown in Fig. 14.14.

Armoured cable (steel wire armour)

The method of terminating s.w.a. cable is best illustrated in Fig. 14.15.

Figure 14.15 Armoured cable.

Lug terminations

There are two types of lugs: crimped lugs and soldered lugs. A crimped lug is fastened to the end of the conductor using the same method as shown in Fig. 14.9. The soldered lug is filled with molten solder and the conductor is pushed into it. Lug terminations are frequently used for connecting a conductor to a bus-bar (Fig. 14.16).

Soldering

For soldering lugs and joints, a tinman's solder is used in conjunction with a flux. The flux keeps the work clean and prevents it from being oxidized during soldering.

For smaller jobs on fine cable, a solder with a flux incorporated in it is used.

Figure 14.16 Lug termination.

Termination into appliances

The methods used to terminate conductors have already been discussed. However, an important aspect of termination of cable into appliances, especially the portable type, is the securing of the cable at or near the point of entry (Fig. 14.17). A good example of this is the cable clamp in a plug top.

Figure 14.17 Cable cord grip.

It is important to note that in any termination and any run of conductor or cable, measures must be taken to avoid any undue strain on that conductor or cable.

Points to Note

1. The type of insulation and sheathing or covering of every cable must be selected so that the cable or conductor is protected against heat, corrosion or mechanical damage, depending on the situation.
2. Cables and conductors must be chosen to suit the voltage and current rating of the circuit.

3. Cables and conductors must be chosen such that the voltage drop between the intake position and any point in the installation does not exceed 3% of the supply voltage for lighting and 5% of the supply voltage for power; that is, for a 230 V installation, the voltage drop must not exceed 3% of

230 V = 6.9 V for lighting and 5% of 230 V = 11.5 V for power.

4. Cables passing through wooden joists should be 50 mm below the top or above the bottom of the joist (Fig. 14.18).

Joist 50 mm Cable

Figure 14.18 Cable through a joist.

5. Where cables pass through holes in metalwork, the hole should be bushed to prevent any abrasion of the cable.
6. Cables should not be installed in a lift shaft unless they are part of the lift installation.
7. Cables – sheathed and unsheathed and/or armoured – must be supported according to the requirements shown in the appropriate IET Regulations On-site Guide tables.
8. Every cable must be selected such that it can withstand the normal ambient temperature of its surroundings.
9. Where cables are to be connected to bus-bars, 150 mm of the insulation should be removed and replaced with heat-resisting insulation. This applies only if the original

insulation is of general-purpose rubber or PVC.

10. Care must be taken to select the correct type of cable for use in flammable or explosive situations.
11. In order to ensure that fire cannot spread, any holes through which cables pass must be made good to the full thickness of the floor, wall, ceiling, etc.
12. Cables and conductors must be chosen carefully in order to avoid damage by corrosion.
13. Every core of a cable used for fixed wiring should preferably be identified throughout its length, or if this is not possible, by sleeves or discs at its terminations.
14. All terminations of cable conductors must be mechanically and electrically sound.
15. Care must be taken to avoid contact between aluminium and copper conductors, unless adequate precautions have been taken to prevent corrosion.
16. Cables buried in walls at a depth less than 50 mm must be enclosed in earthed metallic conduit, metallically protected from penetration by nails or sirens, etc., or protected by a 30 mA RCD.

Plastics

The uses of plastic materials (polymers) in electrical engineering are widespread, the most common being for cable insulation. Other applications include plug tops, socket outlets and motor and transformer winding insulation.

Plastics technology is a vast and complicated subject and hence only the most basic concepts will be discussed here.

Thermoplastic polymers

Thermoplastic polymers soften on heating and solidify to their original state on cooling. Repeated heating and cooling causes no damage.

Thermosetting polymers

Thermosetting polymers become fluid when heated and change permanently to a solid state when cooled. Further heating may cause the polymer to disintegrate.

Polyvinyl chloride

General

Rigid PVC as used in conduits, etc., is a thermoplastic polymer and has the following properties:

1. It has high tensile strength.
2. It can be bent by hand if warmed.
3. It has high electrical resistance.
4. It is weather resistant.
5. It does not crack under stress at normal temperatures.
6. It has a low flammability.
7. It is self-extinguishing when the source of heat is removed.
8. It must be used with special saddles and expansion couplers when used in fluctuating temperatures, as its expansion is five times that of steel.

Flexible PVC (used in cable insulation)

1. is weather resistant;
2. has high electrical resistance;
3. should be kept clear of other plastics to avoid migration of plasticizer.

14

171

The effects of environmental conditions on PVC are discussed at greater length in Chapter 17.

Phenol-formaldehyde

Phenol-formaldehydes are thermosetting polymers and are used with other compounds to manufacture plug tops, socket outlets, etc., and thin insulation for all types of winding. Such equipment can safely be used in temperatures up to 100°C.

Cable management systems

A cable management system is simply a system of supporting and/or supporting cables. The most common are:

▶ Conduit
▶ Trunking
▶ Ducting
▶ Tray
▶ Ladder
▶ Basket
▶ Modular

Conduit

A **conduit** is a tube or pipe in which conductors are run. In effect, the conduit replaces the PVC sheathing of a cable, providing mechanical protection for the insulated conductors.

There are three types of conduits: metal, flexible (metal) and non-metallic (PVC).

Metal conduit

Most metal conduit used nowadays in low-voltage installations is either 'heavy-gauge welded' or 'solid-drawn'. Heavy gauge is made from a single sheet of steel and welded along the seam, while solid-drawn is produced in tubular form and is therefore seamless. This type of conduit is used only for flameproof installations (owing to its expense).

Metal conduit may be threaded and bent, making it a versatile system. It is available covered with black enamel paint or galvanized, the choice depending on the situation in which it is to be used. It is supplied with all the fittings necessary to make it a complete installation system.

When one is preparing conduit, care must be taken to use the correct tools and preparation methods. For example, a proper conduit bending machine and accessories are essential. The machine will have several different sizes of bending wheel and come complete with a pipe vice. The principle on which it is

based is that of a simple second-order lever (Fig. 14.19a–c).

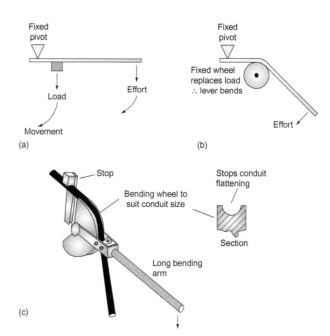

Figure 14.19 Bending metal conduit.

Some typical bends in conduit are illustrated in Fig. 14.20. The **set** is frequently used in order to terminate the conduit at an outlet box (Fig. 14.21).

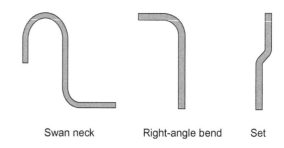

Swan neck   Right-angle bend   Set

Figure 14.20 Conduit bends.

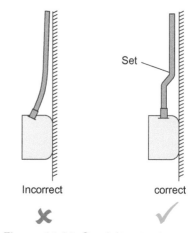

Incorrect   correct

Figure 14.21 Conduit entry to accessories.

There are several ways of joining and terminating conduit as shown in Fig. 14.22. Figure 14.22c shows a 'running coupler'. The conduits to be joined are threaded to a distance greater than the length of the coupler. The coupler is then screwed right on to one thread, the other conduit is butted to the first conduit and the coupler is screwed back over the second thread. This method enables two conduits to be joined without actually turning the conduits themselves. The locknuts are provided to prevent the coupler from moving.

Figure 14.22 Conduit terminations and joints.

Fixing conduit

Figure 14.23 shows the common methods used for securing or fixing conduit.

Figure 14.23 Conduit supports.

Crampets are used for securing conduit in place prior to covering with plaster.

Spacer-bar types are often used to space the conduit out from a wall in order to avoid the need for 'sets'.

Distance saddles space the conduit far enough from the fixing surface to prevent dust from collecting between the conduit and the surface. They were in the past often used in hospitals when they were known as hospital saddles.

Other accessories for use with conduit include elbows (sharp 90° bends), bends (gentle 90° bends) and tees. All of these are available in the inspection and non-inspection varieties. There are also circular boxes with removable lids, to permit ease of wiring a conduit system. Some of these accessories are shown in Fig. 14.24.

Figure 14.24 Conduit accessories.

Flexible conduit

Flexible conduit is of great use when a rigid conduit system is supplying machinery (Fig. 14.25). The use of a short connecting length of flexible conduit ensures that the vibration of the motor does not affect the rigid conduit and enables the motor easily to be aligned, or removed for repair.

Figure 14.25 Flexible conduit.

14

Non-metallic (PVC) conduit

PVC conduit is especially suitable for installation systems in light-industrial premises or offices where surface wiring is required. The fittings are identical to those used for metal conduit with the exception that the system is connected not by screwing but by the use of an adhesive. PVC conduit may be bent by hand using a bending spring. This spring, having the same diameter as the inside of the conduit, is pushed inside the conduit. The conduit may now be bent by hand, the spring ensuring that the conduit keeps its shape. In cold weather a little warmth may need to be applied to achieve a successful bend.

Threading metal conduit

The following procedure should be adopted for the successful threading of metal conduit:

1. Using a pipe vice to hold the conduit, cut the conduit end square (with a hacksaw); if it is not square, then file it until it is correct.
2. Slightly chamfer the edge of the conduit (see Fig. 14.26). This helps to start the thread cutting (see Fig. 14.27).
3. Using a stock and die (see Fig. 14.27), making sure that it is offered square to the end of the conduit, begin cutting the thread. This should not be difficult if the die is sharp, and sufficient 'tallow' (lubricating agent) has been applied.
4. As the thread is cut, a small curl of metal is produced. Do not continue cutting in the same direction all the time, but make two or three turns and then reverse the motion; this will break off the curl of metal and keep the thread clean and even.
5. When the thread is complete, clean away any metal filings and ream out the inside edge

Figure 14.26 Conduit end preparation.

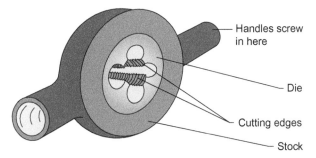

Figure 14.27 Threading tool (stock and die).

to remove any burrs. A little tallow applied to the finished thread will help to screw it into a fitting.

Never cut more thread than is necessary; the finished work will be unsightly, and the exposed thread will need painting to protect it from corrosion.

Pulling a small piece of rag through the conduit with a wire will clean out any particles of metal that have accumulated during the threading process. This ensures, as does reaming (see p. 279), that cable insulation is not damaged when conductors are pulled into the conduit.

Never paint a thread before it is used; the steel conduit itself is usually used as the circuit protective conductor, and it is essential that a good metallic connection is made at every join and termination.

Points to Note

1. The conduit system for each circuit of an installation must be completely erected before any cables are drawn in.
2. Inspection-type fittings must always be accessible for the removal or addition of cables.
3. The number of cables drawn into a conduit may be achieved by selecting the size of conduit using the tables shown in the guidance notes to the regulations. The system used applies to trunking as well as conduit. The IET On-site Guide gives various tables to enable the designer to establish the number of conductors that may be run in conduit or trunking, taking into consideration space factor and the ease of drawing cable in. This system of tables is based on the requirement that, in a conduit run, there should be 40% of cable and 60% of air. In trunking, the cable should only occupy 45% of the available space. So, for example, if eight 2.5 mm² stranded conductors were to be installed in a short straight length of conduit, the factor for the cable (43) wouldbe multiplied by the number of conductors (8) to give an overall cable factor of 43 × 8 = 344.

This overall factor is now used to find a suitable conduit size by comparing it with the conduit factors and selecting a size. In this instance the nearest conduit factor given is (460), giving a conduit size of 20 mm. This method is used for longer lengths with bends, but a different set of tables and factors are used. The same process is also used for trunking capacity.

4. When conduit is bent, the radius of any bend must not be less than 2.5 times the outside diameter of the conduit (Fig. 14.28).

Figure 14.29 Distance between conduit and accessories.

Figure 14.28 Bending radii.

5. If non-inspection elbows or tees are used, they must be installed only at the end of a conduit immediately behind a luminaire (light fitting) or outlet box, etc., of the inspection type. Alternatively one solid elbow may be located not more than 500 mm from a readily accessible outlet box in a conduit not exceeding 10 m between two outlet points providing that all other bends do not add up to more than 90°C (Fig. 14.29).

6. The ends of conduit must be reamed and entries to boxes and accessories must be bushed in order to prevent damage to cables.

7. As we have already seen, a conductor carrying a current has a magnetic field around it, and as a.c. causes the field to change direction continually the field will cut across any metal close to it and in doing so induce e.m.f.'s and eddy currents (heating) in that metal. Running another conductor close to the first with its current flowing in the opposite direction will cancel out the fields. Hence cables of a.c. circuits installed in steel conduit should have all lines and neutral (if any) bunched in the same conduit.

8. Flexible metal conduit must not be used as a circuit-protective conductor; a separate such conductor must be run.

9. Non-metallic conduits must not be installed where extremes of ambient temperature are likely to occur.

10. The supports for rigid PVC conduit must allow for expansion and contraction of the conduit.

11. Drainage points should be provided at the lowest point of a metallic conduit installation to allow moisture due to condensation to drain away.

Trunking

Trunking may be thought of simply as a larger and more accessible conduit system. It is available in two ranges: ordinary wiring trunking and bus-bar trunking. The ordinary system is further available in either metal or PVC.

PVC trunking

The main areas in which PVC trunking is used are domestic and office premises. It is perhaps better described as 'channelling' rather than 'trunking'. All necessary fittings are available to complete any shape of run with little difficulty. Joins in PVC trunking are usually made with adhesive.

Metal trunking

Metal trunking is used extensively in engineering premises. It allows a large number of cables to be drawn into one system with relative ease and because of its accessibility enables changes or additions to circuits to be made with the minimum of effort. As with conduit, it is available either painted or galvanized. A whole range of fittings are available, enabling the most difficult runs to be constructed.

It is available in many different sizes and designs. For example, compartment trunking allows the segregation of certain circuits and support of cables in vertical runs. Figure 14.30 shows several different types of trunking.

Figure 14.30 Trunking types.

Cutting and drilling trunking

It is often necessary to cut or drill trunking. After cutting, all edges should be filed to remove sharp metal. Burrs left after the drilling of a hole should also be filed away. When high-speed twist drills are being used, a lubricant should be applied; this aids the drilling process and prolongs the life of the drill. A lard oil such as tallow is recommended for use with steel (no lubricant is needed when drilling PVC).

It is often necessary to cut large holes in order to connect conduit into the trunking system; this may be achieved by the use of a circular hole saw or a pressure-type hole cutter (Fig. 14.31).

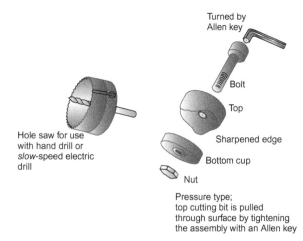

Figure 14.31 Hole cutting.

After cutting, all hole edges must be filed smooth and conduit entries bushed.

Supports for trunking

Trunking may be secured direct to a surface or suspended by means of brackets.

As trunking material is thin, there is no room for countersunk holes. Roundhead screws (Fig. 14.32) are suggested, unless of course the fixing surface requires bolts.

There are occasions when the trunking has to be suspended by means of brackets. Figure 14.33 shows two typical brackets.

Figure 14.32 Screw fixings.

Figure 14.33 Trunking supports.

Bus-bar trunking

There are two main types of bus-bar trunking, the overhead type of bus-bar used for distribution in industrial premises and rising-main trunking (Figs 14.34 and 14.35).

Figure 14.34 Bus-bar trunking.

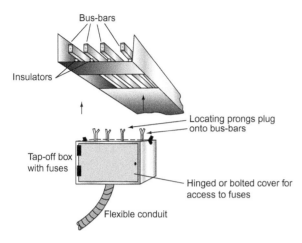

Figure 14.35 Overhead bus-bar trunking.

Figure 14.36 Trunking bends.

Overhead bus-bar trunking is ideal for distribution in factories. It is run at high level, and the tap-off boxes enable machinery to be moved easily. When connected in the form of a ring and incorporating section switches, this system enables parts of the trunking to be isolated when necessary, without all machines losing supply.

Trunking, and in particular steel trunking, can be fabricated into various shapes to allow for bends and tees, etc. This process is, however, laborious and there are many different manufactured shapes available which simply bolt onto straight sections to allow versatility in installation. Figure 14.36 illustrates how such sections are joined together.

Points to Note

1. Where trunking passes through floors, walls or ceilings, a fire barrier must be provided.
2. Many installations contain circuits of different types. These circuits are divided into two voltage bands:
   Band 1: Extra-low voltage (ELV) circuits, including bell and call systems, telecommunication and alarm systems.
   Band 2: Low-voltage circuits (LV), that is typical domestic, commercial and industrial arrangements.
   In trunking systems these bands may be enclosed together provided that either:
   (a) band 1 cables are insulated to the highest voltage present or
   (b) band 1 and 2 circuits are in separate compartments in the trunking.
3. All conductors of bus-bar trunking must be identifiable.
4. Bus-bar trunking must be installed so that the conductors are inaccessible to unauthorized persons.
5. Bus-bars must be free to expand and contract with variations in temperature (Fig. 14.37).

Figure 14.37 Bus-bar expansion strip.

6. All lines and neutral (if any) of a.c. circuits must be contained in the same metal trunking.
7. Entry to trunking must be so placed as to prevent water from entering.
8. The size of trunking may be determined from the guidance notes to the regulations.
9. When metal trunking is used as a circuit protective conductor, copper straps should be fitted across joints to ensure electrical continuity.

177

Traywork

Traywork is another method of supporting several cables along a run. It is simply a perforated metal tray on to which cables can be tied (Fig. 14.38). Its use is restricted normally to situations such as switch rooms, boiler houses and large cable ducts (i.e. places not normally occupied by personnel).

Figure 14.38 Cable tray.

The construction of a cable tray is such that it is easily cut, shaped and bolted, to suit all situations. Support brackets are the same as for trunking.

Ducting

This can come in all different shapes, sizes and materials. It is usually used where many large cables need to be accommodated. Ducts may be earthenware or plastic pipes for use underground or channels in the floor or hollow spaces within a buildings fabric (*appx. 4 BS 7671:2008 gives some examples*).

Ladder

This is similar to tray but in the form of a ladder (Fig. 14.39). It is used in commercial and industrial installations. It is easy and versatile to install and has good ventilation.

Wire basket

This is another variation of tray or ladder and is used generally for carrying many small cables such as would be found in data processing installations, etc. (Fig. 14.40).

Modular

This is a pre-fabricated system used in commercial and industrial installations. It comprises distribution boards, special connection units, outlet boxes and pre-set flexible cable lengths that have plugs on each end.

It requires little effort to erect, the installer just fixes the various connection and outlet units and plugs the cable in at both ends. This clearly avoids the need to cut cable from a reel and terminate the ends (Fig. 14.41).

Fixing and tools

An important aspect of installation work is the fixing of accessories to various surfaces.

Figure 14.42 shows several different methods of fixing.

The main disadvantage of gravity and spring toggles is that it is not possible to remove the bolt without losing the toggle.

Tools

The following is a list of tools used by the electrician; a complete basic tool kit will have all these items.

Figure 14.39 Cable ladder.

Figure 14.40   Wire basket.

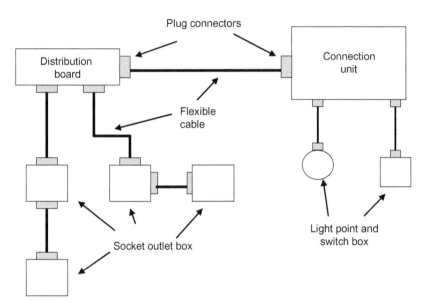

Figure 14.41 Modular wiring system.

FIXING INTO MASONRY
Metal plug

Plastic Rawlplug – use correct drill size and screw size. Do not shorten plug by cutting off.

Rawlbolt – when nut is tightened, metal plug is pulled in direction shown, opening out the main shaft.

Screw fixing

FIXING INTO PARTITIONS

Plasterboard, etc.

Spring-loaded arms

Open when released

Spring toggle

Brolley fixing

Front view

When bolt is tightened whole assembly is distorted

Heavier part drops down when in position

Figure 14.42 Fixing methods.

14

179

First fix

These tools are used for chasing walls, lifting floorboards, etc.

Club hammer	This is for use with cold chisels and bolsters. Ensure that the handle is in good condition. Do not use the hammer if the head is loose.
Bolster (large and small)	This tool is used for chasing walls and brickwork and for making holes for metal boxes. Ensure no mushrooming (see p. 276).
Floorboard chisel	This is similar to a bolster but with a longer handle. The same precautions should be taken as for bolsters.
Scutch hammer and combs	These tools are used for chasing brickwork.
Cold chisels (a selection of various sizes)	These are for use on brickwork and concrete. The same precautions should be taken as for bolsters.
Tenon saw and padsaw	For cutting floorboards, etc. (special floorboard cutters are available). They must be kept sharp and greased when not in use.
Brace and wood bits	These are used for drilling joists, etc. Wood bits should be kept sharp and ratchets oiled.
Claw hammer	For de-nailing joists and floorboards. Precautions to be taken are the same as those for a club hammer.
Rawlplug tool and bits	These are for making holes in brickwork and concrete in order to insert Rawlplugs. Watch for mushrooming of the head.
Hand drill and electric drill	These are for use with metal drills and masonry drills. An electric drill should preferably have two speeds and a hammer action.

Second fix

These tools are used for fitting accessories such as sockets and so on (Fig. 14.43).

Large cabinet-maker's screwdriver	This tool is for use in fixing screws.
Large and small electrician's screwdriver	Used for most screws in accessories. The handles of screwdrivers must be in perfect condition, and the tips must be kept at the correct size. Only the correct size of tip should be used on screws. Do **not** use screwdrivers as chisels.
Wire cutters and strippers	Used for preparing cable ends. They should be kept clean and sharp.

Pliers and sidecutters, insulated; junior hacksaw and blades; Stanley knife and blades	These are used for general work.
Set of tools for termination of m.i.m.s. cable	Cutting tools should be kept sharp and all surplus compounds removed.
Tools for use on conduit and trunking	These include a bending machine with a pipe vice, designed for use with most sizes of conduit. They should be kept clean and free from metal particles. Another tool for use on conduit is a hacksaw frame taking hand blades with 32 teeth per 2.5 cm (Fig. 14.44); the blade must be used in the correct direction.
A selection of files	Files are used for removing sharp edges and burrs. **Always** use a handle on a file.
Reaming tool (Fig. 14.45)	For removing burrs from the inside edge of the conduit.
Stock and dies	These tools are used for cutting threads. A recommended lubricant such as tallow should be used. Dies must be kept clean; do not allow a build-up of tallow and metal.
Adjustable pipe wrench and Stillson (Fig. 14.46)	Used for tightening and untightening conduit.
Bush spanner (Fig. 14.47)	This tool is for use on inaccessible bush nuts.
Draw tape	Used for pulling cables through conduit.
Measuring tape	Steel tapes must *not* be used near live equipment.
Set of twist drills	For use with a hand or electric drill. Cutting edges should be kept sharp. A hot drill must never be quenched in water.
Selection of spanners	For general use.
Spirit level	
Set-square	For use in marking out trunking.
Scriber	Used for marking metal. The points should be kept sharp.
Centre punch	For marking metal prior to drilling.
Hole saw	This type of saw is used for cutting holes in trunking. It should be used only with a hand drill or a low-speed electric drill.
Putty knife and pointing trowel	For making good the plasterwork, etc.

(Continued)

Bending springs	Used for PVC conduit.
Solder, soldering iron, blowlamp	These are used for all general electrical soldering.
Approved test lamp	Used on live circuits. Moisture must be kept away.
First-aid kit	This must be kept stocked at all times.

Figure 14.43 Selection of hand tools.

Figure 14.44
Hacksaw blade teeth.

Figure 14.45 Reamer.

As tools play an important part in installation work, it is perhaps sensible to understand some of the processes involved in their manufacture. This applies particularly to hardened, tipped tools such as chisels, drills, scribes and so on.

Heat treatment

Most good-quality tools are manufactured from carbon steel. When carbon steel is heated, its physical

Figure 14.46 Conduit tools.

Figure 14.47 Conduit termination spanner.

properties change, and these changes may be used to advantage. On being heated, carbon steel increases in temperature until it reaches a point called the **lower critical limit**, when the steel starts to change. It remains at this temperature for a short while before its temperature increases again to the **upper critical limit**, where the change is complete. On slow cooling the reverse process takes place and the steel returns to normal.

Hardening

If a sample of carbon steel is heated to its upper critical limit and then cooled rapidly by plunging it into cold water, it will become very hard. It is usually too brittle for most purposes, however, and has to be tempered.

Tempering

Tempering is carried out by heating the hardened steel to just below its lower critical limit and quenching in water. The steel then loses its brittleness but remains very hard.

Annealing

In order to cold-work to shape and/or machine a sample of steel before it is hardened and tempered, it must be as soft and ductile (able to be permanently deformed without damage) as possible. This is achieved by heating the sample to its hardening temperature and then cooling it very slowly over a long period of time.

Work hardening

Excessive hammering, rolling or bending, etc., of a metal causes it to harden and become brittle. It may be returned to its original condition by annealing.

Comparison of systems

PVC-insulated PVC-sheathed cable

PVC-insulated PVC-sheathed cable is suitable for all types of domestic and commercial wiring installation where there is little risk of mechanical damage, extremes of temperature or corrosion. It is inexpensive and easy to handle, and no specialized tools are needed for working with it.

Mineral-insulated metal-sheathed cable

M.i.m.s. cable is suitable for various applications in commerce and light industry. It has good mechanical strength and can withstand a considerable amount of crushing before it breaks down electrically. It has a high degree of resistance to heat, which makes it useful for fire-alarm circuits. The sheath makes a good circuit protective conductor (cpc). M.i.m.s. cable is not to be used in damp or corrosive situations without an overall PVC covering. It is expensive, although long lasting; it needs specialized tools and skilled craftsmen to install it.

Fire-retardant cable

For fire-alarm circuits and areas of high combustibility, a fire-retardant cable such as Fire-Tuf or FP200 is used. It is a metal (usually aluminium) sheathed cable with high heat-resistant insulation and does not require the same degree of skill for terminations.

Conduit

Steel conduit is used extensively in industrial premises. It has good mechanical strength and may be used as a cpc. Specialized tools are needed to install it, and steel conduit is not easily added to; it is subject to corrosion and is relatively expensive.

PVC conduit is used in commercial and light-industrial premises. It has quite high mechanical strength and is inexpensive compared with steel. It does not corrode, is easily erected and no specialized tools are needed. It cannot stand extremes of temperature and distorts owing to its high degree of expansion unless this is compensated for; it cannot be used as a cpc.

Other types of conduit, including copper and aluminium, are available; these are termed **non-ferrous conduits** (non-magnetic) and are used only in specialized environments. Aluminium conduit is susceptible to a high degree of corrosion.

Trunking

Wirable trunking is used extensively in industry. It is a versatile system allowing the easy addition of further sections. A greater number of cables can be run than in conduit. No specialized tools are needed and there is a range of sizes. It allows different circuits to be segregated. It is, however, more expensive than conduit.

The bus-bar type of trunking is used for rising mains, distribution boards and overhead supplies to machines. The tap-off box system is popular as it gives a measure of control and selective isolation.

System installation

Once a wiring system has been chosen, decisions must be made as to how the system is to be installed, and such decisions will need to take account of the fabric and structure of the building and the aesthetics. For example, PVC flat cables used to rewire a very old dwelling where the walls are 'lath and plaster' (slats of wood covered in a soft lime and sand plaster) are not suitable for chasing out or surface clipping. In this case the best solution is to use an adhesive-backed mini-trunking to enclose the cables.

Another example is that of a traditional brick construction premises with wooden joists, some of which may be load-bearing (i.e. walls are resting on them). These timbers should not be penetrated by drilling or notching in such a way as to weaken them. The Building Regulations indicate where a joist may be penetrated, as shown in Fig. 14.48.

Clearly, all exposed wiring systems will need supporting either vertically or horizontally to ensure that they are not damaged by undue strain. The IET Regulations 'On-site Guide' includes various tables indicating distances between supports for cables, conduit and trunking systems, together with spans and heights of overhead cables, etc.

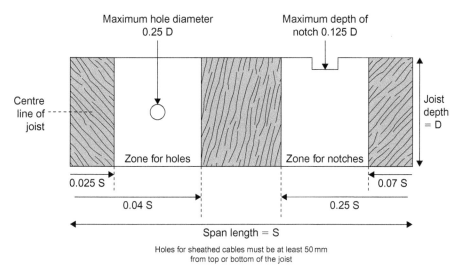

Figure 14.48 Holes and notches in joists.

Self-Assessment Questions

1. Explain the difference between cable insulation and cable sheathing. What is a flexible cord and when is it used?
2. (a) What precautions should be taken when stripping cable sheaths and insulation and why?
   (b) What are the requirements of the IET with respect to identification of conductors?
3. What is a bus-bar and where is it used? Illustrate your answer with sketches.
4. (a) What is m.i.m.s. cable?
   (b) Describe with the aid of sketches one type of termination for m.i.m.s. cable.
5. (a) Describe the correct method of cutting a thread on a conduit end.
   (b) What is an inspection fitting and what are the restrictions on the use of solid (non-inspection) elbows and tees?
6. What is the meaning of the term 'space factor'?
7. Explain, with the aid of sketches, why all lines and neutral (if any) of a.c. circuits must be drawn into the same metallic conduit or trunking.
8. What is meant by 'segregation' of circuits and how is it achieved in trunking?
9. Comment on the advantages and disadvantages of conduit and trunking systems. What special precautions must be taken when using PVC conduit?

14

Electrical Installation Work. 978-1-138-84927-3 © Brian Scaddan.
Published by Taylor & Francis. All rights reserved.

CHAPTER 15

Installation Circuits and Systems

Throughout this chapter there are many references to diagrams/drawings and these may be categorized as follows:

Block diagrams	These show, using squares, rectangles etc., the sequence of a system without too much technical detail (e.g. Figs 15.17 and 15.26).
Layout diagrams	These are very similar to block diagrams, but they indicate more technical detail and tend to show items in their correct geographical location (e.g. Figs 15.18 and 15.19).
Circuit/ schematic	These show how a circuit functions and takes no account of exact locations of terminals or equipment (e.g. Figs 15.28, 15.29, 15.32, 15.34, etc.).
Wiring diagrams	These indicate how a circuit or system is physically wired (e.g. Figs 15.2, 15.4, 15.6, etc.).

Lighting circuits

One-way switching

Figure 15.1 is a circuit diagram showing how the light or lights are controlled, while Fig. 15.2 shows how the point would be wired in practice.

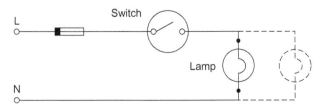

Figure 15.1 Single switch controlling one or more lamps. *Note*: Switch in line conductor.

Points to Note

1. The ends of black, blue or grey switch wires have brown sleeving to denote line conductor. (This is not required for conduit wiring as the cable will be brown.)
2. The earth wire terminations have green and yellow sleeving. (This is not required for conduit.)
3. The light-switch point has an earth terminal.
4. The ceiling rose has an earth terminal.
5. The flexible cord from the rose to the lampholder is capable of withstanding the maximum likely temperature.
6. If a batten holder is used instead of a ceiling rose, then the cable entries should be sleeved with heat-resistant sleeving.
7. The maximum mass suspended by flexible cord shall not exceed:
   2 kg for a 0.5 mm² cord;
   3 kg for a 0.75 mm² cord;
   5 kg for a 1.0 mm² cord.
8. The line terminal in a ceiling rose must be shrouded.
9. For the purpose of calculating the cable size supplying a lighting circuit, each lighting point must be rated at a minimum of 100 W.
10. A ceiling rose, unless otherwise designed, must accommodate only one flexible cord.

Figure 15.2 Wiring diagram.

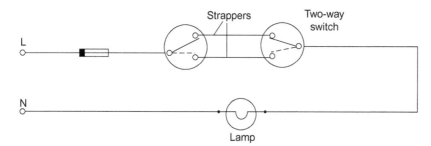

Figure 15.3 Circuit diagram – two-way switching.

Two-way switching (Figs 15.3 and 15.4)

Points 1–10 are as for Fig. 15.2. A typical application is for stairway lighting.

Two-way and intermediate switching

The circuit diagram for two-way and intermediate switching is shown in Fig. 15.5.

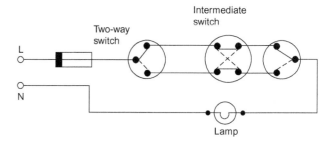

Figure 15.5 Circuit diagram – two-way and intermediate switching.

> ### Note
>
> The earth is omitted from the diagram, for clarity.

Brown sleeving or discs are used on grey, black and blue cables from two-way and intermediate switches to the junction box in order to denote line cables (Fig. 15.6).

The applications of this type of switching are for stairs and landings and in long corridors.

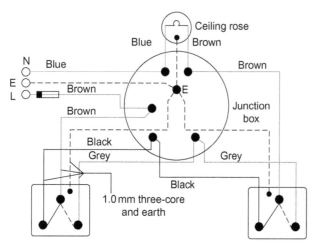

Figure 15.4 Wiring diagram – two-way switching.

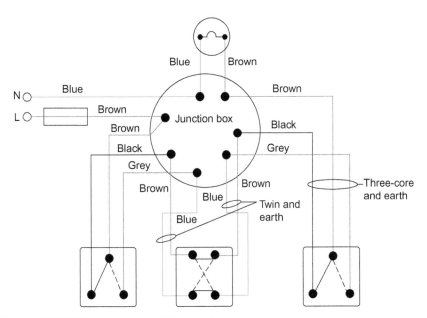

Figure 15.6 Wiring diagram (PVC sheathed).

Lighting layouts

There are two main methods of wiring a lighting installation:

1. Each ceiling rose or junction box is fed from the previous one in the form of a chain.
2. The main feed is brought into a central junction box and each point is fed from it (like the spokes of a wheel).

There are of course variations involving combinations of these two methods depending on the shape and size of the installation.

Power circuits

Radial circuits

Radial circuits are arranged in the same way as point 1, above, in lighting layouts, in that each socket outlet is supplied via the previous one (Fig. 15.7).

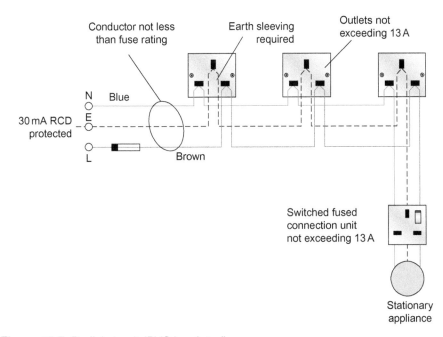

Figure 15.7 Radial circuit (PVC insulated).

187

Points to Note

1. The number of socket outlets, complying with BS 1363, is unlimited for a floor area of up to 50 m² if the circuit protection is a 30 or 32 A cartridge fuse or circuit breaker, the cable being 4.0 mm² PVC copper or 2.5 mm² mineral insulated (m.i.).
2. The number of socket outlets is unlimited for a floor area up to 20 m² with any type of circuit protection of 20 A using 2.5 mm² PVC copper cable or 1.5 mm² m.i.
3. The total number of fused spurs is unlimited.

of all points supplied by the fused spur must not exceed 13 A.

4. Non-fused spurs must supply *no more* than one single or one double socket or one stationary appliance.
5. For domestic premises of area less than 100 m², a ring circuit may serve an unlimited number of points.
6. Cable sizes for ring circuits using copper conductors are:
   2.5 mm² PVC insulated or
   1.5 mm² m.i.
7. Both ring and radial socket outlet up to 20 A for general use by ordinary persons must be 30 mA RCD protected.

Ring final circuits

These circuits are the same as radial circuits except that the final socket outlet is wired back to the supply position. In effect, any outlet is supplied from two directions (Fig. 15.2).

Points to Note

1. Every twin-socket outlet counts as two single-socket outlets.
2. The number of non-fused spurs must not exceed the total number of points on the ring. Figure 15.8 shows seven points on the ring and five spurs.
3. The fuse rating of a fused connection unit must not exceed 13 A and the current rating

Note

When using some 2.5 mm² flat twin with earth cable, a 1.0 mm² circuit protective conductor (cpc) is too small to comply with the regulations unless a circuit breaker or high breaking capacity (HBC) fuse is used.

Water heater circuits

Water heaters of the immersion type (i.e. having an uninsulated element completely immersed in water) are available in several different forms, two of the more popular being the large 140 l storage type, and

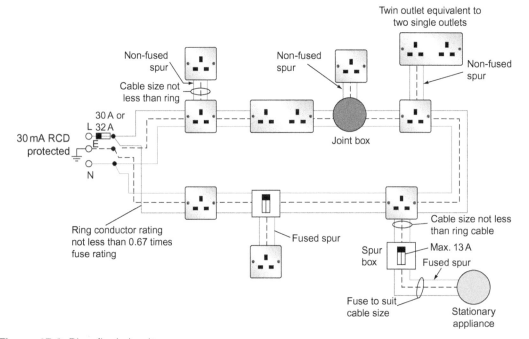

Figure 15.8 Ring final circuit.

the smaller open-outlet type (10–12 l) used for small quantities of instant hot water.

Both types are stationary appliances and could therefore be connected into a ring circuit. However, it is recommended that heaters over 15 l have their own circuit.

> **Points to Note**
>
> 1. The heater shall be effectively connected to earth, and the inlet and outlet pipes must be made of metal. An earthing terminal must be provided.
> 2. The heater must be supplied from a double-pole linked switch (*not* a plug and socket).

Cooker circuits

A cooker exceeding 3 kW should be supplied on its own circuit. As it is rare to have every heating element working at once, *diversity* is applied to calculate the assumed current demand as follows.

The first 10 A of the total rated current of the connected cooking appliance, plus 30% of the remainder of the total rated current of the connected appliance, plus 5 A if there is a socket outlet in the control unit.

> **Example**

The full-load rating of a 230 V cooker is 11.5 kW. Calculate, using diversity, the assumed current demand. The cooker control unit has a socket outlet.

Actual full-load demand:

$$P = I \times V$$

$$I = \frac{P}{V}$$

$$= \frac{11500}{230}$$

$$I = 50\,A$$

Assumed demand using diversity:

$$I = 10 + \frac{30 \times (50 - 10)}{100} + 5$$

$$= 10 + \frac{30 \times 40}{100} + 5$$

$$= 10 + 12 + 5$$

$$I = 27\,A$$

This means that although not advisable, a 27 A cable could be used rather than a 50 A cable. But, the use of diversity is mainly to size the main incoming tails,

by adding together all the other assumed current demands of other circuits.

> **Points to Note**
>
> 1. If the assumed current demand of a cooker circuit exceeds 15 A but does not exceed 50 A, two or more cooking appliances may be fed, if they are in the same room.
> 2. Every cooker must have a control switch within 2 m. If two cookers are installed, one switch may be used, provided that neither cooker is more than 2 m from the switch.

Space heating systems

There are three methods of transferring heat from one place to another: conduction, convection and radiation.

Conduction of heat occurs when the source of heat and the object it is heating are in direct contact; for example, a metal rod held in a flame will become hot at the holder's end as the heat is conducted along the rod. Conduction of heat does not really concern us in the subject of space heating.

Convection heating is a method by which cold air or liquid is heated, rises, cools and falls, and is reheated again, the operation continuing until the required temperature is reached.

Radiation transfers the heat from the source directly to the object to be heated through the surrounding air.

Radiant or direct heating

Direct-heating appliances include all apparatus that gives a person or persons warmth by the direct transfer of heat from the source to the person (e.g. an electric fire (Fig. 15.9)).

The infrared or halogen type of heater found in bathrooms also gives out heat by radiation. These types are permitted in zone 2 of a room containing a

Figure 15.9 Infrared fire.

fixed bath tub or shower, its control being by means of a pull-cord switch.

Convection heating

Convectors work on the principle of circulating warm air in a space (Fig. 15.10). Figure 15.11 shows a typical domestic convector heater.

Figure 15.10 Circulation of a convection heater.

Figure 15.11 Floor or wall-mounted convector.

Storage heaters

Another type of heating appliance which uses the convection principle is the block storage heater.

In this case the elements are embedded in special fireproof heat-retaining bricks. The bricks collect the heat from the elements while they are energized and slowly give off the heat to the surrounding air over a period of time. These heaters are designed to be used on a special 'off-peak' tariff, switching on at 11.00 p.m. and switching off at 7.00 a.m. During the day the stored heaters are sometimes installed to give a boost to the room temperature. The fan blows hot air from around the bricks into the room.

Other variations of this system are underfloor heating installations and ducted warm-air systems.

Floor warming

Mineral-insulated single-core heating cable can be laid in the solid floor of a dwelling for heating purposes.

The floor composition is usually such that it readily retains heat, giving off the required warmth during the day.

Ducted warm-air heating

There are many different types of ducted warm-air heating systems on the market. All, however, have the same basic approach, that is a large, centrally located storage heater, with ducts leading from it to separate rooms. Air is forced from the unit to the rooms by means of one or more fans and released into the rooms via adjustable grills.

Thermostats

In order to control a heating appliance automatically, a method of detecting the temperature of the air, or water, or element being heated, must be found. Such a device is a *thermostat*. It detects changes in temperature and switches the heating appliance either on or off. The different types of thermostat are as follows.

The bimetallic strip

Different metals expand at different rates when heated, and this is the basis of the bimetallic-strip type of thermostat (Fig. 15.12a and b). Metal A expands faster than B when subjected to a rise in temperature, causing the strip to distort (Fig. 15.12b) and makes the contact.

Figure 15.12 Bimetal strip.

The Invar rod

In a variation of the bimetallic-strip type of thermostat, a length of nickel–steel alloy, Invar, is secured inside a copper tube. With a rise in temperature the copper expands but the Invar rod does not (Fig. 15.13a and b).

Both of these types of thermostat, the Invar rod and the bimetal strip, are used with water heaters. The bimetal type is used for the control of irons.

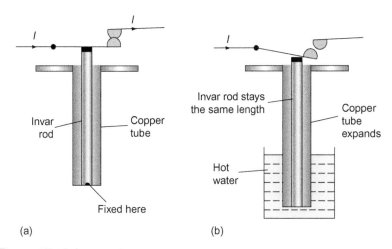

Figure 15.13 Invar rod.

Air thermostats

Air thermostats are commonly used to control central heating systems. Figure 15.14a and b illustrates the two different types used.

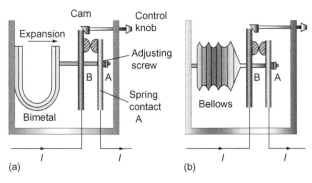

(a)   (b)

Figure 15.14 Air thermostat: (a) bimetal type; (b) bellows type.

In Fig. 15.14a, increase in air temperature causes the bimetal to expand and the spring contact A to move away from B, opening the circuit. Heat settings can be achieved by the control knob, which alters the distance between A and B.

Figure 15.14b shows a bellows type. In this case the air inside the bellows expands and contracts with a change in temperature, so causing contacts A and B to open or close.

Simmerstat

A simmerstat is used mainly for heat control on cooker plates and is shown in Fig. 15.15. Once the control knob has been selected to a particular value, the bimetal at A is the only part that can move, point B being fixed by the position of the cam.

Figure 15.15 Simmerstat.

Ovenstat

This type of thermostat works on the bellows principle. In this case, however, the bellows are expanded by the pressure from a bulb and tube filled with liquid, which expands as the temperature rises (Fig. 15.16).

Figure 15.16 Ovenstat.

Installation systems

The IET Regulations recommend that every consumer's installation should have a means of isolation, a means of overcurrent protection

191

Figure 15.17 (a) Single-phase control; (b) three-phase control. TP and N: triple pole and neutral.

and a means of earth leakage protection. This recommendation applies whatever the size or type of installation, and the sequence of this equipment will be as shown in Fig. 15.17a and b.

Industrial installations

Industrial installations differ basically from domestic and commercial ones only in the size and type of equipment used. The supplies are three-phase four-wire, and switchgear is usually metal clad. For extremely heavy loads, switch-fuses are replaced by circuit breakers, and much use is made of overhead bus-bar trunking systems. Figures 15.18 and 15.19 show typical layouts.

With the larger types of installation, an alphanumeric system is very useful for cross-reference between block diagrams and floor plans showing architectural symbols.

Figures 15.20–15.22 illustrate a simple but complete scheme for a small garage/workshop. Figure 15.20 is an isometric drawing of the garage and the installation, from which direct measurements for materials may be taken.

Figure 15.21 is the associated floor plan, which cross-references with the DB schedule and interconnection details shown in Fig. 15.22.

Figure 15.18 Layout of industrial installation.

Multi-storey commercial or domestic installations

In order to supply each floor or individual flat in a block, it is necessary to run cables from the main intake position. These cables are called *risers,* and the distribution cables which run from these to each individual supply point are called *laterals.*

Figure 15.19 Distribution system.

Figure 15.20 Isometric drawing for garage/workshop. 1, three-phase supply to ramp: 20 mm² conduit; 2, single-phase supply to double sockets: 20 mm² conduit; also 3, 5, 6, 9, 11, 13; 4, single-phase supply to light switch in store: 20 mm² conduit; 7, single-phase supply to light switch in compressor: 20 mm² conduit; 8. three-phase supply to compressor: 20 mm² conduit; 10, single-phase supply to heater in WC: 20 mm² conduit; 12, single-phase supply to light switch in WC: 20 mm² conduit; 14, single-phase supply to light switch in office: 20 mm² conduit; 15, main intake position; 16, single-phase supplies to switches for workshop lights: 20 mm² conduit; 17, 50 mm × 50 mm steel trunking; 18, supplies to fluorescent fittings: 20 mm² conduit.

Figure 15.21 Floor plan for garage/workshop.

Figure 15.22 Details of connection diagram for garage/workshop.

The majority of rising mains are in the form of bus-bar trunking with either rectangular or circular conductors; this enables easy tapping off of submain cables. Figure 15.23 shows a typical system.

The rising-main system is similar to the ordinary radial circuit in that one cable run supplies several points. Hence the current flowing in the cable at the far end will be less than that at the supply end and the voltage drop will be greater at the far end with all loads connected (Figs 15.24 and 15.25).

Hence, the currents may be found in any part of a radial distribution cable. Also, if the resistance per metre of the cable is available and the position of the loads along the cable is known, then the voltage drop at points along the cable may be calculated.

Example

A 230 V radial distributor is 70 m long and has a resistance of 0.0008 Ω/m supply and return. Four loads

Figure 15.23 Rising main in a block of flats.

Figure 15.24 Radial circuit.

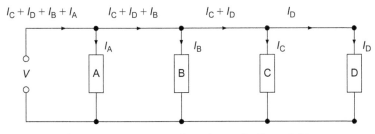

Figure 15.25 Circuit equivalent to that shown in Fig. 15.24.

A, B, C and D rated at 30, 45, 60 and 80 A are fed from the cable at distances of 20, 10, 15 and 25 m, respectively. Calculate the total current drawn from the supply, the current in the cable between each of the loads and the voltage at load D if all the loads are connected.

Total load $I = 30 + 45 + 60 + 80$

$= 215\,A$

$I_{CD} = 80\,A$

$I_{BC} = 60 + 80$

$= 140\,A$

$A_{AB} = 45 + 60 + 80$

$= 185\,A$

Resistance between supply and A $= 20 \times 0.0008$

$= 0.016\,\Omega$

∴ Voltage drop between supply and A $= 0.016 \times I$

$= 0.016 \times 215$

$= 3.44\,V$

Resistance between A and B $= 10 \times 0.0008$

$= 0.008\,\Omega$

∴ Voltage drop between A and B $= 0.008 \times I_{AB}$

$= 0.008 \times 185$

$= 1.48\,V$

Resistance between B and C $= 15 \times 0.0008$

$= 0.012\,\Omega$

∴ Voltage drop between B and C $= 0.012 \times I_{BC}$

$= 0.012 \times 140$

$= 1.68\,V$

Resistance between C and D $= 25 \times 0.0008$

$= 0.02\,\Omega$

∴ Voltage drop between C and D $= 0.02 \times I_{CD}$

$= 0.02 \times 80$

$= 1.6\,V$

Total voltage drop $= 3.44 + 1.48$

$+ 1.68 + 1.6$

$= 8.2\,V$

∴ Voltage at load D $= 230 - 8.2$

$= 221.8\,V$

Off-peak supplies

As the name implies, off-peak electricity is supplied to the consumer at a time, usually between 11 p.m. and 7 a.m., when demand is not at a peak. This ensures a greater economy in the use of generators and hence the cost per unit to the consumer is low.

These supplies are used mainly for space and water heating; however, some intake arrangements allow all energy-using devices to be used off-peak.

Standard off-peak arrangement

Figure 15.26 shows the arrangement at the supply intake position. In this arrangement the time clock controls the contractor coil. In installations with only a light off-peak load, the time clock contacts are able to control the load directly.

Figure 15.26 Arrangement for off-peak supplies (earths omitted for clarity).

The white meter

A white meter has two recording dials, one for normal supplies, and the other for off-peak. Two time clocks and a contactor are used to change over to off-peak. With this system the whole installation can be run during off-peak hours. Figure 15.27 illustrates this system.

During normal hours, only the normal-rate dial will record. The consumer can in fact use off-peak appliances during this period, by overriding the time clock. Of course, any energy used by these appliances is charged at the normal rate.

At a preset time, say 11 p.m., the authority's time clock automatically changes the connections to the dials in the meter, and energy used by any appliance will be metered at off-peak rates.

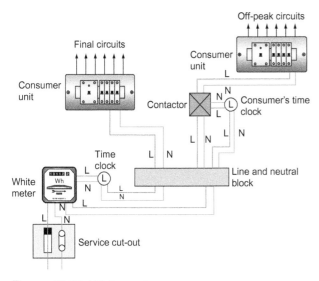

Figure 15.27 White meter.

Alarm and emergency systems

These types of systems are usually supplied by an extra-low voltage (up to 50 V) although the operating voltage may be supplied via a transformer, whose primary may be at a low voltage.

Before each of these systems is discussed, it is perhaps best to establish a convention with regard to relays. A *relay* is an electromagnet which causes pairs of contacts to make or break, when it is energized. All diagrams should show the relay *de-energized;* the contacts are then said to be in their *normal* position (Fig. 15.28).

Another important point to note is the difference between a circuit or schematic diagram and a wiring diagram. A circuit diagram shows how the system *functions*, and relay contacts, switches, and accessories are shown on a diagram in a position most convenient for drawing and understanding. A wiring diagram shows how the system is to be *wired*, and all components of the circuit should be shown in their correct places (Fig. 15.29a and b).

The 'hold-on' circuit shown in Fig. 15.29a is most important especially in fire- and burglar-alarm systems. It operates as follows. With switch S closed R is still not energized. By pushing the button B coil R is energized and the normally open contact C will close giving another route for the supply to reach the relay coil. When the button is released the relay will 'hold-on' through its own contact. If switch S were opened, the relay would de-energize and contact C would open, but closing S again would not re-energize the relay. Button B would need to be pushed again to re-energize the relay.

Both security and fire-alarm systems are basically the same, in that various sensors are wired to a control panel, which in turn will activate an alarm in the event of sensor detection (Figs 15.30 and 15.31). Some modern panels have the facility for incorporating both systems in the same enclosure.

Open-circuit system

In this system the call points (sensors, detectors, etc.) are wired in parallel such that the operation of any one will give supply to the relay RA and the sounder via the reset button. N/O contacts RA1 will then close, holding on the relay and keeping to the sounder via these contacts. This hold-on facility is most important as it ensures that the sounder is not interrupted if an attempt is made to return the activated call point to its original 'off' position.

Fire-alarm systems are usually wired on an open-circuit basis, with a two-wire system looped from one detector to the next, terminating across an end-of-line resistor (EOLR). This provides a circuit cable monitoring facility; the EOLR is of sufficiently high value to prevent operation of the alarm. Figure 15.33 shows a typical connection diagram.

Closed-circuit system

This system has the call points wired in series, and the operation of the reset button energizes relay RA.

Figure 15.28 (a) De-energized; (b) energized.

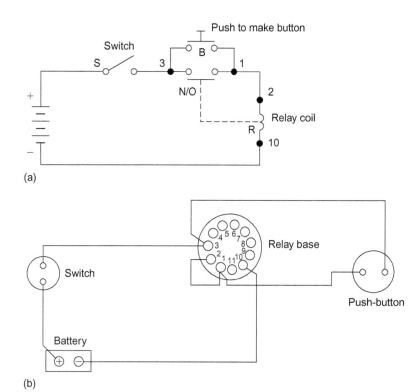

Figure 15.29 (a) Simple 'hold-on' circuit; (b) wiring diagram.

Figure 15.30 Block diagram for fire-alarm system.

Figure 15.31 Block diagram for security alarm system.

Figure 15.32 Open circuit.

N/O contacts RA1 close and N/C contacts RA2 open, the relay RA remaining energized via contacts RA1 when the reset button is released. The alarm sounder mute switch is then closed, and the whole system is now set up.

An interruption of the supply to the relay RA, by operation of any call point, will de-energize the relay, open RA1 and close RA2, thus actuating the alarm sounder. The system can only be cancelled and reset by use of the reset button (Fig. 15.34).

The closed-circuit system is quite popular, as it is self-monitoring in that any malfunction of the relay or break in the call point wiring will cause operation of the system as if a call point had been activated.

Intruder alarm systems tend, in the main, to be based on the closed-circuit type. Figure 15.35 shows the connection diagram for a simple two-zone system with tamper loop and personal attack button.

The tamper loop is simply a continuous conductor wired to a terminal in each detector in the system. It is continuously monitored irrespective of whether the alarm system is switched on or off, and if interrupted will cause immediate operation of the alarm.

The entry/exit circuit is usually confined to the front and/or back doors. The facility exists to alter the time delay between setting the system and exiting, and between entering and switching the system off. This adjustment is made inside the control panel.

Figure 15.33 Fire-alarm system.

Figure 15.34 Closed circuit.

All security and fire-alarm systems should have battery back-up with charging facilities.

Call systems

Once again these fall into different categories, such as telephone systems and page and bleeper systems. However, the nurse-call variety which uses push-buttons and lamp indication is probably the most popular.

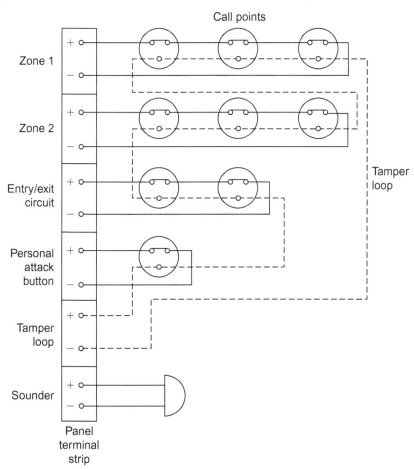

Figure 15.35 Security alarm system.

199

With this type, each room is equipped with a call button of some description, a patient's reassurance light and a cancel button. Outside each room, there is an indicator light, and at strategic points in the building there are zone buzzers. Centrally located is a display panel which incorporates a buzzer and an indication of which room is calling.

Figure 15.36 illustrates, in a simple form, the principle of operation of such a system. This system should by now be quite familiar to the reader; it is simply another variation of the hold-on circuit. Any patient pushing a call button energizes his or her corresponding relay in the main control panel, which is held on by a pair of N/O contacts. At the same time the reassurance, the room and the panel lights 1, 2 and 3 are all illuminated. The zone and panel buzzers are energized via the relay's other pair of N/O contacts.

It is usual to locate the cancel button only in the patient's room, as this ensures that staff visit the patient in question.

Emergency lighting systems

These fall into two categories: maintained and non-maintained. Both of these systems may be utilized by individual units or by a centralized source.

Maintained system

In this system the emergency lighting unit is energized continuously via a step-down transformer, and in the event of a mains failure it remains illuminated via a battery (Fig. 15.37).

Non-maintained system

Here the lighting units remain de-energized until a mains failure occurs, at that time they are illuminated by a battery supply (Fig. 15.38).

It should be noted that modern systems use electronic means to provide the changeover from mains to

Figure 15.37 Maintained system.

Figure 15.36 Nurse-call system.

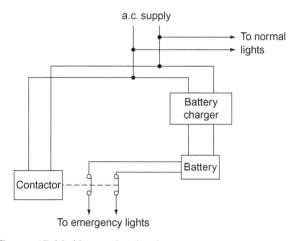

Figure 15.38 Non-maintained system.

battery supply. The contactor method, however, serves to illustrate the principle of operation.

Central heating systems

Let us take a look at the two most basic arrangements: the pumped central heating (CH) and gravity-fed hot-water (HW) system, and the fully pumped system with mid-position valve. It must be remembered that, whatever the system, it is imperative that the wiring installer has knowledge of the function of the system in order to do a competent job.

Pumped CH and gravity HW

This system comprises a boiler with its own thermostat to regulate the water temperature, a pump, a hot-water storage tank, a room thermostat and some form of timed programmer. The water for the HW (i.e. the taps, etc.) is separate from the CH water, but the boiler heats both the systems.

Figure 15.39 shows such a system. From the diagram it might appear that when the requirement for HW is switched off at the programmer, the CH cannot be called for as the boiler has lost its feed. In fact, such programmers have a mechanical linkage between the switches: HW is allowed without CH, but selection of CH automatically selects HW also.

Note the little heating element in the room thermostat; this is known as an accelerator. Its purpose is to increase the sensitivity of the thermostat; manufacturers claim that it increases the accuracy of the unit to within 1°C. The inclusion of an accelerator (if required) does mean an extra conductor for connection to neutral.

Fully pumped system

Two additional items are required for this system: a cylinder thermostat and a mid-position valve. In this system HW and CH can be selected independently. The mid-position valve has three ports: a motor will drive the valve to either HW only, CH only, or HW and CH combined. With this system the boiler and pump always work together. Figure 15.40 illustrates the system, and Fig. 15.41 shows the internal connections of a mid-position valve.

Some difficulties may be experienced in wiring when the component parts of the system are produced by different manufacturers. In this case it is probably best to draw one's own wiring diagram from the various details available.

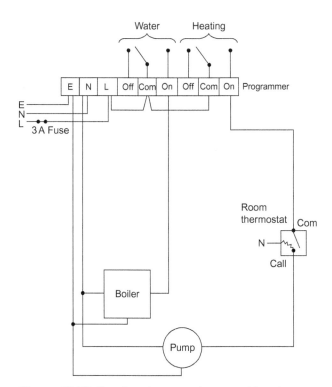

Figure 15.39 Gravity primary and pumped heating.

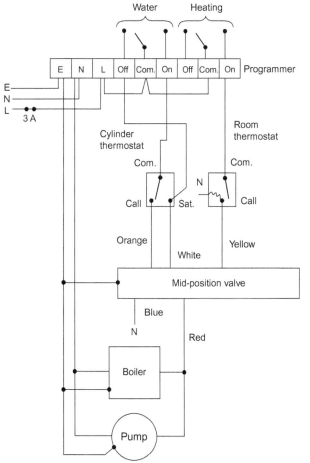

Figure 15.40 Fully pumped system.

Figure 15.41 Internal connection of mid-position valve.

Extra-low-voltage lighting

These systems, incorrectly referred to as low-voltage lighting (low voltage is 50–1000 V a.c.), operate at 12 V a.c. They employ tungsten–halogen dichroic lamps, which have a very high performance in comparison with 230 V halogen lamps. For example, a 50 W dichroic lamp has the same light intensity as a 150 W PAR lamp.

Extra-low-voltage (ELV) lighting is becoming very popular, especially for bathrooms and kitchens. There is very little heat emission, the colour rendering is excellent, and energy consumption is very low.

The 12 V a.c. to supply the lamps is derived from a 230 V/12 V transformer specially designed to cater for the high starting surges, and only these types should be used. The voltage at each lamp is critical: 0.7 V overvoltage can cause premature ageing of the lamp, and 0.7 V undervoltage will reduce the light output by 30 %. Hence variation in voltage must be avoided.

Figure 15.42

To achieve this, leads and cables must be kept as short as possible, and the correct size must be used to avoid excessive voltage drop. When several lamps are to be run from one transformer, it is advisable to use a fused splitter unit rather than to wire them in a parallel chain (Fig. 15.42).

It is important to remember that, for example, a 50 W ELV lamp will draw 4.17 A from the 12 V secondary of the transformer ($I = P/V$). Although a 1.0 mm2 cable will carry the current, the voltage drop for only 3 m of this cable will be 0.55 V.

Choice of system

The choice of any particular wiring system and its accessories will depend on the environment in which it is to be installed. Under normal conditions, typical wiring systems would include:

PVC-insulated, PVC-sheathed	domestic premises; small shops and offices, etc.
PVC conduit or trunking	offices; light industry
Metal conduit; trunking or armoured cable	any situation where there is a serious risk of mechanical damage
m.i.m.s.	fire-alarm systems; boilerhouses; earthed concentric wiring, etc.

There are, however, certain environments which need particular attention, and these are detailed Part 7 of the IET Regulations, special locations.

Special Locations

The bulk of BS 7671 relates to typical, single-and three-phase, installations. There are, however, some special installations or locations that have particular requirements. Such locations may present the user/occupant with an increased risk of death or injuries from electric shock.

BS 7671 categorizes these special locations in Part 7 and they comprise the following:

Section 701	Bathrooms, shower rooms, etc.
Section 702	Swimming pools and other basins
Section 703	Rooms containing sauna heaters
Section 704	Construction and demolition sites
Section 705	Agricultural and horticultural premises
Section 706	Conducting locations with restrictive movement
Section 708	Caravan/camping parks
Section 709	Marinas and similar locations
Section 711	Exhibition shows and stands
Section 712	Solar photovoltaic power supply systems
Section 714	Outdoor lighting installations
Section 715	Extra low voltage lighting installations
Section 717	Mobile or transportable units
Section 721	Caravans and motor caravans
Section 722	Electric vehicle charging
Section 740	Amusement devices, fairgrounds, circuses, etc.
Section 753	Floor and ceiling heating systems

Let us now briefly investigate the main requirements for each of these special locations.

BS 7671 Section 701: Bathrooms, etc.

This section deals with rooms containing bathtubs, shower basins or areas where showers exist but with tiled floors (e.g. leisure/recreational centres, sports complexes, etc.).

Each of these locations is divided into zones 0, 1 and 2 which give an indication of their extent and the equipment/wiring, etc., that can be installed in order to reduce the risk of electric shock.

So! Out with the tape measure, only to find that in a one-bedroom flat, there may be no zone 2. How can you conform to BS 7671?

The stark answer (mine) is that you may not be able to conform exactly. You do the very best you can in each particular circumstance to ensure safety. Let us not forget that the requirements of BS 7671 are based on reasonableness.

Zone 0

This is the interior of the bathtub or shower basin or, in the case of a shower area without a try, it is the space having a depth of 100 mm above the floor out to a radius of 1200 mm from a fixed shower head (Fig. 15.43).

Points to Note

1 Only SELV (12 V) or ripple-free d.c. may be used as a measure against electric shock, the safety source being outside zones 0, 1 and 2.
2 Other than current using equipment specifically designed for use in this zone, **no** switchgear or accessories are permitted.
3 Equipment designed for use in this zone must be to at least IPX7.
4 Only wiring associated with equipment in this zone may be installed.

Zone 1

This extends above zone 0 around the perimeter of the bathtub or shower basin to 2.25 m above the floor level, and includes any space below the bath or basin that is accessible without the use of a key or tool. For showers without basins, zone 1 extends out to a radius of 1200 mm from a fixed shower head (Fig. 15.44).

Points to Note

1 Other than switches and controls of equipment specifically designed for use in this zone, and cord operated switches, only SELV switches are permitted.
2 Provided they are suitable, fixed items of current using equipment such as:
Showers,
Shower pumps,

Figure 15.43 Zone.

Figure 15.44 Zone 1.

> Towel rails,
> Luminaires, etc.
> **3** Equipment designed for use in this zone must be to at least IPX4, or IPX5, where water jets are likely to be used for cleaning purposes.

Zone 2

This extends 600 mm beyond zone 1 and to a height of 2.25 m above the floor level (Fig. 15.45).

Points to Note

Other than switches and controls of equipment specifically designed for use in this zone, and cord operated switches, only SELV switches are permitted.

> **1** Equipment designed for use in this zone must be to at least IPX4, or IPX5 where water jets are likely to be used for cleaning purposes.
> **2** For showers without basins there is no zone 2, just an extended zone 1.
> **3** Socket outlets other than SELV may **not** be installed within 3 m of the boundary of zone 1.

Supplementary equipotential bonding

Supplementary bonding may be established connecting together the cpc's, exposed and extraneous conductive parts within the location.

Such extraneous conductive parts will include:

▶ metallic gas, water, waste and central heating pipes
▶ metallic structural parts that are accessible to touch
▶ metal baths and shower basins.

This bonding may be carried out inside or outside the location preferably close to the entry of the extraneous conductive parts to the location.

However, this bonding may be omitted if the premises has a protective earthing and automatic disconnection system in place, all extraneous conductive parts of the locations are connected to the protective bonding and all circuits are RCD protected (which they have to be anyway!).

Electric floor units may be installed below any zone provided that they are covered with an earthed metal grid or metallic sheath and connected to the protective conductor of the supply circuit.

Figure 15.45 Zone 2 and extended zone 1.

BS 7671 Section 702: Swimming pools

In a similar fashion to bathrooms and shower rooms, etc. swimming pool locations are also divided into zones 0, 1 and 2:

Zone 0 is in the pool/basin or fountain.
Zone 1 extends 2.0 m horizontally from the rim of zone 0 and 2.5 m vertically above it regardless of the pool being above or below ground level. If there are diving boards, shutes or viewing galleries, etc., the height extends to a point of 2.5 m from their top surface and 1.5 m horizontally either side of such shutes, etc.
Zone 2 extends a further 1.5 m horizontally from the edge of zone 1 and 2.5 m above ground level.

Now, what can we install in these zones?

Zones 0 and 1

Protection against shock

Only SELV to be used.

Wiring systems

Only systems supplying equipment in these zones are permitted. Metal cable sheaths or metallic covering of wiring systems shall be connected to the supplementary equipotential bonding. Cables should preferably be enclosed in PVC conduit.

Switchgear, controlgear and socket outlets

None permitted except for locations where there is no zone 2. In this case, a switch or socket outlet with an insulated cap or cover may be installed beyond 1.25 m from the edge of zone 0 at a height of no less than 300 mm. Additionally, the circuits must be protected by:

1. SELV or
2. Automatic disconnection using a 30 mA RCD or
3. Electrical separation.

Equipment

Only that which is designed for these locations.

Other equipment may be used when the pool/basin is not in use (cleaning, maintenance, etc.) provided the circuits are protected by:

1. SELV or
2. Automatic disconnection using a 30 mA RCD or
3. Electrical separation.

Socket outlets and control devices should have a warning notice indicating to the user that they should not be used unless the location is unoccupied by persons.

Zone 2 (there is no zone 2 for fountains)

Switchgear and controlgear

Socket outlets and switches, provided they are protected by:

1. SELV or
2. Automatic disconnection using a 30 mA RCD or
3. Electrical separation.

IP Rating of enclosures

Zone 0 IPX8 (submersion)
Zone 1 IPX4 (splashproof) or IPX5 (where water jets are used for cleaning)
Zone 2 IPX2 (drip proof) indoor pools
    IPX4 (splashproof) outdoor pools
    IPX5 (where water jets are used for cleaning).

Supplementary bonding

All extraneous conductive parts in zones 0, 1 and 2 must be connected by supplementary bonding conductors to the protective conductors of exposed conductive parts in these zones.

BS 7671 Section 703: Hot air saunas

Once again a zonal system, that is, 1, 2 and 3, has been used as per Fig. 15.46. In this case the zones are based on temperature.

Additional protection

All circuits in the location should have additional protection against shock by 30 mA RCDs except sauna heater circuits unless recommended by the manufacturer.

Wiring systems

It is preferred that the wiring systems for the sauna will be installed outside. However, any wiring inside should be heat resistant and any metal sheaths or conduit must be inaccessible in normal use.

Equipment

All should be at least IPX4 and IPX5 if water jets are to be used for cleaning.

Zone 1 only the sauna equipment
Zone 2 no restriction regarding temperature resistance
Zone 3 must be suitable for 125°C and cable sheaths for 175°C

15

Sauna room

Figure 15.46

Switchgear, controlgear and accessories

Only that which is associated with the sauna heater equipment may be installed in zone 2 and in accordance with the manufacturer's instructions. All other should be outside.

BS 7671 Section 704: Construction sites

Not as complicated as one may think. The only areas that require special consideration are where construction work is being carried out, not site huts, etc.

So, let us keep all this as simple as possible. Clearly, construction sites are hazardous areas and in consequence the shock risk is greater.

Protection

For socket outlet circuits, of rating up to and including 32 A and circuits supplying equipment of rating up to and including 32 A, the means of protection shall be:

1. Reduced low voltage (preferred for portable hand tools and hand lamps up to 2 kw).
2. Automatic disconnection of supply with additional protection by 30 mA RCDs.
3. Electrical separation.
4. SELV or PELV (SELV being preferred for portable hand lamps in damp locations).

For socket outlet circuits rated above 32 A, a 500 mA RCD is required.

External influences

These are not addressed in BS 7671, presumably as there are so many different possibilities. So common sense must prevail and equipment must be used with an appropriate degree of protection in accordance with the severity of the influence.

Wiring systems

Apart from some requirements for flexible cables, the only comment relates to ensuring the cables that pass under site roads, etc., are protected against mechanical damage.

Isolation and switching

An assembly for construction sites (ACS) which is basically the main intake supply board, should comprise a lockable isolator and, for current using equipment:

1. Overcurrent devices.
2. Fault protective devices.
3. Socket outlets if required.

Plugs and sockets/cable couplers

All should be to BS EN 60309-2.

BS 7671 Section 705: Agricultural and horticultural locations

The requirements apply only to locations that do not include the main farmhouse outside of which the environment is hazardous and where, of course, livestock is present (animals are susceptible to lethal shock levels at 25 V a.c.).

Protection

Protection against shock may be provided by:

1. Automatic disconnection of supply with additional RCD protection for
   (a) final circuits supplying socket outlets rated at 32 A or less (30 mA),
   (b) final circuits supplying socket outlets more than 32 A (100 mA),
   (c) all other circuits (300 mA).
2. SELV or PELV.

Protection against thermal effects:

1. Heating appliances should be mounted at appropriate distances from combustible materials and livestock, with radiant heaters at a minimum distance of 0.5 m.
2. For fire protection an RCD rated at 300 mA or less should be used.

Supplementary bonding

Wherever livestock is housed, supplementary bonding must be carried out connecting all exposed and extraneous conductive parts that can be touched by livestock. All metal grids in floors must be connected to the supplementary equipotential bonding.

External influences

1. All equipment must be to at least IP44, and luminaires exposed to dust and moisture ingress, IP54.
2. Appropriate protection for socket outlets where influences are greater than AD4 AE3 and/or AG1.
3. Appropriate protection where corrosive substances are present.

Diagrams

The user of the installation should be provided with plans and diagrams showing the location of all equipment, concealed cable routes, distribution and the equi-potential bonding system.

Wiring systems

Any! Just as long as it is suitable for the environment and fulfils the required minimum degrees of protection.

A high impact PVC conduit/trunking system would be appropriate in many cases as it is not affected by corrosion, is rodent proof and has no exposed conductive parts. However, the system would

be designed to suit the particular environmental conditions.

Wiring systems should be erected so as to be, where possible, inaccessible to livestock. Overhead lines should be insulated and where vehicles/mobile equipment are used, underground cables should be at least 0.6 m deep and mechanically protected and 1.0 m deep in arable land. Self-supporting suspended cables should be at a height of at least 6 m.

Switchgear and controlgear

Whatever! As long as it is suitable for the conditions and that emergency switching is placed in a position inaccessible to livestock and can be accessed in the event of livestock panic (stampede!).

BS 7671 Section 706: Restrictive conductive locations

These are very rare locations which could include metal tanks, boilers, ventilation ducts, etc., where access is required for maintenance, repair or inspection. Body movement will be severely restricted and in consequence such areas are extremely dangerous.

This section deals with the installation inside the location and the requirements for bringing in accessories/equipment from outside.

For fixed equipment in the location, one of the following methods of protection shall be used:

1. Automatic disconnection of supply but with additional supplementary bonding.
2. The use of Class II equipment backed up by a 30 mA RCD.
3. Electrical separation.
4. SELV.

For hand-held lamps and tools and mobile equipment, SELV or electrical separation should be used.

BS 7671 Section 708: Caravan and camping parks

We drive into a caravan/camping park for our holiday and need to connect to a supply of electricity for all our usual needs. This is accommodated by the provision of suitably placed socket outlets supplied via distribution circuits.

External influences

Equipment should have at least the following protection codes:

1. IPX4 for the presence of splashes (AD4).
2. IP3X for presence of small objects (AE2).

15

3. IK08 for presence of high severity mechanical stress (AG3).
(The IK codes are for impact and 08 is an impact of 5 J.)

Wiring systems

The distribution circuits are erected either underground or overhead:

1. Underground cable (preferred) should be suitably protected against mechanical damage, tent pegs, steel spikes, etc., and at a depth of not less than 0.6 m.
2. If overhead, then 6 m above ground where there is vehicle movement and 3.5 m elsewhere.

Switchgear and socket outlets

1. Supply equipment should be adjacent to or within 20 m of the pitch.
2. Socket outlets should be: to BS EN 60309-2; IP44, at between 0.5 and 1.5 m above ground, rated not less than 16 A and have individual overcurrent and 30 mA RCD protection.
3. If the supply is TNC-S (PME) the protective conductor of each socket needs to be connected to an earth rod.

BS 7671 Section 709: Marinas

This location is basically a camping park for boats and has similar requirements to those of caravans and camping parks.

It is where you arrive in your 40 ft 8 berth cruiser (some hope) looking for a place to park!

However, the environment is a little more harsh than the caravan park due to the possibilities of corrosion, mechanical damage, structural movement and flammable fuels together with the increased risk of electric shock.

External influences

Due to the harsh conditions mentioned, the classification of influences would include:

▶ AD water
▶ AE solid foreign bodies
▶ AF corrosion and
▶ AG impact.

Wiring systems

Distribution circuits, like those in caravan parks, can be either underground or overhead as well as PVC covered, mineral insulated, cables in cable management systems, etc.

However, overhead cables on or incorporating a support wire, cables with aluminium conductors or mineral insulated cables shall not be installed above a jetty or pontoon, etc.

Underground cables should have additional mechanical protection and be installed 0.5 m deep. Overhead cables should be at the same heights as in caravan parks.

Isolation, switching and socket outlets

Generally the same as caravan parks.

Socket outlets should be installed not less than 1 m above the highest water level except that for floating pontoons, walkways, etc., this height may be reduced to 300 mm.

BS 7671 Section 711: Exhibitions, shows and stands

This section deals with the protection of the users of temporary structures erected in or out of doors and is typical of antique fairs, motorbike shows, arts and craft exhibitions, etc.

It does not cover public or private events that form part of entertainment activities which are the subject of BS 7909.

External influences

None particularly specified. Clearly they must be considered and addressed accordingly.

Wiring

Armoured or mechanically protected cables where there is a risk of mechanical damage. Cables shall have a minimum conductor size of 1.5 mm^2.

Protection

Against shock:

▶ Supply cables to a stand or unit, etc., must be protected at the cable origin by a time delayed RCD of residual current rating not exceeding 300 mA.
▶ All socket outlet circuits not exceeding 32 A and all other final circuits, excepting emergency lighting, shall have additional protection by 30 mA RCDs.
▶ Any metallic structural parts accessible from within the unit stand, etc., shall be connected by a main protective bonding conductor to the main earthing terminal of the unit.

Against thermal effects:

▶ Clearly in this case all luminaires, spot lights, etc., should be placed in such positions as not to cause a build-up of excessive heat that could result in fire or burns.

Isolation

Every unit, etc., should have a readily accessible and identifiable means of isolation.

Inspection and testing

Tongue in cheek here! Every installation **should** be inspected and tested on site in accordance with Part 6 of BS 7671.

BS 7671 Section 712: Solar photovoltaic (PV) supply systems

These are basically solar panels generating d.c. which is then converted to a.c. via an inverter. Those dealt within BS 7671 relate to those systems that are used to 'top up' the normal supply. There is a need for consideration of the external influences that may affect cabling from the solar units outside to control gear inside.

There must be protection against overcurrent and a provision made for isolation on both the d.c. and a.c. side of the inverter.

As the systems can be used in parallel with or as a switched alternative to the public supply, reference should be made to Chapter 55 of BS 7671.

BS 7671 Section 714: Outdoor lighting installations.

This section deals with outdoor lighting such as found on highways, sports stadiums, stately homes, telephone kiosks and town plans. It does not apply to lighting fixed to the outside wall of a building and which is supplied from inside the premises.

The disconnection time for all circuits for highway power supplies is 5 seconds and certain installations in places such as telephone kiosks and town plans should be protected by an RCD not exceeding 30mA.

BS 7671 Section 715: Extra low voltage lighting installations.

This is typical of the ELV downlighters for bathrooms. The system is usually SELV and derived from an isolating transformer.

BS 7671 Section 717: Mobile or transportable units

Medical facility units, mobile workshops, canteens, etc., are the subject of this section. They are self-contained with their own installation and designed to be connected to a supply by, for instance, a plug and socket.

The standard installation protective measures against shock are required with the added requirement that the automatic disconnection of the supply should be by means of an RCD. Also all socket outlets for the use of equipment outside the unit should have additional protection by 30 mA RCDs.

The supply cable should be HO7RN-F, oil- and flame-resistant heavy duty rubber with a minimum copper conductor size of 2.5mm^2.

Socket outlets outside should be to a minimum of IP44.

BS 7671 Section 721: Caravans and motor caravans

These are the little homes that people tow behind their cars or that are motorized, not those that tend to be located on a fixed site. It would be unusual for the general electrical contractor to wire new, or even rewire old, units. How many of us ever rewire our cars? Consequently, only the very basic requirements are considered here.

Protection

These units are small houses on wheels and subject to the basic requirements of protection against shock and overcurrent. Where automatic disconnection of supply is used, this must be provided by a 30 mA RCD.

Wiring systems

The wiring systems should take into account the fact that the structure of the unit is subject to flexible/mechanical stresses and, therefore, our common flat twin and three core cables should not be used.

Inlets

Unless the caravan demand exceeds 16 A, the inlet should conform to the following:

(a) BS EN 60309-1 or 2 if interchangeability is required
(b) no more than 1.8 m above ground level
(c) readily accessible and in a suitable enclosure outside the caravan
(d) identified by a notice that details the nominal voltage, frequency and rated current of the unit.

Also, inside the caravan, there should be an isolating

15

switch and a notice detailing the instructions for the connection and the disconnection of the electricity supply and the period of time between inspection and testing (3 years).

General

Accessories and luminaires should be arranged such that no damage can occur due to movement, etc.

There should be no compatibility between sockets of low and extra-low voltage. Any accessory exposed to moisture should be IP55 rated (jet proof and dust proof).

BS 7671 Section 722: Electric vehicle (EV) charging.

This covers the requirements for socket outlets used to charge electric vehicles and relates to various modes of charging.

> *Mode 1 charging.* This is the connection of the EV to a standard socket outlet, single or three phase, not exceeding 16A.
> *Mode 2 charging.* This is the connection of the EV to a standard socket outlet, single or three phase, not exceeding 32A but utilizing a special cable that ensures that current only flows if the cpc is connected.
> *Mode 3 charging.* This is similar to Mode 2 but the supply is from dedicated supply equipment and feeds control equipment in the EV.
> *Mode 4 Charging.* Like Mode 3 but with an 'off board' charger.

BS 7671 Section 740: Amusement devices, fairgrounds, circuses, etc.

This is not an area that is familiar to most installation electricians and hence will only be dealt with very briefly.

The requirements of this section are very similar to those of Section 711 exhibitions, shows, etc., and parts of Section 706 agricultural locations (because of animals) regarding supplementary bonding.

For example, additional protection by 30 mA is required for:

1. Lighting circuits.
2. All socket outlet circuits rated up to 32 A.
3. Mobile equipment supplied by a flexible cable rated up to 32 A.

Automatic disconnection of supply must be by an RCD.

Equipment should be rated to at least IP44.

The installation between the origin and any equipment should be inspected and tested after each assembly on site.

BS 7671 Section 753: Floor and ceiling heating systems

Systems referred to in this section are those used for thermal storage heating or direct heating.

Protection

Against shock:

▶ Automatic disconnection of supply with disconnection achieved by 30 mA RCD.
▶ Additional protection for Class II equipment by 30 mA RCDs.
▶ Heating systems provided without exposed conductive parts shall have a metallic grid of spacing not more than 300 mm installed on site above a floor system or below a ceiling system and connected to the protective conductor of the system.

Against thermal effects:

▶ Where skin or footware may come into contact with floors the temperature shall be limited, for example to 30°C.
▶ To protect against overheating of these systems the temperature of any zone should be limited to a maximum of 80°C.

External influences

Minimum of IPX1 for ceilings and IPX7 for floors.

The designer must provide a comprehensive and detailed plan of the installation which should be fixed on or adjacent to the system distribution board.

Installations in flammable and/or explosive situations

In premises such as petrol stations, gas works, flour mills, etc., using electricity involves an obvious risk. Serious arcing at contacts or in faulty conductors or equipment could cause an explosion and/or fire. It is therefore important to ensure that such a dangerous situation does not arise, by installing suitable cable and fittings. An outline of the hazards and suitable wiring systems is given in CP 1003. The following paragraphs are extracts from that Code of Practice, reproduced by permission of the BSI, 389 Chiswick High Rd, London, W4 4AL, from whom complete copies of the Code of Practice may be obtained.

Extracts from BS Code of Practice 1003 Part 1, 1964

'Electrical apparatus and associated equipment for use in explosive atmospheres of gas or vapour (other than mining)' Part 1: 'Choice, installation and maintenance of flameproof and intrinsically safe equipment'.

Foreword

In dealing with the risk of fire or explosion from the presence of flammable liquids, gases or vapours, three sets of conditions are recognized.

Division 0

An area or enclosed space within which any flammable or explosive substance, whether gas, vapour or volatile liquid, is continuously present in concentrations within the lower and upper limits of flammability.

Division 1

An area within which any flammable or explosive substance, whether gas, vapour or volatile liquid, is processed, handled or stored, and where during normal operations an explosive or ignitable concentration is likely to occur in sufficient quantity to produce a hazard.

Division 2

An area within which any flammable or explosive substance, whether gas, vapour or volatile liquid, although processed or stored, is so well under conditions of control that the production (or release) of an explosive or ignitable concentration in sufficient quantity to constitute a hazard is only likely under abnormal conditions.

The conditions described as appertaining to Division 0 are such as normally to require the total exclusion of any electrical equipment, except in such special circumstances as to render this impracticable, in which case recourse may be possible to special measures such as pressurization or the use of intrinsically safe equipment.

A risk of the nature described under Division 1 can be met by the use of flameproof or intrinsically safe equipment, with which Part 1 is concerned, or by the use of the means described in Part 2 of the Code which is concerned with methods, other than the use of flameproof or intrinsically safe equipment, of securing safety in flammable and explosive atmospheres.

The certifying authority for flameproof apparatus is the Ministry of Power. The certifying authority for intrinsically safe electrical apparatus for use in factories coming within the scope of the Factories Act is the Ministry of Labour. The recognized testing authority for flameproof enclosures and intrinsically safe circuits and apparatus is the Ministry of Power[1].

Types of hazard

Two main types to be considered: gases and vapours or flammable liquids.

1. *Explosive gases* or vapours are grouped according to the grade of risk and four groups are recognized.

Group I	Gas encountered in coal mining.
Group II	Various gases commonly met within industry.
Group III	Ethylene, diethyl ether, ethylene oxide, town gas and coke oven gas.
Group IV	Acetylene, carbon disulphide, ethyl nitrate, hydrogen and water-gas.

(i) Flameproof apparatus. Apparatus with flameproof enclosures, certified appropriately for the gas group which constitutes the risk, should be used. It should be noted that no apparatus is certified for Group IV; other techniques (e.g. pressurization) must therefore be applied if electrical apparatus has to be installed where gases in this group may be present in dangerous concentrations.

In general, apparatus certified for the higher groups will cover situations where gases from the lower groups are present.

(ii) Selection of apparatus for diverse risks. If, in an installation capable of subdivision, Group III apparatus is required for some parts while Group II would suffice for other parts, it is recommended that the former, which could cover the risks in a lower group, should be used throughout test apparatus of the latter group should inadvertently be transferred to a place where Group III is required.

2. *Flammable liquids*. Flammable liquids give rise, in a greater or lesser degree according to their flashpoints and the temperature to which they are subjected, to explosive vapours which should be treated as under (1) above.

The liquid, however, constitutes a further

---

[1] The Ministry of Power is now part of the Department of Trade and Industry, and the Factories Act is now overseen by the Health and Safety Executive.

15

risk in which fires may occur as a result of unvaporized liquid in the form of spray or otherwise, coming into contact with electrical equipment and then being ignited by a spark or other agency.

Types of wiring

Danger areas:

(a) Cables drawn into screwed solid-drawn steel conduit.

(b) Lead-sheathed, steel-armoured cable.

(c) Mineral-insulated, metal-sheathed cable.

(d) PVC insulated and armoured cable with an outer sheath of PVC.

(e) Polyethylene-insulated, PVC-covered overall and armoured.

(f) Cables enclosed in a seamless aluminium sheath with or without armour.

Automatic electrical protection

(a) All circuits and apparatus within a danger area should be adequately protected against overcurrent, short circuit and earth-leakage current.

(b) Circuit breakers should be of the free-handle trip-free type to preclude misuse, such as tying-in or holding-in under fault conditions against the persistence of which they are designed to afford protection, and an indicator should be provided in all cases to show clearly whether the circuit breaker is open or closed.

Portable and transportable apparatus and its connections

Portable electrical apparatus should only be permitted in any hazardous area in the most exceptional circumstances which make any other alternative extremely impracticable, and only then if it is of a certified type.

CHAPTER 16

Earthing and Bonding

Earth: What it is, and why and how we connect to it

The thin layer of material which covers our planet, be it rock, clay, chalk or whatever, is what we in the world of electricity refer to as earth. So, why do we need to connect anything to it? After all, it is not as if earth is a good conductor.

Perhaps it would be wise at this stage to revise potential difference (p.d.). A p.d. is exactly what it says it is: a difference in potential (volts). Hence two conductors having p.d.'s of, say, 20 and 26 V have a p.d. between them of 26 − 20 = 6 V. The original p.d.'s, that is, 20 and 26 V, are the p.d.'s between 20 and 0 V and 26 and 0 V.

So where does this 0 V or zero potential come from? The simple answer is, in our case, the earth. The definition of earth is therefore the conductive mass of earth, whose electric potential at any point is conventionally taken as zero.

Hence if we connect a voltmeter between a live part (e.g. the line conductor of, say, a socket outlet) and earth, we would probably read 230 V; the conductor is at 230 V, the earth at zero. Of course it must be remembered that we are discussing the supply industry in the United Kingdom, where earth potential is very important. We would measure nothing at all if we connected our voltmeter between, say, the positive 12 V terminal of a car battery and earth, as in this case the earth plays no part in any circuit. Figure 16.1 illustrates this difference.

Note the connection of the supply neutral in Fig. 16.1a to earth, which makes it possible to have a complete circuit via the earth. Supply authority neutrals should be at around 0 V, and in order to maintain this condition they are connected to the zero potential of earth.

This also means that a person in an installation touching a live part whilst standing on the earth would take the place of the voltmeter in Fig. 16.1a, and could suffer a severe electric shock. Remember that the accepted **lethal** level of shock current passing through a person is only 50 mA or 1/20 A. The same situation would arise if the person was touching, say, a faulty appliance and a gas or water pipe (Fig. 16.2).

One method of providing some measure of protection against these effects is to join together (bond) all metallic parts and connect them to earth. This ensures that all metalwork in a healthy situation is at or near 0 V, and under fault conditions all metalwork will rise to the same potential. So, simultaneous contact with two such metal parts would not result in a shock, as there will be no p.d. between them. This method is known as earthed equipotential bonding.

Unfortunately, as previously mentioned, earth itself is not a good conductor unless it is very wet, and therefore it presents a high resistance to the flow of fault current. This resistance is usually enough to restrict fault current to a level well below that of the rating of the protective device, leaving a faulty circuit uninterrupted. Clearly this is an unhealthy situation. The methods of overcoming this problem will be dealt with later.

In all but the most rural areas, consumers can be connected to a metallic earth return conductor which is ultimately connected to the earthed neutral of the

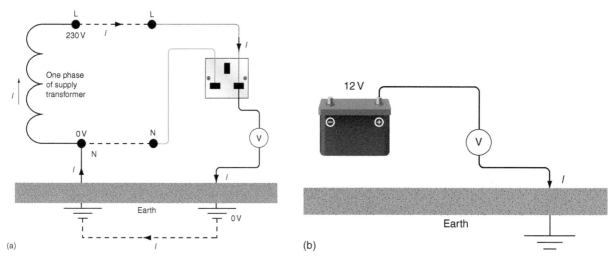

Figure 16.1 (a) Earth path, (b) no earth path.

supply. This, of course, presents a low-resistance path for fault currents to operate the protection.

Summarizing, then, connecting metalwork to earth places that metal at or near zero potential, and bonding between metallic parts puts such parts at the same potential even under fault conditions.

The Faraday cage

In one of his many experiments, Michael Faraday (1791–1867) placed himself in an open-sided cube which was then covered in a conducting material and insulated from the floor. When this cage arrangement was charged to a high voltage, he found that he could move freely within it touching any of the sides, with no adverse effects. He had in fact created an

equipotential zone, and of course in a correctly bonded installation we live and/or work in Faraday cages!

Connecting to earth

In the light of previous comments, it is obviously necessary to have as low an earth path resistance as possible, and the point of connection to earth is one place where such resistance may be reduced. When two conducting surfaces are placed in contact with each other, there will be a resistance to the flow of current depending on the surface areas in contact. It is clear, then, that the greater surface contact area with earth that can be achieved, the better.

There are several methods of making a connection to earth, including the use of rods, plates and tapes. By far the most popular method in everyday use is the rod

Figure 16.2 Shock path.

earth electrode. The plate type needs to be buried at a sufficient depth to be effective and, as such plates may be 1 or 2 m², considerable excavation may be necessary. The tape type is predominantly used in the earthing of large electricity substations, where the tape is laid in trenches in a mesh formation over the whole site. Items of plant are then earthed to this mesh.

Rod electrodes

These are usually of solid copper or copper-clad carbon steel, the latter being used for the larger-diameter rods with extension facilities. These facilities comprise a thread at each end of the rod to enable a coupler to be used for connecting the next rod; a steel cap to protect the thread from damage when the rod is being driven in; a steel driving tip; and a clamp for the connection of an earth tape or conductor (Fig. 16.3).

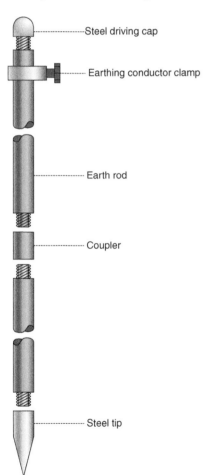

Figure 16.3 Earth rod.

The choice of length and diameter of such a rod will, as previously mentioned, depend on the soil conditions. For example, a long thick electrode is used for earth with little moisture retention. Generally, a 1–2 m rod, 16 mm in diameter, will give a relatively low resistance.

Earth electrode resistance

If we were to place an electrode in the earth and then measure the resistance between the electrode and points at increasingly larger distances from it, we would notice that the resistance increased with distance until a point was reached (usually around 2.5 m) beyond which no increase in resistance was seen (Fig. 16.4).

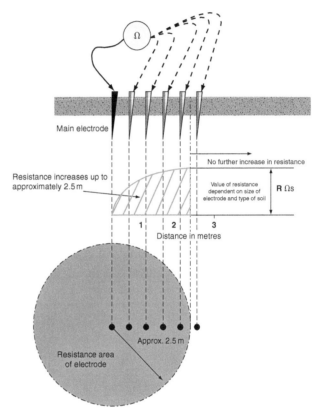

Figure 16.4 Resistance area of electrode.

The value of this **electrode resistance** will depend on the length and cross-sectional area (c.s.a.) of the electrode and the type of soil.

This resistance area is particularly important with regard to voltage at the surface of the ground (Fig. 16.5).

Figure 16.5 Electrode resistance.

Figure 16.6 Earth electrode installation.

For a 2 m earth rod, with its top at ground level, 80% to 90% of the voltage appearing at the electrode under fault conditions is dropped across the earth in the first 2.5 to 3 m. This is particularly dangerous where livestock are present as the hind and fore legs of an animal can be, respectively, inside and outside the resistance area: 25 V can be lethal. This problem can be overcome by ensuring that the whole of the electrode is well below ground level and by providing protection that will operate in a fraction of a second (earth leakage circuit breaker) (Fig. 16.6).

Of course, values of resistance will depend on the type of soil and its moisture content. Any soil that holds moisture such as clay or marshy ground has a relatively low resistivity, whereas gravel or rock has a high resistivity. Typical values for soil resistivity in ohms are as follows:

Garden soil	5–50
Clay	10–100
Sand	250–500
Rock	1000–10 000

Earthing systems

Contact with metalwork made live by a fault is clearly undesirable. One popular method of providing some measure of protection against such contact is by protective earthing, protective equipotential bonding and automatic disconnection of supply. This entails the bonding together and connection to earth of the following:

1. All metalwork associated with electrical apparatus and systems, termed exposed conductive parts. Examples include conduit, trunking and the metal cases of apparatus.
2. All metalwork liable to introduce a potential including earth potential, termed extraneous conductive parts. Examples are gas, oil and water pipes, structural steelwork, radiators, sinks and baths.

The conductors used in such connections are called **protective conductors**, and they can be further subdivided into the following:

1. Circuit protective conductors (cpc), for connecting exposed conductive parts to the main earthing terminal.
2. Main protective bonding conductors, for bonding together main incoming services, structural steelwork, etc.
3. Supplementary protective bonding conductors, for bonding together sinks, baths, taps, radiators, etc., and exposed conductive parts in bathrooms and swimming pools.

The effect of all this bonding is to create a zone in which all metalwork of different services and systems will, even under fault conditions, be at a substantially equal potential. If, added to this, there is a low-resistance earth return path, the protection should operate fast enough to prevent danger.

The resistance of such an earth return path will depend upon the system. These systems have been designated in the IET Regulations using the letters T, N, C and S. These letters stand for:

T   Terre (French for earth) and meaning a direct connection to earth

N   Neutral

C   Combined

S   Separate

When these letters are grouped they form the classification of a type of system. The first letter in such a classification denotes how the supply source is earthed. The second denotes how the metalwork of an installation is earthed. The third and fourth indicate the functions of neutral and protective conductors. Hence:

1. A TT system has a direct connection of the supply source to earth and a direct connection of the installation metalwork to earth. An example is an overhead line supply with earth electrodes, and the mass of earth as a return path (Fig. 16.7).
2. A TN–S system has the supply source directly connected to earth, the installation metalwork connected to the earthed neutral of the supply source via the metal sheath of the supply cable, and the neutral and protective conductors

throughout the whole system performing separate functions (Fig. 16.8).

3. A TN–C–S system is as the TN–S system but the supply cable sheath is also the neutral, that is, it forms a combined earth/neutral conductor known as a PEN (Protective Earthed Neutral) conductor (Fig. 16.9). The installation earth and neutral are separate conductors. This system is also known as PME (Protective Multiple Earthing).

Note that only single-phase systems have been shown, for simplicity.

With this system (PME system), it is important to ensure that the neutral is kept at earth potential by earthing it at many points along its length (hence 'multiple' earthing). If this is not done, a fault to neutral in one installation could cause a shock risk in all the other installations connected to that system (Fig. 16.10).

The cable usually used for such a system is concentric cable which consists of a single-core cable (for single phase) surrounded by armouring which is the earth and neutral conductor. Three core with concentric

Figure 16.7 TT system.

Figure 16.8 TN–S system.

16

Figure 16.9 TN–C–S system.

Figure 16.10 PME or TN–C–S system.

neutral would be used for three-phase, four-wire cable (Fig. 16.11).

There are, however, several hazards associated with the use of a PME system. These include:

1. Shock risk if neutral is broken. In this case, a fault on a PME system with a broken neutral would result in the neutral becoming live to earth either side of the break. This situation is more likely to occur with overhead supplies.

2. Fire risk. As heavy currents are encouraged to flow, there is a risk of fire starting during the time it takes for the protective devices to operate.

The chance of a broken neutral is lessened in underground cable to some extent by the use of concentric cable, as it is unlikely that the neutral conductor in such a cable could be broken without breaking the live conductor.

In view of the hazards of such a system there are strict regulations for its use, and approval from the Department of Trade and Industry must be obtained before it can be installed.

Summary

In order to avoid the risk of serious electric shock, it is important to provide a path for earth leakage currents to operate the circuit protection, and to endeavour to maintain all metalwork at the same potential. This is achieved by bonding together metalwork of electrical and non-electrical systems to earth. The path for leakage currents would then be via the earth itself in TT systems or by a metallic return path in TN–S or TN–C–S systems.

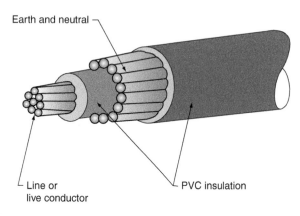

Figure 16.11 Single-core concentric cable.

Earth fault loop impedance

As we have seen, circuit protection should operate in the event of a direct fault from line to earth. The speed of operation of the protection is of extreme importance and will depend on the magnitude of the fault current, which in turn will depend on the impedance of the earth fault loop path.

Figure 16.12 shows this path. Starting at the fault, the path comprises:

1. The circuit protective conductor (cpc).
2. The consumer's earthing terminal and earth conductor.
3. The return path, either metallic or earth.
4. The earthed neutral of the supply transformer.

5. The transformer winding.
6. The line conductor from the transformer to the fault.

Figure 16.13 is a simplified version of this path. We have

$$Z_s = Z_e + R_1 + R_2$$

where Z_s is the actual total loop impedance, Z_e is the impedance external to the installation, R_1 is the resistance of the line conductor and R_2 is the resistance of the cpc. We also have

$$I = U_0 / Z_s$$

where I is the fault current and U_0 is the nominal voltage to earth (usually 230 V).

Figure 16.12 Earth fault loop path.

Figure 16.13 Simplified earth fault loop path.

Determining the value of total loop impedance Z_s

The IET Regulations require that when the general characteristics of an installation are assessed, the loop impedance Z_e external to the installation shall be ascertained. This may be measured in existing installations using a line-to-earth loop impedance tester. However, when a building is only at the drawing board stage it is clearly impossible to make such a measurement. In this case, we have three methods available to assess the value of Z_e:

1. Determine it from details (if available) of the supply transformer, the main distribution cable and the proposed service cable.
2. Measure it from the supply intake position of an adjacent building which has service cable of similar size and length to that proposed.
3. Use maximum likely values issued by the supply authority as follows:

   (a) TT system: $21\,\Omega$ maximum
   (b) TN–S system: $0.8\,\Omega$ maximum
   (c) TN–C–S system: $0.35\,\Omega$ maximum.

Method 1 will be difficult for anyone except engineers. Method 3 can, in some cases, result in pessimistically large cable sizes. Method 2, if it is possible to be used, will give a closer and more realistic estimation of Z_e. However, if in any doubt, use method 3.

Having established a value for Z_e, it is now necessary to determine the impedance of that part of the loop path internal to the installation. This is, as we have seen, the resistance of the line conductor plus the resistance of the cpc, that is, $R_1 + R_2$. Resistances of copper conductors may be found from manufacturers' information, which gives values of resistance/metre for copper and aluminium conductors at 20°C in $m\Omega/m$. Table 16.1 gives resistance values for copper conductors up to $35\,mm^2$.

Table 16.1 Resistance of copper conductors in $m\Omega/m$ at 20°C

Conductor (mm²)	Resistance (Ω)
1.0	18.1
1.5	12.1
2.5	7.41
4.0	4.61
6.0	3.08
10.0	1.83
16.0	1.15
25.0	0.727
35.0	0.524

So, a $25.0\,mm^2$ line conductor with a $4.0\,mm^2$ cpc has $R_1 = 0.727\,m\Omega$ and $R_2 = 4.61\,\Omega$, giving $R_1 + R_2 = 0.727 + 4.61 = 5.337\,m\Omega/m$. So having established a value for $R_1 + R_2$, we must now multiply it by the length of the run and divide by 1000 (the values given are in milliohms per metre). However, this final value is based on a temperature of 20°C, but when the conductor is fully loaded, its temperature will increase. In order to determine the value of resistance at conductor-operating temperature, a multiplier is used. This multiplier, applied to the 20°C value of resistance, is 1.2 for PVC cables.

Hence, for a 20 m length of 70°C PVC-insulated $16.0\,mm^2$ line conductor with a $4\,mm^2$ cpc, the value of $R_1 + R_2$ would be

$$R_1 + R_2 = (1.15 + 4.61) \times 20 \times 1.2/1000 = 0.138\,\Omega$$

We are now in a position to determine the total earth fault loop impedance Z_s from

$$Z_s = Z_e + R_1 + R_2$$

As mentioned, this value of Z_s should be as low as possible to allow enough fault current to flow to operate the protection as quickly as possible. The IET Regulations give maximum values of loop impedance for different sizes and types of protection for final circuits not exceeding 32 A and distribution circuits. Provided that the actual calculated values do not exceed those tabulated, final circuits will disconnect under earth fault conditions in 0.4 s or less, and distribution circuits in 5 s or less. The reasoning behind these different times is based on the time that a faulty circuit can reasonably be left uninterrupted and are based on the probable chances of someone being in contact with exposed or extraneous conductive parts at the precise moment that a fault develops.

Example

Let's have a look at a typical example of a shower circuit run in an 18 m length of $6.0\,mm^2$ (6242 Y) twin cable with cpc, and protected by a 30 A BS 3036 semi-enclosed rewirable fuse. A $6.0\,mm^2$ twin cable has a $2.5\,mm^2$ cpc. We will also assume that the external loop impedance Z_e is measured as $0.27\,\Omega$. Will there be a shock risk if a line-to-earth fault occurs?

The total loop impedance $Z_s = Z_e + R_1 + R_2$ and we are given $Z_e = 0.27\,\Omega$. For a $6.0\,mm^2$ line conductor with a $2.5\,mm^2$ cpc, $R_1 + R_2$ is $10.49\,m\Omega/m$. Hence, with a multiplier of 1.2 for 70°C PVC, the total $R_1 + R_2 = 18 \times 10.49 \times 1.2/1000 = 0.23\,\Omega$. Therefore, $Z_s = 0.27 + 0.23 = 0.53\,\Omega$. This is less than the $0.95\,\Omega$ maximum given in the IET Regulations for a 30 A BS 3036 fuse. Consequently, the protection will disconnect the circuit in less than 0.4 s.

Residual current devices

We have seen how very important the total earth loop impedance Z_s is in the reduction of shock risk. However, in TT systems where the mass of earth is part of the fault path, the maximum values of Z_s given in the IET Regulations may be hard to satisfy. Added to this, climatic conditions will alter the resistance of the earth in such a way that Z_e may be satisfactory in wet weather but not in very dry.

The regulations recommend therefore that the preferred method of earth fault protection for installations in a TT systems be achieved by a residual current device (RCD), such that the product of its residual operating current and the loop impedance will not exceed a figure of 50 V. Residual current breakers (RCBs), residual current circuit breakers (RCCBs) and RCDs are one and the same thing. Modern developments in circuit breaker (cb), RCD and consumer unit design now make it easy to protect any individual circuit with a combined cb/RCD (RCBO), making the use of split-load boards unnecessary.

In domestic premises the use of 30 mA RCDs is required for the protection of all socket outlets rated at not more than 20 A, for all circuits in a bath or shower room and for cables embedded in walls and partitions at a depth less than 50 mm. Socket outlets not intended for general use, for example a those provided for non-portable equipment such as freezers, etc., are exempt from this requirement, provided they are suitably labelled or identified.

Principle of operation of an RCD

Figure 16.14 illustrates the construction of an RCD. In a healthy circuit the same current passes through the line coil, the load and back through the neutral coil. Hence the magnetic effects of line and neutral currents cancel out.

In a faulty circuit, either line to earth or neutral to earth, these currents are no longer equal; therefore the out-of-balance current produces some residual magnetism in the core. As this magnetism is alternating, it links with the turns of the search coil, inducing an electromotive force (e.m.f.) in it. This e.m.f. in turn drives a current through the trip coil, causing operation of the tripping mechanism.

Figure 16.14

It should be noted that a line-to-neutral fault will appear as a load, and hence the RCD will not operate for this fault.

The test switch creates an out-of-balance condition which tips the breaker. Its only purpose is to indicate that the breaker is in working order. **It does not check the condition of any part of the earth system.**

Out-of-balance currents as low as 5 to 30 mA will be detected, and therefore a person touching unearthed live metalwork would cause the breaker to operate before the lower lethal limit of 50 mA was reached (Fig. 16.15). It is still necessary, however, to ensure that the earth system of an installation is connected to a suitable earth electrode.

Nuisance tripping

Certain appliances such as cookers, water heaters and freezers tend to have, by the nature of their construction and use, some leakage currents to earth. These are quite normal, but could cause the operation of an RCD protecting an entire installation. This can be overcome by using split-load consumer units, where socket outlet circuits are protected by a 30 mA RCD, leaving all other circuits controlled by a normal mains

Figure 16.15 RCD operation via a human body.

switch. Better still, especially in TT systems, is the use of a 100 mA RCD for protecting circuits other than socket outlets.

Requirements for RCD protection

30 mA

▶ All socket outlets rated at not more than 20 A.
▶ Mobile **equipment** rated at not more than 32 A for use outdoors.
▶ All circuits in a bath/shower room.
▶ Preferred for all circuits in a TT system.
▶ All cables installed less than 50 mm from the surface of a wall or partition unless they are protected by other means and also at any depth if the construction of the wall or partition includes metallic parts.
▶ In zones 0, 1 and 2 of swimming pool locations.
▶ All circuits in a location containing saunas, etc.
▶ Socket outlet final circuits not exceeding 32 A in agricultural locations.
▶ Circuits supplying Class II equipment in restrictive conductive locations.
▶ Each socket outlet in caravan parks and marinas and final circuit for houseboats.
▶ All socket outlet circuits rated not more than 32 A for show stands, etc.
▶ All socket outlet circuits rated not more than 32 A for construction sites (where reduced low voltage, etc. is not used).
▶ All socket outlets supplying equipment outside mobile or transportable units.
▶ All circuits in caravans.
▶ All circuits in circuses, etc.
▶ A circuit supplying Class II heating equipment for floor and ceiling heating systems.

100 mA

▶ Socket outlets of rating exceeding 32 A in agricultural locations.

300 mA

▶ At the origin of a temporary supply to circuses, etc.
▶ Where there is a risk of fire due to storage of combustible materials.
▶ All circuits (except socket outlets) in agricultural locations.

500 mA

▶ Any circuit supplying one or more socket outlets of rating exceeding 32 A, on a construction site.

Where loop impedances are too high, RCD ratings can be calculated.

Bonding: Questions and Answers

By now, we should know why bonding is necessary. The next question is to what extent bonding should be carried out? This is, perhaps, answered best by means of question and answer examples:

1. **Do I need to bond the kitchen hot and cold taps and a metal sink together?**

   The IET Regulations do not require this bonding to be carried out in domestic kitchens/utility rooms, etc.

2. **Do I have to bond radiators in a premises to, say, metal clad switches or socket outlets?**

   Supplementary bonding is only necessary when extraneous conductive parts are simultaneously accessible with exposed conductive parts and when the disconnection time for the circuit concerned cannot be achieved. In these circumstances, the bonding conductor should have a resistance $R < 50/I_a$, where,

   R = resistance of supplementary bonding conductor

   50 = touch voltage of 50 V

   I_a = current causing operation of protection.

3. **Do I need to bond metal window frames?**

   In general, no. Apart from the fact that most window frames will not introduce a potential from anywhere, the part of the window most likely to be touched is the opening portion, to which it would not be practicable to bond. There may be a case for the bonding of patio doors, which could be considered earthy with rain running from the lower portion to the earth. However, once again, the part most likely to be touched is the sliding section, to which it is not possible to bond. In any case, there would need to be another simultaneously accessible part to warrant considering any bonding.

4. **What size of bonding conductors should I use?**

   Main protective bonding conductors should be not less than half the size of the main earthing conductor, subject to a minimum of 6.0 mm² or, where PME (TN-C-S) conditions are present, 10.0 mm². For example, most new domestic installations now have a 16.00 mm² earthing conductor, so all main bonding will be in 10.0 mm². Supplementary bonding conductors are subject to a minimum of 2.5 mm² if mechanically protected or 4.0 mm² if not. However, if these bonding conductors are connected to exposed conductive parts, they must be the same size as the cpc connected to the exposed conductive part (once again subject to the minimum sizes mentioned). It is sometimes difficult to protect a bonding conductor mechanically throughout its length, especially at terminations, so it is perhaps better to use 4.0 mm² as the minimum size.

5. Do I have to bond free-standing metal cabinets, screens, work benches, etc.?

No. These items will not introduce a potential into the equipotential zone from outside and cannot,

therefore, be regarded as extraneous conductive parts.

6. What do I do when the pipework is plastic or a mixture of metal and plastic?

No bonding is required.

7. What about supplementary bonding in a bathroom?

Provided that the premises has a protective bonding and automatic disconnection system in place and all circuits are RCD protected, then NO supplementary bonding is required (see page 309).

Points to Note

1. An earthing terminal must be provided adjacent to the consumer's terminals. This is usually in the form of a rectangular metal block with cable entries and screws (Fig. 16.16). The earthing conductor from the main earthing terminal is connected to the cable sheath (if this system is used for earth return) by means of an earth clamp usually of the type shown in Fig. 16.17.

Fixing holes

Figure 16.16 Consumer's main earthing terminal.

carry current shall be connected to the appropriate circuit protective conductors. This includes things such as metal boxes for socket outlets and metal casings of fires. There are some exceptions to these requirements, as follows:

(a) Short isolated lengths of metal used for the mechanical protection of cables (conduit used to carry cables overhead between buildings is **not** exempt).
(b) Metal cable clips.
(c) Metal lamp caps.
(d) Metal screws, rivets or nameplates isolated by insulating material.
(e) Metal chains used to suspend luminaires.

Connecting terminal

SAFETY ELECTRICAL CONNECTION

DO NOT REMOVE

Label to be fixed on clamp letters NO LESS than 4.75 mm high

Figure 16.17 Earth clamp.

2. All the exposed conductive parts of wiring systems and apparatus not intended to

(f) Metal luminaires (such as lampholders) using filament lamps, provided that they are installed above a non-conducting floor and are screened, or positioned so that they cannot be touched by a person able to come into contact with earthed metal.

(g) Catenary wires where insulated hangers are used. If there is any metalwork in an installation that is likely to come into contact with earthed metal accidentally, then it must be either effectively bonded to, or segregated from, that earthed metal. This includes metal baths, sinks, exposed pipes, radiators, tanks, any structural steelwork that is accessible, and the framework of any mobile equipment such as cranes and lifts which have electrical apparatus fitted. The minimum size of copper bonding lead is 2.5 mm² with mechanical protection or 4.0 mm² without.

3. Gas and water services must be bonded to the consumer's earthing terminal as near as possible to the point of entry into the building, and on the consumer's side of the service. It is not permissible to use a gas or water pipe as an earth electrode, the minimum size of copper bonding lead is 6.00 mm². The bonding may be achieved by using the clamp illustrated in Fig. 16.17.

4. A circuit protective conductor other than copper strip must be insulated throughout its length, and where insulation is removed at terminations (i.e. stripping back twin with earth cable) the resulting bare cpc must be sleeved.

Self-Assessment Questions

1. What is the resistance of a 10 m length of 6.0 mm² copper line conductor if the associated cpc is 1.5 mm²?

2. What is the length of a 6.0 mm² copper line conductor with a 2.5 mm² cpc if the overall resistance is 0.189 Ω?

3. If the total loop impedance of a circuit is 0.96 Ω and the cable is a 20 m length of 4.0 mm² copper with a 1.5 mm² cpc, what is the external loop impedance?

4. Will there be a shock risk if a double-socket outlet, fed by a 23 m length of 2.5 mm² copper conductor with a 1.5 mm² cpc, is protected by a 20 A BS 3036 rewirable fuse and the external loop impedance is measured as 0.5 Ω?

5. A cooker circuit is protected by a 32 A BS EN 60898 Type C cb, and wired in 6.0 mm² copper with a 2.5 mm² cpc. The run is some 30 m and the external loop impedance of the TN–S system cannot be measured. Is there a shock risk, and if so how could it be rectified?

CHAPTER 17

Protection

Protection

What is protection?

The meaning of the word 'protection' as used in the electrical industry is no different from that in everyday usage. People protect themselves against personal or financial loss by means of insurance and from injury or discomfort by the use of the correct protective clothing. They further protect their property by the installation of security measures such as locks and/or alarm systems. In the same way, electrical systems need the following:

1. To be protected against mechanical damage, the effects of the environment and electrical overcurrents.
2. To be installed in such a fashion that persons and/or livestock are protected from the dangers that can be caused by such an electrical installation.

Let us now look at these protective measures in more detail.

Protection against mechanical damage

The word 'mechanical' is somewhat misleading in the sense that most of us associate it with machinery of some sort. In fact, a serious electrical overcurrent left uninterrupted for too long can cause distortion of conductors and degradation of insulation; both of these effects are considered to be mechanical damage.

However, let us start by considering the ways of preventing mechanical damage caused by physical impact and the like.

Cable construction

A cable comprises one or more conductors, each covered with an insulating mater-ial. This insulation provides protection from shock and prevents the passage of leakage currents between conductors.

Clearly, insulation is very important and should be protected from damage itself.This may be achieved by covering the insulated conductors with a protective sheathing during manufacture, or by enclosing them in conduit or trunking at the installation stage.

The type of sheathing chosen and/or the installation method will depend on the environment in which the cable is to be installed. For example, metal conduit with polyvinyl chloride (PVC) singles or mineral-insulated (m.i.) cable would be used in preference to PVC-sheathed cable clipped direct, in an industrial environment. Figure 17.1 shows the effect of physical impact on m.i. cable.

Protection against corrosion

Mechanical damage to cable sheaths and metalwork of wiring systems can occur through corrosion, and hence care must be taken to choose corrosion-resistant materials and to avoid contact between dissimilar metals in damp situations.

Protection against thermal effects

This is the subject of Chapter 42 of the IET Regulations. It basically requires common-sense decisions regarding the placing of fixed equipment, such that surrounding materials are not at risk from damage by heat.

Figure 17.1 Mineral-insulated cable. On impact, all parts including the conductors are flattened, and a proportionate thickness of insulation remains between conductors and conductors and sheath, without impairing the performance of the cable at normal working voltages.

In addition to these requirements is the need to protect persons from burns by guarding parts of equipment that are liable to excessive temperatures.

Polyvinyl chloride

PVC is a thermoplastic polymer widely used in electrical installation work for cable insulation, conduit and trunking. General-purpose PVC is manufactured to the British Standard BS 6746.

PVC in its raw state is a white powder; it is only after the addition of plasticizers and stabilizers that it acquires the form that we are familiar with.

Degradation

All PVC polymers are degraded or reduced in quality by heat and light. Special stabilizers added during manufacture help to retard this degradation at high temperatures. However, it is recommended in the IET Regulations that PVC-sheathed cables or thermoplastic fittings for luminaires (light fittings) should not be installed where the temperature is likely to rise above 60°C. Cables insulated with high-temperature PVC (up to 80°C) should be used for drops to lampholders and entries into batten holders. PVC conduit and trunking should not be used in temperatures above 60°C.

Embrittlement and cracking

PVC exposed to low temperatures becomes brittle and will easily crack if stressed. Although both rigid and flexible PVC used in cables and conduit can reach as low as −5°C without becoming brittle, the Regulations recommend that general-purpose PVC-insulated cables should not be installed in areas where the temperature is likely to be consistently below 0°C. They further recommend that PVC-insulated cable should not be handled unless the ambient temperature is above 0°C and unless the cable temperature has been above 0°C for at least 24 h.

When rigid PVC conduit is to be installed in areas where the ambient temperature is below −5°C but not lower than −25°C, Type B conduit manufactured to BS 4607 should be used.

When PVC-insulated cables are installed in loft spaces insulated with polystyrene granules, contact between the two polymers can cause the plasticizer in the PVC to migrate to the granules. This causes the PVC to harden and although there is no change in the electrical properties, the insulation may crack if disturbed.

External influences

Appendix 5 of the IET Regulations classifies external influences that may affect an installation. This classification is divided into three sections, the environment (A), how that environment is utilized (B) and construction of buildings (C). The nature of any influence within each section is also represented by a letter, and the level of influence is represented by a number. The following table gives examples of the classification:

Environment	Utilization	Building
Water	Capability	Materials
AD6 waves	BA3 handicapped	CA1 non-combustible

With external influences included in drawings and specifications, installations and materials used can be designed accordingly.

Protection against ingress of solid objects, liquid and impact

In order to protect equipment from damage by foreign bodies, liquid or impact, and also to prevent persons from coming into contact with live or moving parts, such equipment is housed inside enclosures or cable management systems such as conduit, trunking ducts, etc.

The degree of protection offered by such an enclosure is the subject of BS EN 60529 and BS EN 62262, commonly known as the IP and IK codes, parts of which are as shown in the accompanying tables. It will be seen from the IP table (Table 17.1) that,

Table 17.1 IP codes

First numeral: Mechanical protection

0. No protection of persons against contact with live or moving parts inside the enclosure. No protection of equipment against ingress of solid foreign bodies.
1. Protection against accidental or inadvertent contact with live or moving parts inside the enclosure by a large surface of the human body, for example a hand, not for protection against deliberate access to such parts. Protection against ingress of large solid foreign bodies.
2. Protection against contact with live or moving parts inside the enclosure by fingers. Protection against ingress of medium-sized solid foreign bodies.
3. Protection against contact with live or moving parts inside the enclosure by tools, wires or such objects of thickness greater than 2.5 mm. Protection against ingress of small foreign bodies.
4. Protection against contact with live or moving parts inside the enclosure by tools, wires or such objects of thickness greater than 1 mm. Protection against ingress of small foreign bodies.
5. Complete protection against contact with live or moving parts inside the enclosures. Protection against harmful deposits of dust. The ingress of dust is not totally prevented, but dust cannot enter in an amount sufficient to interfere with satisfactory operation of the equipment enclosed.
6. Complete protection against contact with live or moving parts inside the enclosures. Protection against ingress of dust.

Second numeral: Liquid protection

0. No protection.
1. Protection against drops of condensed water. Drops of condensed water falling on the enclosure shall have no effect.
2. Protection against drops of liquid. Drops of falling liquid shall have no harmful effect when the enclosure is tilted at any angle up to 15° from the vertical.
3. Protection against rain. Water falling in rain at an angle equal to or smaller than 60° with respect to the vertical shall have no harmful effect.
4. Protection against splashing. Liquid splashed from any direction shall have no harmful effect.
5. Protection against water jets. Water projected by a nozzle from any direction under stated conditions shall have no harmful effect.
6. Protection against conditions on ships' decks (deck with watertight equipment). Water from heavy seas shall not enter the enclosures under prescribed conditions.
7. Protection against immersion in water. It must not be possible for water to enter the enclosure under stated conditions of pressure and time.
8. Protection against indefinite immersion in water under specified pressure. It must not be possible for water to enter the enclosure.
X Indicates no *specified* protections.

for instance, an enclosure to IP56 is dustproof and waterproof.

IK codes

Protection against mechanical impact.

The most commonly quoted IP codes in the Regulations are IPXXB and IP2X (the X denotes that no protection is specified, **not** that no protection exists).

Hence, IP2X means that an enclosure can withstand the ingress of medium-sized solid foreign bodies (12.5 mm diameter), and a jointed test finger, known affectionately as the British Standard finger! IPXXB denotes protection against the test finger only.

For accessible horizontal top surfaces of enclosures, the IP code is IPXXD and IP4X. This indicates protection against small foreign bodies and a 1 mm diameter test wire. IPXXD is the l mm diameter wire only.

IET Regulations, Section 522, give details of the types of equipment, cables and enclosure that, for example, may be selected for certain environmental conditions; for example, an enclosure housing equipment in an AD8 environment (under water) would need to be IPX8.

Protection against electric shock

There are two ways of receiving an electric shock: by contact with intentionally live parts and by contact with conductive parts made live due to a fault. It is obvious that we need to provide protection against both of these conditions.

Basic protection

Clearly, it is not satisfactory to have live parts accessible to touch by persons or livestock. The IET Regulations recommend five ways of minimizing this danger:

1. By covering the live part or parts with insulation, which can only be removed by destruction, for example cable insulation.
2. By placing the live part or parts behind a barrier or inside an enclosure, providing protection to at least IPXXB or IP2X. In most cases, during the life of an installation it becomes necessary to open an enclosure or remove a barrier. Under these circumstances, this action should only be possible by the use of a key or tool, for example by using a screwdriver to open a junction box. Alternatively, access should only be gained after the supply to the live parts has been disconnected, for example by isolation on the front of a control panel where the cover cannot be removed until the isolator is in the 'off' position. An intermediate barrier of at least IP2X or IPXXB will give protection when an enclosure is opened: a good example of this is the barrier inside distribution fuseboards, which prevents accidental contact with incoming live feeds.
3. By placing obstacles to prevent unintentional approach to or contact with live parts. This method must only be used where skilled persons are working.

4. By placing out of arm's reach; for example, the high level of the bare conductors of travelling cranes.
5. By using a residual current device (RCD) as additional protection. Whilst not permitted as the sole means of protection, this is considered to reduce the risk associated with contact with live parts, provided that one of the other methods just mentioned is applied, and that the RCD has a rated operating current, $I_{\Delta n}$, of not more than 30 mA and an operating time not exceeding 40 ms at five times $I_{\Delta n}$, that is, 150 mA.

Fault protection

The IET Regulations suggest five ways of fault protection against contact with parts made live due to a fault. Of these, protective earthing, protective equipotential bonding and automatic disconnection in the event of a fault have already been discussed in Chapter 16. The other methods are as follows.

Reinforced insulation

Often referred to as double-insulated equipment, this is typical of modern appliances where there is no provision for the connection of a circuit protective conductor (cpc). This does not mean that there should be no exposed conductive parts and that the casing of equipment should be of an insulating material; it simply indicates that live parts are so well insulated that faults from live to conductive parts cannot occur.

Non-conducting location

This is basically an area in which the floor, walls and ceiling are all insulated. There must be no protective conductors within such an area, and socket outlets should have no earthing connections.

It must not be possible simultaneously to touch two exposed conductive parts, or to touch an exposed conductive part and an extraneous conductive part. This requirement clearly prevents shock current passing through a person in the event of an earth fault, and the insulated construction prevents shock current passing to earth.

Earth-free local equipotential bonding

This is, in essence, a Faraday cage, where all metals are bonded together but not to earth. Obviously great care must be taken while entering such a zone in order to avoid differences in potential between inside and outside.

The areas mentioned in this method and the previous method are very uncommon. Where they do exist, they should be under constant supervision to ensure that no additions or alterations can reduce the protection intended.

Electrical separation

This method relies on a supply from a safety source such as an isolating transformer to BS EN 60742, which has no earth connection on the secondary side. In the event of a circuit that is supplied from a source developing a live fault to an exposed conductive part, there would be no path for shock current to flow (see Fig. 17.2).

Figure 17.2 Electrical separation.

Once again, great care must be taken to maintain the integrity of this type of system as an inadvertent connection to earth or interconnection with other circuits would render the protection useless.

Exemptions

As with most sets of rules and regulations, there are certain areas that are exempt from the requirements. These areas are listed quite clearly in IET Regulations 410.3.9, and there is no point in repeating them all here. However, one example is the dispensing of the need to earth exposed conductive parts such as small fixings, screws and rivets, provided that they cannot be touched or gripped by a major part of the human body (not less than 50 mm × 50 mm), and that it is difficult to make and maintain an earth connection.

Protection by automatic disconnection of supply

This measure is a combination of basic and fault protection.

Separated extra low voltage

This is simply extra low voltage (less than 50 V a.c.) derived from a safety source such as a class II safety isolating transformer to BS EN 61558-2-6, or

a motor generator, which has the same degree of isolation as the transformer, or a battery or diesel generator, or an electronic device such as a signal generator.

Live or exposed conductive parts of separated extra low voltage (SELV) circuits should not be connected to earth or protective conductors of other circuits, and SELV circuit conductors should ideally be kept separate from those of other circuits. If this is not possible, then the SELV conductors should be insulated to the highest voltage present.

Obviously, plugs and sockets of SELV circuits should not be interchangeable with those of other circuits.

SELV circuits supplying socket outlets are mainly used for hand lamps or soldering irons, for example in schools and colleges. Perhaps a more common example of a SELV circuit is a domestic bell installation, where the transformer is to BS EN 60742. Note that bell wire is usually only suitable for 50–60 V, which means that it should not be run together with circuit cables of higher voltages.

Reduced low voltage systems

The Health and Safety Executive accepts that a voltage of 63.5 V to earth, three phase, or 55 V to earth, single phase, will give protection against severe electric shock. They therefore recommend that portable tools used on construction sites, etc., be fed from a 110 V centre-tapped transformer to BS 4343. Figure 17.3 shows how 55 V is derived. Earth fault loop impedance values for these systems may be taken from Table 41.6 of the Regulations.

Figure 17.3 Reduced low voltage system.

Protection against overcurrent

An overcurrent is a current greater than the rated current of a circuit. It may occur in two ways:

1. As an overload current; or
2. As a fault current, which may be subdivided into
   (a) A short-circuit current
   (b) An earth fault current.

These conditions need to be protected against in order to avoid damage to circuit conductors and equipment. In practice, fuses and circuit breakers will fulfil both of these needs.

Overloads

Overloads are overcurrents occurring in healthy circuits. They may be caused, for example, by faulty appliances or by surges due to motors starting or by plugging in too many appliances in a socket outlet circuit.

Short circuits and earth faults

A short-circuit current is the current that will flow when a 'dead short' occurs between live conductors (line to neutral for single phase; line to line for three phase). Earth fault current flows when there is a short between a line conductor and earth. Prospective short-circuit current (PSCC) and prospective earth fault current (PEFC) are collectively known as prospective fault current (PFC). The term is usually used to signify the value of fault current at fuse or circuit breaker positions.

Prospective fault current is of great importance. However, before discussing it or any other overcurrent further, it is perhaps wise to refresh our memories with regard to fuses and circuit breakers and their characteristics.

Fuses and circuit breakers

As we all know, a fuse is the weak link in a circuit, which will break when too much current flows, thus protecting the circuit conductors from damage.

It must be remembered that the priority of the fuse is to protect the **circuit conductors**, not the appliance or the user. Calculation of cable size therefore automatic-ally involves the correct selection of protective devices.

There are many different types and sizes of fuse, all designed to perform a certain function. The IET Regulations refer to only four of these types: BS 3036, BS 88, BS 1361 and BS 1362 fuses. It is perhaps sensible to include, at this point, circuit breakers to BS 3871, BS EN 60898 and RCBOs to BS EN 61009, although the BS 3871 miniature circuit breakers (mcb's) are no longer included in the IET Regulations.

Fuses

A fuse is simply a device that carries a metal element, usually tinned copper, which will melt and break the circuit when excessive current flows. There are three types of fuses:

1. The rewirable or semi-enclosed fuse.
2. The cartridge fuse and fuse link.
3. The high-rupturing-capacity (HRC) fuse.

The rewirable fuse (BS 3036)

A rewirable fuse consists of a fuse, a holder, a fuse element and a fuse carrier, the holder and carrier being made of porcelain or Bakelite (Fig. 17.4). The circuits, for which this type of fuse is designed, have colour codes that are marked on the fuse holder and are as follows:

45 A – Green
30 A – Red
20 A – Yellow
15 A – Blue
5 A – White

Figure 17.4 Typical rewirable fuse assembly.

Although this type of fuse was very popular in domestic installations, as it is cheap and easy to repair, it has serious disadvantages:

1. The fact that it is repairable enables the wrong size of fuse wire (element) to be used.
2. The elements become weak after long usage and may break under normal conditions.
3. Normal starting-current surges (e.g. when motors, etc., are switched on) are 'seen' by the fuse as an overload and will therefore break the circuit.

Figure 17.5 Cartridge fuse.

Figure 17.6 HRC fuse.

4. The fuse holder and carrier can become damaged as a result of arcing in the event of a heavy overload.

Cartridge fuse (BS 1362 and BS88-3)

A cartridge fuse consists of a porcelain tube with metal and caps to which the elem-ent is attached. The tube is filled with silica (Fig. 17.5).

These fuses are found generally in modern plug tops used with 13 A socket outlets (BS1362). The BS 88-3 are found in some distribution boards and at mains intake positions (DNOs fuse). They have some advantages over the rewirable fuse as they: do not deteriorate; maintain accuracy in breaking at rated values; and do not arc when interrupting faults. Cartridge fuses are, however, expensive to replace.

HRC fuses

The HRC fuse is a sophisticated variation of the cartridge fuse and is normally found protecting motor circuits and industrial installations.

It consists of a porcelain body filled with silica with a silver element and lug type and caps. Another feature of HRC fuse is the indicating element, which shows when the fuse has blown.

It is very fast-acting and can discriminate between a starting surge and an overload (Fig. 17.6).

Circuit breakers

These protective devices have two elements, one thermal and one electro-magnetic. The first element looks after overloads and the second element looks after short circuits.

Circuit breakers have one great advantage over the fuse in that, once having operated, they can be reset. They are also very accurate (tripping current) and fast, and therefore provide a high degree of discrimination.

Class of protection

It will be evident that each of the protective devices just discussed provides a different level of protection, that is, rewirable fuses are slower to operate and less accur-ate than mcb's. In order to classify these devices, it is important to have some means of knowing their circuit-breaking and 'fusing' performance. This is achieved for fuses by the use of a fusing factor:

$$\text{Fusing factor} = \frac{\text{fusing current}}{\text{current rating}}$$

where the **fusing current** is the minimum current causing the fuse to blow, and the **current rating** is the maximum current, which the fuse can sustain without blowing. For example, a 5 A fuse, which blows only when 9 A flows, will have a fusing factor of 9/5 = 1.8:

    Rewirable fuses have a fusing factor of about 1.8
    Cartridge fuses have a fusing factor of between
        1.25 and 1.75
    HRC fuses have a fusing factor of up to 1.25
        (maximum)
    Circuit breakers are designed to operate at no
        more than 1.5 times their rating

Breaking capacity of fuses and circuit breakers

When a fault occurs, the current may, for a fraction of a second, reach hundreds or even thousands of amperes. The protective device must be able to break, and in the case of circuit breakers, make such a current without damage to its surroundings by arcing, overheating or the scattering of hot particles.

Tables 17.2 and 17.3 indicate the performance of circuit breakers and the more commonly used British Standard fuse links.

17

Table 17.2

Circuit Breakers	Breaking	Capacity (kA)
BS 3871 Types 1, 2, 3, etc.	1	(M1)
	1.5	(M1.5)
	3	(M3)
	4.5	(M4.5)
	6	(M6)
	9	(M9)
BS EN 60898 Types B, C, D	Icn 1.5	Ics 1.5
	3	3
	6	6
	10	7.5
	15	7.5
	25	10

Icn is the rated ultimate breaking capacity.
Ics is the maximum breaking capacity operation after which the breaker may still be used without loss of performance.

Although all reference to BS 3871 mcb's have been removed from BS 7671, they are still in existence and therefore worthy of mention.

Fuse and circuit breaker operation

Let us consider a protective device rated at, say, 10 A. This value of current can be carried indefinitely by the device, and is known as its nominal setting I_n. The value of the current, which will cause operation of the device, I_2, will be larger than I_n, and will be dependent on the device's **fusing factor**. This is a figure which, when multiplied by the nominal setting I_n, will indicate the value of operating current I_2.

For fuses to BS 88 and BS 1361 and circuit breakers to BS 3871, this fusing factor is approximately 1.45; hence our 10 A device would not operate until the current reached $1.45 \times 10 = 145$ A.

The IET Regulations require coordination between conductors and protection when an overload occurs, such that:

1. The nominal setting of the device I_n is greater than or equal to the design current of the circuit I_b ($I_n \geq I_b$).
2. The nominal setting I_n is less than or equal to the lowest current carrying capacity I_z of any of the circuit conductors ($I_n \leq I_z$).

Table 17.3 British Standards for fuse links

Standard	Current Rating	Voltage Rating	Breaking Capacity	Notes
1. BS 2950	Range 0.05 to 25 A	Range 1000 V (0.05 A) to 32 V (25 A) a.c. and d.c.	Two or three times current rating	Cartridge fuse links for telecommunication and light electrical apparatus. Very low breaking capacity.
2. BS 646	1, 2, 3 and 5 A	Up to 250 V a.c. and d.c.	1000 A	Cartridge fuse intended for fused plugs and adaptors to BS 546: 'round-pin' plugs.
3. BS 1362 cartridge	1, 2, 3, 5, 7, 10 and 13 A	Up to 250 V a.c.	6000 A	Cartridge fuse primarily intended for BS 1363: 'flat-pin' plugs.
4. BS 88-3 HRC cut-out fuses	5, 15, 20, 30, 45 and 60 A	Up to 250 V d.c.	16 500 A	Cartridge fuse intended for use in domestic consumer units. The dimensions prevent interchangeability of fuse links which are not of the same current rating.
5. BS 88 motors	Four ranges, 2 to 1200 A	Up to 600 V, but normally 250 or 400 V a.c. and 250 or 500 V d.c.	Ranges from 10 000 to 80 000 A in four a.c. and three d.c. categories	Part 1 of Standard gives performance and dimensions of cartridge fuse links, whilst Part 2 gives performance and requirements of fuse carriers and fuse bases designed to accommodate fuse links complying with Part 1.
6. BS 2692	Main range from 5 to 200 A; 0.5 to 3 A for voltage transformer protective fuses	Range from 2.2 to 132 kV	Ranges from 25 to 750 MVA (main range) 50 to 2500 MVA (VT fuses)	Fuses for a.c. power circuits above 660 V.
7. BS 3036	5, 15, 20, 30, 45, 60, 100, 150 and 200 A	Up to 250 V to earth	Ranges from 1000 to 12 000 A	Semi-enclosed fuses (the element is a replacement wire) for a.c. and d.c. circuits.
8. BS 4265	500 mA to 6.3 A; 32 mA to 2 A	Up to 250 V a.c.	1500 A (high breaking capacity) 35 A (low breaking capacity)	Miniature fuse links for protection of appliances of up to 250 V (metric standard).

3. The operating current of the device I_2 is less than or equal to 1.45 I_z ($I_2 \leq 145I_z$).

So, for our 10 A device, if the cable is rated at 10 A then condition 2 is satisfied. Since the fusing factor is 1.45, condition 3 is also satisfied: $I_2 = I_n \times 1.45 = 10 \times 1.45$, which is also 1.45 times the 10 A cable rating.

The problem arises when a BS 3036 semi-enclosed rewirable fuse is used, as it may have a fusing factor of as much as 2. In order to comply with condition 3, I_n should be less than or equal to $0.725I_z$.

This figure is derived from 1.45/2 = 0.725. For example, if a cable is rated at 10 A, then I_n for a BS 3036 should be $0.725 \times 10 = 7.25$ A. As the fusing factor is 2, the operating current is $I_2 = 2 \times 7.25 = 14.5$, which conforms to condition 3, that is, $I_2 \leq 1.45 \times 10 = 14.5$.

All of these foregoing requirements ensure that conductor insulation is undamaged when an overload occurs.

Under fault conditions it is the conductor itself, which is susceptible to damage and must be protected. Figure 17.7 shows one half-cycle of short-circuit current if there were no protection. The RMS value (0.7071 × maximum value) is called the prospective fault current. The cut-off point is where the fault current is interrupted and an arc is formed; the time t_1 taken to reach this point is called the pre-arcing time. After the current has been cut off, it falls to zero as the arc is being extinguished. The time t_1 is the total time taken to disconnect the fault.

Figure 17.7

During the time t_1, the protective device is allowing energy to pass through to the load side of the circuit. This energy is known as the pre-arcing let-through energy and is given by $I_f^2 t_1$ where I_f is the short-circuit current. The total let-through energy from start to disconnection of the fault is given by $I_f^2 t_2$ (see Fig. 17.8).

Figure 17.8 Let through energy.

For faults of up to 5 s duration, the amount of heat energy that a cable can withstand is given by $k^2 s^2$, where s is the cross-sectional area (c.s.a.) of the conductor and k is a factor dependent on the conductor material. Hence the let-through energy should not exceed $k^2 s^2$, that is, $I_f^2 t = k^2 s^2$. If we transpose this formula for t, we get $t = k^2 s^2 / I_f^2$, which is the maximum disconnection time in seconds.

Remember that these requirements refer to short-circuit currents only. If in fact the protective device has been selected to protect against overloads and has a breaking capacity not less than the prospective short-circuit current I_p at the point of installation, it will also protect against short-circuit currents. However, if there is any doubt the formula should be used.

For example, in Fig. 17.9, if I_n has been selected for overload protection, the questions to be asked are as follows:

1. Is $I_n \geq I_b$? Yes.
2. Is $I_n \leq I_z$? Yes.
3. Is $I_2 \geq 1.45I_z$? Yes.

Figure 17.9

Then if the device has a rated breaking capacity not less than I_p, it can be con-sidered to give protection against short-circuit current also.

When an installation is being designed, the prospective short-circuit current at every relevant point must be determined, by either calculation or measurement. The value will decrease as we move farther away from the intake position (resistance

17

increases with length). Thus, if the breaking capacity of the lowest rated fuse in the installation is greater than the prospective short-circuit current at the origin of the supply, there is no need to determine the value except at the origin.

Discrimination

Where more than one fuse protects a circuit (Fig. 17.10), it is clearly sensible that the correct fuse should blow under fault conditions. A fault on the appliance should cause fuse C to blow. If fuse B blew, although it would break the circuit to the faulty appliance, it would unnecessarily render the whole radial circuit dead. If fuse A blew, instead of B or C, all circuits from the distribution would be pointlessly disconnected.

Figure 17.10 Discrimination.

The arrangement of fuses to protect the correct part of a circuit is called **discrimination**.

It cannot be assumed that discrimination is achieved just because protective devices have different ratings. This is especially the case where a mixture of different types of device is used. However, as a general rule a 2:1 ratio with the lower-rated devices will be satisfactory. Table 17.4 shows how fuse links may be chosen to ensure discrimination.

Fuses will give discrimination if the figure in column 3 does not exceed the figure in column 2. Hence:

A 2 A fuse will discriminate with a 4 A fuse
A 4 A fuse will discriminate with a 6 A fuse
A 6 A fuse will *not* discriminate with a 10 A fuse
A 10 A fuse will discriminate with a 16 A fuse

All other fuses will **not** discriminate with the next highest fuse, and in some cases, several sizes higher are needed, for example, a 250 A fuse will only discriminate with a 400 A fuse.

Position of protective devices

When there is a reduction in the current-carrying capacity of a conductor, a protective device is required. There are, however, some exceptions to this requirement; these are listed quite clearly in the IET Regulations. As an example, protection is not needed

Table 17.4 I_f^2t characteristics: 2–800 A fuse links. Discrimination is achieved if the total I_f^2t of the minor fuse does not exceed the pre-arcing I_f^2t of the major fuse

Rating (A)	I_f^2t Pre-arcing	I_f^2t Total at 400 V
2	0.9	1.7
4	4	12
6	16	59
10	56	170
16	190	580
20	310	810
25	630	1700
32	1200	2800
40	2000	6000
50	3600	11 000
63	6500	14 000
80	13 000	36 000
100	24 000	66 000
125	34 000	120 000
160	80 000	260 000
200	140 000	400 000
250	230 000	560 000
315	360 000	920 000
350	550 000	1 300 000
400	800 000	2 300 000
450	700 000	1 400 000
500	900 000	1 800 000
630	2 200 000	4 500 000
700	2 500 000	5 000 000
800	4 300 000	10 000 000

in a ceiling rose where the cable size changes from 1.0 mm² to, say, 0.5 mm² for the lamp-holder flex. This is permitted, as it is not expected that lamps will cause overloads.

Protection against undervoltage

From the point of view of danger in the event of a drop or loss of voltage, the protection should prevent automatic restarting of machinery, etc. In fact, such protection is an integral part of motor starters in the form of the control circuit.

The essential part of a motor control circuit that will ensure undervoltage protection is the 'hold-on' circuit (Fig. 17.11) (see also p. 304).

Figure 17.11 Hold-on circuit.

When the start button is pushed, the coil becomes energized and its normally open (N/O) contacts close. When the start button is released the coil remains energized via its own N/O contacts. These are known as the 'hold-on' contacts.

The coil can only be de-energized by opening the circuit by the use of the stop button or by a considerable reduction or loss of voltage. When this happens, the N/O contacts open, and, even if the voltage is restored or the circuit is made complete again, the coil will remain de-energized until the start button is pushed again. Figure 17.12 shows how this 'hold-on' facility is built into a typical single-phase starter.

Figure 17.12 Single-phase motor starter.

Control

After deciding how we are going to earth an installation, settling on the method of protecting persons and livestock from electric shock and

preventing conductors and insulation from damage, we must now investigate the means of controlling the installation. In simple terms, this means the switching of the installation or any part of it 'on' or 'off'. The IET Regulations refer to this topic as 'isolation and switching'.

Isolation and switching

By definition, isolation is the cutting off of electrical energy from every source of supply, and this function is performed by a switch, a switch fuse or a fuse switch.

Isolator

This is simply a double- or triple-pole switch in which the moving switch contacts are mechanically linked. In this way, both line and neutral or all phases in a three-phase system are disconnected from the supply (Fig. 17.13).

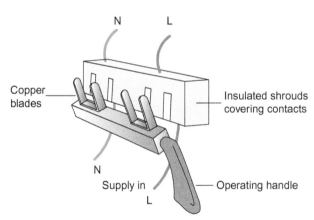

Figure 17.13 Internal arrangement of single-phase isolator.

Switch fuse

This is an extension of the isolator, in a way that the load side of the supply is interrupted by a fuse in the line conductor.

Fuse switch

Fuse switches are used on three-phase systems. Hence, the fuse forms part of the moving operating blade.

Consumer unit

The consumer unit is found in nearly all domestic installations. It consists of a double-pole isolator and a distribution board in one assembly.

17

Circuit breaker

The means of isolation must be double or triple pole. Some modern consumer units have a residual current device installed in place of the usual isolator.

Sequence of control

Figure 17.14a–d illustrates some typical control sequences.

Figure 17.14 Typical control sequence.

With a domestic installation, the main switch in a consumer unit is considered to be a means of isolation for the whole installation, and each fuse or circuit breaker is considered to be an isolator for the individual circuits. Ideally all of these devices should have some means of preventing unintentional re-energization, either by locks or by interlocks. In the case of fuses and circuit breakers, these can be removed and kept in a safe place.

In many cases, isolating and locking off come under the requirements for switching off for mechanical maintenance. Hence a switch controlling a motor circuit should have, especially if it is remote from the motor, a means of locking in the 'off' position (Fig. 17.15).

Figure 17.15 Switching off for mechanical maintenance

A one-way switch controlling a lighting point is a functional switch, but could be considered as a means of isolation, or a means of switching off for mechanical maintenance (changing a lamp). A two-way switching system, however, does not provide a means of isolation, as neither switch cuts off electrical energy from all sources of supply.

In an industrial or workshop environment, it is important to have a means of cutting off the supply to the whole or parts of the installation in the event of an emergency. The most common method is the provision of stop buttons suitably located and used in conjunction with a contactor or relay (Fig. 17.16).

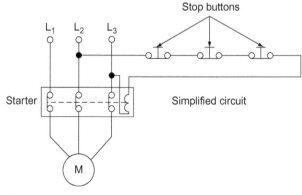

Figure 17.16 Emergency switching.

Pulling a plug from a socket to remove a hazard is not permitted as a means of emergency switching. It is, however, allowed as a means of functional switching, for example switching off a hand lamp by unplugging.

Whilst we are on the subject of switching, it should be noted that a switch controlling discharge lighting (this includes fluorescent fittings) should, unless it is specially designed for the purpose, be capable of carrying at least twice the steady load of the circuit. The reason for this is that discharge lighting contains chokes that are highly inductive and cause arcing at switch contacts. The higher rating of the switch enables it to cope with such arcing.

Points to Note

1. Every consumer's installation must incorporate a means of overcurrent protection, which should comprise either a fuse or circuit breaker inserted in each line conductor of the supply.
2. Every method of overcurrent protection should be based on the prospective short-circuit current (i.e. the current that would flow in the event of a dead short between lines on the load side of the protection).
3. Fuses and circuit breaker must be arranged in an installation such that discrimination is ensured.
4. Protective devices must be inserted only in the line conductor of a two-wire system in which one wire is connected to earth (i.e. neutral is earthed at the supply transformer).
5. Every circuit conductor must be protected by a fuse or circuit breaker installed at the origin of the circuit. The protective device must have a setting not less than the design current of the circuit (e.g. a 5A lighting circuit is protected by no less than 5A protection). Also, the setting should not exceed the lowest current-carrying capacity of any conductor in the circuit, and the value at which the protection operates must not exceed 1.45 times the lowest current-carrying capacity of any circuit conductor.
6. Effective and accessible means of isolation must be provided at the origin of the installation to cut off all voltage as may be necessary to prevent danger.
7. When an installation serves two or more detached buildings (i.e. house and detached garage) a means of isolation must be provided in each building.
8. If the purpose of a switch or circuit breaker is not clear, it should be labelled to show which apparatus it controls.

17

237

CHAPTER 18

Circuit and Design

Design procedure

The requirements of IET Regulations make it clear that circuits must be designed and the design data made readily available. In fact this has always been the case with previous editions of the IET Regulations, but it has not been so clearly indicated.

How then do we begin to design? Clearly, plunging into calculations of cable size is of little value unless the type of cable and its method of installation is known. This in turn will depend on the installation's environment. At the same time, we would need to know whether the supply was single or three phase, the type of earthing arrangements and so on. Here then is our starting point, and it is referred to in the IET Regulations, chapter 3, as 'Assessment of general characteristics'.

Having ascertained all the necessary details, we can decide on an installation method, the type of cable, and how we will protect against electric shock and overcurrents. We would now be ready to begin the calculation part of the design procedure.

Basically there are eight stages in such a procedure. These are the same whatever the type of installation, be it a cooker circuit or a sub-main cable feeding a distribution board in a factory. Here then are the eight basic steps in a simplified form:

1. Determine the design current I_b.
2. Select the rating of the protection I_n.
3. Select the relevant rating factors.
4. Divide I_n by the relevant rating factors (CFs) to give tabulated cable current-carrying capacity I_t.
5. Choose a cable size to suit I_t.
6. Check the voltage drop.
7. Check for shock risk constraints.
8. Check for thermal constraints.

Let us now examine each stage in detail.

Design current

In many instances the design current I_b is quoted by the manufacturer, but there are times when it has to be calculated. In that case there are two formulae involved, one for single phase and one for three phase:

Single phase:

$$I_b = \frac{P}{V} \quad (V \text{ usually } 230\,\text{V})$$

Three phase:

$$I_b = \frac{P}{\sqrt{3} \times V_L} \quad (V_L \text{ usually } 400\,\text{V})$$

Current is in amperes, and power P in watts.

If an item of equipment has a power factor (PF) and/or has moving parts, efficiency (eff) will have to be taken into account. Hence:

Single phase:

$$I_b = \frac{P \times 100}{V \times PF \times eff}$$

Three phase:

$$I_b = \frac{P \times 100}{\sqrt{3} \times V_L \times PF \times eff}$$

Nominal setting of protection

Having determined I_b we must now select the nominal setting of the protection I_n such that $I_n \geq I_b$. This value may be taken from IET Regulations, Tables 41.2, 41.3 or 41.4, or from manufacturers' charts. The choice of fuse or circuit breaker (cb) type is also important and may have to be changed if cable sizes or loop impedances are too high.

Rating factors

When a cable carries its full-load current it can become warm. This is not a problem unless its temperature rises further due to other influences, in which case overheating could damage the insulation. These other influences are: high ambient temperature; cables grouped together closely; uncleared overcurrents; and contact with thermal insulation.

For each of these conditions there is a rating factor which will respectively be called C_a, C_g, C_f and C_i, respectively, and which derates cable current-carrying capacity or conversely increases cable size (IET Regulations Table 52.2).

Ambient temperature C_a

The cable ratings in the IET Regulations are based on an ambient temperature of 30°C, and hence it is only above this temperature that an adverse correction is needed. Table 4B1 of the Regulations gives factors for all types of protection.

Grouping C_g

When cables are grouped together they impart heat to each other. Therefore the more cables there are, the more heat they will generate, thus increasing the temperature of each cable. Table 4C1 of the Regulations gives factors for such groups of cables or circuits. It should be noted that the figures given are for cables of the same size, and hence **no** correction need to be made for cables grouped at the outlet of a domestic consumer unit, for example where there is a mixture of different sizes.

A typical situation where rating factors need to be applied would be in the calculation of cable sizes for a lighting system in a large factory. Here many cables of the same size and loading may be grouped together in trunking and could be expected to be fully loaded all at the same time.

Protection by BS 3036 fuse C_f

As we have already discussed in Chapter 17, because of the high fusing factor of BS 3036 fuses, the rating of the fuse I_n should be less than or equal to $0.725 I_z$. Hence 0.725 is the rating factor which is to be used when BS 3036 fuses are used.

Thermal insulation C_i

With the modern trend towards energy saving and the installation of thermal insulation, there may be a need to derate cables to account for heat retention.

The values of cable current-carrying capacity given in appendix 4 of the IET Regulations have been adjusted for situations when thermal insulation touches one side of a cable. However, if a cable is totally surrounded by thermal insulation for more than 0.5 m, a factor of 0.5 must be applied to the tabulated clipped direct ratings. For less than 0.5 m, derating factors (Table 52.2) should be applied.

Application of rating factors

Some or all of the onerous conditions just outlined may affect a cable along its whole length or parts of it, but not all may affect it at the same time. So, consider the following:

1. If the cable in Fig. 18.1 ran for the whole of its length, grouped with others of the same size in a high ambient temperature, and was totally surrounded with thermal insulation, it would seem logical to apply all the CFs, as they all affect the whole cable run. Certainly the factors for the BS 3036 fuse, grouping and thermal insulation should be used. However, it is doubtful if the ambient temperature will have any effect on the cable, as the thermal insulation, if it is efficient, will prevent heat from reaching the cable. Hence apply C_g, C_f and C_i.

2. In Fig. 18.2a the cable first runs grouped, then leaves the group and runs in high ambient temperature, and finally is enclosed in thermal insulation. We therefore have three different conditions, each affecting the cable in different areas. The BS 3036 fuse affects the whole cable run and therefore C_f must be used. However there is no need to apply all of the remaining factors as the worse one will automatically compensate for the others. The relevant factors are shown in Fig. 18.2b: apply only $C_f = 0.725$ and $C_i = 0.5$. If protection was **not** by BS 3036 fuse, then apply only $C_i = 0.5$.

3. In Figs 18.3 and 18.4, a combination of cases 1 and 2 is considered. The effect of grouping and ambient temperature is $0.7 \times 0.97 = 0.69$. The factor for thermal insulation is still worse than this combination, and therefore C_i is the only one to be used.

Figure 18.1

Figure 18.2

18

Figure 18.3

Figure 18.4

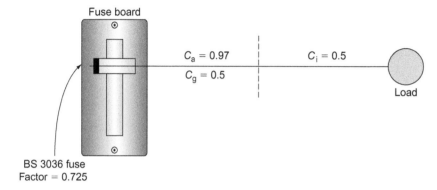

Figure 18.5

Having chosen the **relevant** rating factors, we now apply them to the nominal rating of the protection I_n as divisors in order to calculate the tabulated current-carrying capacity I_t of the cable.

Current-carrying capacity

The required formula for tabulated current-carrying capacity I_t is

$$I_t \geq \frac{I_n}{\text{relevant CFs}}$$

In Fig. 18.4, the current-carrying capacity is given by

$$I_t \geq \frac{I_n}{C_f C_i} \geq \frac{30}{0.725 \times 0.5} \geq 82.75\,\text{A}$$

or, without the BS 3036 fuse,

$$I_t \geq \frac{30}{0.5} \geq 60\,\text{A}$$

In Fig. 18.5, $C_a C_g = 0.97 \times 0.5 = 0.485$, which is worse than C_i (0.5). Hence

$$I_t \geq \frac{I_n}{C_f C_a C_g} \geq \frac{30}{0.725 \times 0.485} \geq 85.3\,\text{A}$$

or, without the BS 3036 fuse,

$$I_t \geq \frac{30}{0.485} \geq 61.85\,\text{A}$$

Choice of cable size

Having established the tabulated current-carrying capacity I_t of the cable to be used, it is now essential to choose a cable to suit that value. The tables in appendix 4 of the IET Regulations list all the cable sizes, current-carrying capacities and voltage drops of the various types of cable. For example, in the case of single core 70°C thermoplastic insulated cables, which are single phase, in conduit and have a current-carrying capacity of 45 A, the installation is by reference method B (Table 4A2), the cable table is 4DIA and the column is 4. Hence the cable size is 10.0 mm² (column 1).

Voltage drop

The resistance of a conductor increases as the length increases and/or the cross-sectional area (c.s.a.) decreases. Associated with an increased resistance is a drop in voltage, which means that a load at the end

Figure 18.6 Voltage drop

of a long thin cable will not have the full supply voltage available (Fig. 18.6).

The IET Regulations require that the voltage drop V should not be so excessive that equipment does not function safely. They further indicate that the following percentages of the nominal voltage at the **origin** of the circuit will satisfy. This means that:

	LV lighting 3% (V)	LV power 5% (V)
230 V single phase	6.9	11.5
400 V three phase	12	20

For example, the voltage drop on a power circuit supplied from a 230 V source by a 16.0 mm² two core copper cable 23 m long, clipped direct and carrying a design current of 33 A, will be

$$V_c = \frac{mV \times I_n \times L}{1000} \quad \text{(mV from Table 4D2B)}$$

$$= \frac{28 \times 33 \times 23}{1000} = 2.125 \, V$$

As we know that the maximum voltage drop in this instance (230 V) is 11.5 V, we can determine the maximum length by transposing the formula:

$$\text{Maximum length} = \frac{V_c \times 1000}{mV \times I_b}$$

$$= \frac{11.5 \times 1000}{28 \times 23} = 178 \, m$$

There are other constraints, however, which may not permit such a length.

Shock risk

This topic has already been discussed in full in Chapter 16. To recap, however, the actual loop impedance Z_s should not exceed those values given in Tables 41.2, 41.3 and 41.4 of the IET Regulations. This ensures that circuits feeding final and distribution circuits will

be disconnected, in the event of an earth fault, in the required time.

Remember

$$Z_s = Z_e + R_1 + R_2$$

Thermal constraints

The IET Regulations require that we either select or check the size of a circuit protective conductor (cpc) against Table 54.7 of the Regulations, or calculate its size using an adiabatic equation.

Selection of cpc using table 54.7

Table 54.7 of the Regulations simply tells us that:

1. For line conductors up to and including 16 mm², the cpc should be at least the same size.
2. For sizes between 16 and 35 mm², the cpc should be at least 16 mm².
3. For sizes of line conductor over 35 mm², the cpc should be at least half this size.

This is all very well, but for large sizes of line conductor the cpc is also large and hence costly to supply and install. Also, composite cables such as the typical twin with cpc 6242Y type have cpc's smaller than the line conductor and hence do not comply with Table 54.7.

Calculation of cpc using an adiabatic equation

The adiabatic equation

$$s = \sqrt{I^2 t}/k$$

enables us to check on a selected size of cable, or on an actual size in a multicore cable. In order to apply the equation we need first to calculate the earth fault current from

$$I = U_0/Z_s$$

18

where U_0 is the nominal line voltage to earth (usually 230 V) and Z_s is the actual earth fault loop impedance. Next we select a k factor from Tables 54.2 to 54.7 of the Regulations, and then determine the disconnection time t from the relevant curve.

For those unfamiliar with such curves, using them may appear a daunting task. A brief explanation may help to dispel any fears. Referring to any of the curves for fuses in Appendix 3 of the IET Regulations, we can see that the current scale goes from 1 to 10 000 A, and the time scale from 0.01 to 10 000 s. One can

imagine the difficulty in drawing a scale between 1 and 10 000 A in divisions of 1 A, and so a logarithmic scale is used. This cramps the large scale into a small area. All the subdivisions between the major divisions increase in equal amounts depending on the major division boundaries; for example, all the subdivisions between 100 and 1000 are in amounts of 100 (Fig. 18.7).

Figures 18.8 and 18.9 give the IET Regulations time/current curves for BS 88 fuses. Referring to the appropriate curve for a 32 A fuse (Fig. 18.8), we find

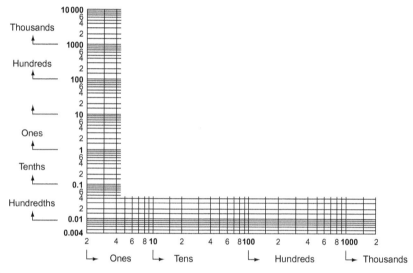

Figure 18.7 Log–Log scale graphs.

Figure 18.8 Time/current characteristics.

Figure 18.9 Time/current characteristics.

Figure 18.10 Determining current values.

that a fault current of 200 A will cause disconnection of the supply in 0.6 s.

Where a value falls between two subdivisions, for example 150 A, an estimate of its position must be made. Remember that even if the scale is not visible, it would be cramped at one end; so 150 A would not fall half-way between 100 and 200 A (Fig. 18.10).

It will be noted in Appendix 3 of the Regulations that each set of curves is accompanied by a table which indicates the current that causes operation of the protective device for disconnection times of 0.1, 0.4 and 5 s.

The IET Regulations curves for cb's to BS EN 60898 Type C and RCBO's are shown in Fig. 18.11.

Having found a disconnection time, we can now apply the formula.

Example of the use of an adiabatic equation

Suppose that in a design the protection was by 40 A BS 88 fuse; we had chosen a 4.0 mm² copper cpc running with our line conductor; and the loop impedance Z_s was 1.15 Ω. Would the chosen cpc size be large enough to withstand damage in the event of an earth fault? We have

$$I = U_0/Z_s = 230/1.15 = 200 \text{ A}$$

From the appropriate curve for the 40 A BS 88 fuse (Fig. 18.12), we obtain a disconnection time t of 2 s. From table 54.3 of the Regulations, $k = 115$. Therefore the minimum size of cpc is given by

$$s = \sqrt{I^2 t}/k = \sqrt{200^2 \times 2}/115 = 2.46 \text{ mm}^2$$

18

245

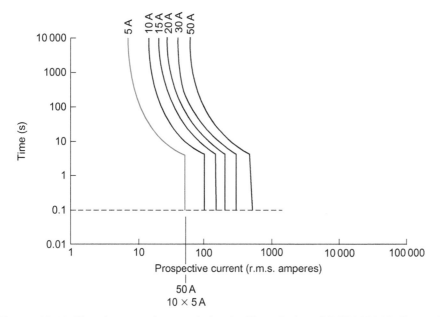

Figure 18.11 Time/current characteristics for Type C cb to BS EN 60898. Example for 50 A superimposed. For times less than 20 ms, the manufacturer should be consulted.

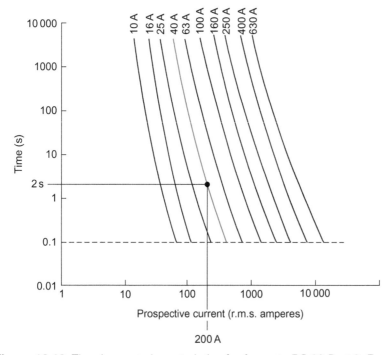

Figure 18.12 Time/current characteristics for fuses to BS 88 Part 2. Example for 40 A superimposed.

So our 4.0 mm² cpc is acceptable. Beware of thinking that the answer means that we could change the 4.0 mm². If we did, the loop impedance would be different and hence *I* and *t* would change; the answer for *s* would probably tell us to use a 4.0 mm².

In the example shown, '*s*' is merely a check on the actual size chosen.

Installation reference methods

Method 100

Touching insulation one side (Fig. 18.13).

Method 101

Touching insulation one side (Fig. 18.14).

Method 102

Touching inner wall surface (Fig. 18.15).

Method 103

Not touching inner wall surface (Fig. 18.16).

Method A

Single- or multi-core cables in conduit in an insulated wall (Fig. 18.17).

Method B

Enclosed in conduit or trunking on or in a wall (Fig. 18.18).

Method C

Single- or multi-core cable clipped direct or in masonry (Fig. 18.19).

Figure 18.13 Method 100.

Figure 18.14 Method 101.

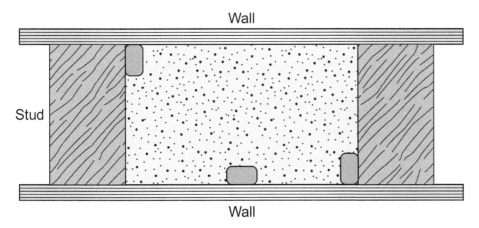

Figure 18.15 Method 102.

18

Figure 18.16 Method 103.

Figure 18.17 Method A.

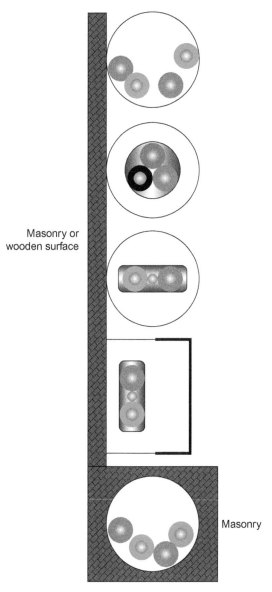

Figure 18.18 Method B.

Installation methods

Figures 18.13–18.19 illustrate some of the common methods of cable installation.

Having discussed each component of the design procedure, we can now put them all together to form a complete design.

Example of circuit design

A consumer lives in a bungalow with a detached garage and workshop, as shown in Fig. 18.20. The building method is traditional brick and timber.

Rating factors
  C_a – 0.87 or 0.94 if fuse is BS 3036
  C_g – not applicable

  C_f – 0.725 **only** if fuse is BS 3036
  C_i – 0.5 if cable is totally surrounded in thermal insulation.

The mains intake position is at high level and comprises an 80 A BS 1361 230 V main fuse, an 80 A rated meter and a six-way 80 A consumer unit housing BS EN 60898 Type B cb's which are as follows:

Ring circuit	32 A
Lighting circuit	6 A
Immersion heater circuit	16 A
Cooker circuit	32 A
Shower circuit	32 A
Spare way	–

Figure 18.19 Method C.

Figure 18.20

The cooker is rated at 30 A, with no socket in the cooker unit. The main tails are 16.0 mm² double-insulated PVC, with a 6.0 mm² earthing conductor. There is no main protective bonding. The earthing system is TN–S, with an external loop impedance Z of 0.3 Ω. The prospective short-circuit current (PSCC) at the origin has been measured as 800 A. The roof space is insulated to full depth of the ceiling joists and the temperature in the roof space is not expected to exceed over 35°C.

The consumer wishes to convert the workshop into a pottery room and install an 8.6 kW/230 V electric kiln. The design procedure is as follows.

Assessment of general characteristics

The present maximum demand, applying diversity, is

| Ring | 32 A |
| Lighting (66% of 6 A) | 3.96 A |

Immersion heater	16 A
Cooker (10 A + 30% of 20 A)	16 A
Shower	32 A
Total	**100 A**

Reference to the current rating tables in the IET Regulations will show that the existing main tails are too small and should be uprated. So, the addition of another 8.6 kW of load is not possible with the present arrangement.

The current taken by the kiln is 8600/230 = 37.4 A. Therefore, the new maximum demand is 100 + 37.4 = 137.4 A.

Supply details are:

Single-phase 230 V, 50 Hz earthing: TN–S PSCC at origin (measured): 800 A.

Decisions must now be made as to the type of cable, the installation method and the type of protective device. As the existing arrangement is not satisfactory, the supply authority must be informed of the new maximum demand, as a larger main fuse and service cable may be required.

Sizing the main tails

1. The new load on the existing consumer unit will be 137.4 A. From the IET Regulations, the cable size is 25.0 mm².

18

249

2. The earthing conductor size, from the IET Regulations, will be 16.0 mm². The main equipotential bonding conductor size, from the IET Regulations, will be 10.0 mm².

For a domestic installation such as this, a PVC flat twin-cable, clipped direct (avoiding any thermal insulation) through the loft space and the garage, etc. would be the most appropriate.

Sizing the kiln circuit cable

Design current I_b

$$I_b = \frac{P}{V} = \frac{8600}{230} = 37.4\,A$$

Rating and type of protection I_n

As we have seen, the requirement for the rating I_n is that $I_n \geq I_b$. Therefore, using the tables in the IET Regulations, I_n will be 40 A.

Rating factors

C_a – 0.94
C_g – not applicable
C_f – 0.725 **only** if the fuse is BS 3036 (not applicable here)
C_i – 0.5 if the cable is totally surrounded in thermal insulation (not applicable here)

Tabulated current-carrying capacity of cable

$$I_t = \frac{I_n}{CF} = \frac{40}{0.94} = 42.5\,A$$

Cable size based on tabulated current-carrying capacity

Table 4D5A IET Regulations give a size of 6.0 mm² for this value of I_t (method C).

Check on voltage drop

The actual voltage drop is given by

$$\frac{mV \times I_b \times 1}{1000} = \frac{7.3 \times 37.4 \times 24.5}{1000}$$
$$= 6.7\,V \text{ (well below the maximum of 11.5 V)}$$

This voltage drop, whilst not causing the kiln to work unsafely, may mean inefficiency, and it is perhaps better to use a 10.0 mm² cable.

For a 10.0 mm² cable, the voltage drop is checked as

$$\frac{4.4 \times 37.4 \times 24.5}{1000} = 4.04\,V$$

Shock risk

The cpc associated with a 10.0 mm² twin 6242Y cable is 4.0 mm². Hence, the total loop impedance will be

$$Z_s = Z_c + \frac{\{(R_1 + R_2) \times L \times 1.2\}}{1000}$$
$$= 0.3 + \frac{(6.44 \times 24.5 \times 1.2)}{1000} = 0.489\,\Omega$$

> **Note**
>
> 6.44 is the tabulated $(R_1 + R_2)$ value and the multiplier 1.2 takes account of the conductor resistance at its operating temperature.
>
> It is likely that the chosen cb will be a Type B.

Thermal constraints

We still need to check that the 4.0 mm² cpc is large enough to withstand damage under earth fault conditions. So, the fault current would be

$$I = \frac{U_0}{Z_s} = \frac{230}{0.489} = 470\,A$$

The disconnection time t for this current for this type of protection (from the relevant curve in the IET Regulations) is as follows.

40 A cb Type B = 0.1 s (the actual time is less than this but 0.1 s is the instantaneous time)

From the regulations, the factor for $k = 115$. We can now apply the adiabatic equation

$$S = \frac{\sqrt{I^2 \times t}}{k} = \frac{\sqrt{470^2 \times 0.1}}{115} = 1.29\,mm^2$$

Hence, our 4.0 mm² cpc is of adequate size.

Summarizing:

1. The kiln circuit would be protected by a 40 A BS EN 60898 Type B cb and supplied from a spare way in the consumer unit using the main fuse which would then need to be uprated to 100 A.
2. The main tails would be changed to 25.0 mm².
3. The earthing conductor would be changed to 16.0 mm².
4. Main protective bonding conductors would need to be installed 10.0 mm² twin with earth PVC cable.

CHAPTER 19

Testing

Measurement of electrical quantities

Instruments

Instruments play an important role in installation work enabling the measurement of the current, voltage, resistance, power and power factor.

The basic ammeter and voltmeter work on either the moving-iron or the moving-coil principle, whereas the modern digital instrument uses complex electronics.

The moving-iron instrument (repulsion type)

Figure 19.1 illustrates a moving-iron instrument of the repulsion type.

It comprises a coil, with a fixed iron, a pointer with an iron vane attached (moving iron) and a sampling device inside it.

When a supply, either a.c. or d.c., is applied to the coil, both fixed and moving irons are magnetized to the same polarity and will therefore repel each other.

The design of the irons ensures that the repulsion is always in the same direction.

The damper ensures a slow and even movement of the pointer. It consists of a cylinder closed at one end and a light piston inside it. The pointer, which is attached to the piston, is slowed down by the air pressure that builds up in the cylinder, resisting the movement of the piston. A spring returns the pointer to zero when the current supply is removed.

As the amount of movement depends on the square of the supply current, a small current produces a small movement and a large current produces a larger movement. Hence the scale tends to be cramped at the lower values of the current.

The moving-coil instrument

Moving-coil instrument works on the motor principle of a current-carrying coil in a magnetic field. Figure 19.2a and b shows two variations of this type of instrument.

The basic components of both the systems are a magnetic field, a core or shell of soft iron and a coil wound on an aluminium former. The connection to the coil is made via the control springs.

Damping is achieved by eddy currents in the aluminium coil former. These currents cause small magnetic fields to flow, which in turn interact with the main field and slow down the movement of the coil.

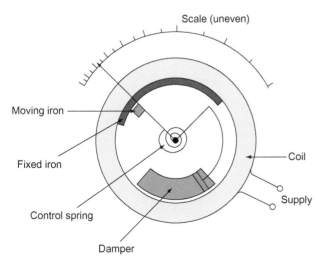

Figure 19.1 Moving iron instrument.

Electrical Installation Work. 978-1-138-84927-3 © Brian Scaddan.
Published by Taylor & Francis. All rights reserved.

The digital instrument

This type of instrument has generally replaced the analogue types discussed in the previous section, but the theory of operation is too complex to deal with in this chapter and hence only basic details will be considered.

A digital instrument is basically an electronic voltmeter with four sections:

1. The power supply and reference generators.
2. The signal-conditioning circuitry (current, resistance, voltage, etc.).
3. The analogue-to-digital (A/D) converter.
4. The count/display module.

Figure 19.3 shows a block diagram of a digital multimeter.

A comparison between the different types of instruments is shown in Table 19.1.

Measurement of current

It is often necessary to extend the range of an ammeter to read values of current higher than the instrument's movement is designed for, and for this purpose, shunts or current transformers (CTs) are used.

Ammeter shunts

As shown in Fig. 19.4, a shunt is simply a low-value resistor connected in parallel with the instrument.

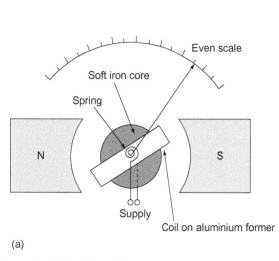

(a)

(b)

Figure 19.2 Moving-coil instruments.

Figure 19.3 Block diagram of digital instrument.

Table 19.1 Comparison of instrument types

	Advantages	Disadvantages
Moving iron	Cheap, strong, can be used on a.c. and d.c.	Uneven scale, affected by heat and stray magnetic fields
Moving coil	Even scale, very accurate, unaffected by stray magnetic fields	Fragile, expensive, can only be used on d.c.
Digital	Robust, no moving parts, accurate, easy reading	Requires regular battery changes

Figure 19.4 Ammeter shunt.

Example

A moving-coil ammeter gives full-scale deflection (f.s.d.) at 15 mA. If the instrument resistance is $5\,\Omega$, calculate the value of shunt required to enable the instrument to read currents up to 3 A.

For f.s.d., Fig. 19.5 gives:

$$\text{Potential difference (p.d.) across meter} = I_A R_A$$
$$= 15 \times 10^{-3} \times 5$$
$$= 75\,\text{mV}$$
$$\therefore \text{p.d. across shunt} = 75\,\text{mV}$$
$$\text{Shunt current} = I - I_A$$
$$= 3 - (15 \times 10^{-3})$$
$$= 3 - 0.015$$
$$= 2.985\,\text{A}$$
$$\therefore \text{Shunt resistance} = \frac{V_s}{I_s}$$
$$= \frac{75 \times 10^{-3}}{2.985}$$
$$= 0.025\,\Omega$$

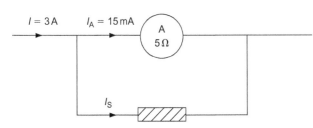

Figure 19.5 Ammeter shunt.

The shunt may be used in conjunction with either a.c. or d.c. instruments.

For measuring high a.c. currents, however, a CT is used.

Current transformer

CTs are usually of the wound or bar type transformers as shown in Fig. 19.6. As in any transformer, the secondary current will depend on the transformer ratio, that is,

$$\frac{I_p}{I_s} = \frac{N_s}{N_p}$$

Example

An ammeter capable of taking 2.5 A is to be used in conjunction with a CT to measure a bus-bar current of up to 2000 A. Calculate the number of turns on the transformer as follows:

$$\frac{I_p}{I_s} = \frac{N_s}{N_p}$$
$$\frac{2000}{2.5} = \frac{N_s}{1}$$
$$\therefore N_s = \frac{2000}{2.5}$$
$$= 800\,\text{turns}$$

Great care must be taken when using CTs, as high voltages normally associated with high currents will be stepped up on the secondary side, creating a potentially dangerous situation.

Before removing an ammeter or load (burden) from a CT, the secondary terminals must be shorted out.

19

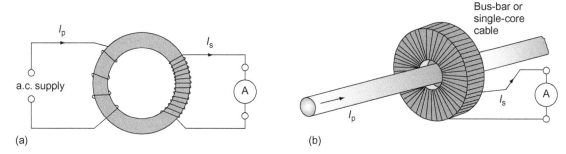

Figure 19.6 CTs: (a) wound-type primary; (b) bar-type primary.

Measurement of voltage

As with current measurement, moving-iron and moving-coil instruments are also used while measuring the voltage.

The extension of the range of a voltmeter is achieved by using either a multiplier or a voltage transformer (VT) in case of high a.c. voltages.

Voltmeter multiplier

A voltmeter multiplier is simply a resistance in series with the instrument as shown in Fig 19.7.

Figure 19.7 Voltmeter multiplier.

Example

A moving-coil instrument of resistance $5\,\Omega$ and f.s.d. at $20\,mA$ is to be used to measure voltages up to $100\,V$. Calculate the value of the series multiplier required.

$$\text{Instrument voltage at f.s.d.} = I_V \times R_V$$
$$= 20 \times 10^{-3} \times 5$$
$$= 0.1\,V$$
$$\therefore \text{Voltage to be dropped across multiplier} = V - V_V$$
$$= 100 - 0.1$$
$$= 99.9\,V$$
$$\therefore \text{Value of resistance} = \frac{V_m}{I_m}$$
$$= \frac{99.9}{20 \times 10^{-3}}$$
$$= 4995\,\Omega$$

Voltage transformer (VT)

A VT is simply a typical double-wound step-down transformer with many turns on the primary and a few on the secondary.

Instruments in general

Multimeters

There are many types of multimeter now available, the more expensive ones usually rendering greater accuracy. They all work on the moving-coil principle and use rectifiers for the d.c. ranges. Shunts, multipliers, VTs and CTs are switched in or out when ranges and scales are changed by the operator.

Wattmeters

A wattmeter is simply a combination of an ammeter and a voltmeter in a single instrument, usually a dynamometer.

Figure 19.9 shows how high-voltage connections can be made to a wattmeter.

Tong tester

The tong tester or clip-on ammeter is a variation of the bar primary CT. It consists of an insulated iron core in two parts that can be separated like tongs, on one end of which the secondary winding and an ammeter are placed.

The iron core is clipped round a bus-bar or single core cable and the current is registered on the ammeter (Fig. 19.10).

Phase-rotation indicator

The phase-rotation indicator is a simple three-phase induction motor. When connected to a three-phase supply, a disc that is connected to the motor rotates in the direction of the supply sequence. It is also used when two three-phase systems are to be paralleled together (Fig. 19.11).

Figure 19.8 Wattmeter connection.

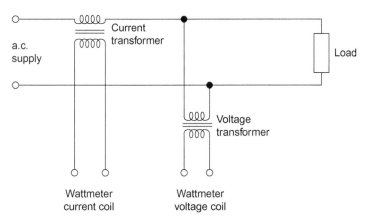

Figure 19.9 Use of CT and VT.

Figure 19.10 Clip-on ammeter.

Figure 19.11 Phase-rotation indicator.

the more commonly used tests prescribed by the IET Wiring Regulations and assign typical values to them, we can at least provide a basis for the choice of the most suitable instruments. It is clear from Table 19.2 that all that is required is an ohmmeter of one sort or another, a residual current device (RCD) tester and an instrument for measuring prospective short-circuit current.

It is clearly the most sensible thing to purchase instruments from one of the established manufacturers rather than to attempt to save money by buying cheap, less known brands. Also, since the instruments used in the world of electrical installation

Selection of test instruments

Let us now look at the range of quantities that an electrician is likely to encounter in the normal practice of his or her profession. If we take the sequence of

255

Table 19.2 Installation test instruments

Test	Range	Type of Instrument
1. Continuity of protective conductors	2–0.005 Ω or less	Low-reading ohmmeter
2. Continuity of ring final conductors	0.05–0.08 Ω	Low-reading ohmmeter
3. Insulation resistance	Infinity to <1 MΩ	High-reading ohmmeter
4. Polarity	None	Ohmmeter, bell, etc.
5. Earth fault loop impedance	0–2000 Ω	Special ohmmeter
6. Earth electrode resistance	Any value over about 3 or 4 Ω	Special ohmmeter
7. Operation of RCD	5–500 mA	Special instrument
8. Prospective short-circuit current	2–20 kA	Special instrument

Figure 19.12 Approved test lamp.

are bound to be subjected to harsh treatment, a robust construction is most important.

Many of the well-known instrument companies provide a dual facility in a single instrument, for example prospective fault current (PFC) and loop impedance together, or insulation resistance and continuity. Hence it is likely that only three or four instruments would be required, together with an approved test lamp.

Approved test lamps and indicators

If you search your toolboxes, you will have little difficulty in finding your 'neon screwdriver' or 'testascope'. The next thing you need to do is to locate a very deep pond and drop it in.

Can you actually imagine allowing electric current at low voltage (LV) (50–1000 V a.c.) to pass through one's body in order to activate a test lamp? It only takes around 10–15 mA to cause a severe electric shock, and about 50 mA (1/20th of an ampere) to kill.

Apart from the fact that such a device will register any voltage from about 5 V upwards, the safety of the user depends entirely on the integrity of the current-limiting resistor in the unit. In one of the instances of such usage, an electrician received a considerable shock while using such an instrument after his apprentice had dropped it in a sink of water, simply wiped it dry and replaced it in the tool box. The water had seeped into the device and shorted out the resistor.

An approved test lamp should be of similar construction to that shown in Fig. 19.12.

Accidental rcd operation

Using a test lamp to probe between line and earth for indication of a live supply on the line terminal has long been in practice. However, this can present a problem where RCDs exist in the circuit, as the test applies a deliberate line to earth fault.

Some test lamps have LED indicators, and the internal circuitry of such test lamps limits the current to the earth to a level below which the RCD can operate. The same limiting effect applies to multimeters. However, it is always advisable to check that the testing device does not have any effect on RCDs.

Calibration, zeroing and care of instruments

Precise calibration of instruments is usually outside the province of the electricians and would normally be carried out by the manufacturer or a local service representative. A check, however, can be made by the user to determine whether calibration is necessary by comparing readings with an instrument known to be accurate, or by measurement of known values of voltage, resistance, etc.

Sometimes the readings can be incorrect simply because either the instrument is not zeroed before use, or because the internal battery needs to be replaced. Most modern instruments have battery condition indication, and of course this should never be ignored.

Always turn off the selection switches after testing. A lot of instrument fuses are blown when, for example, a multimeter is inadvertently left on the ohms range and then used to check for mains voltage.

The following set procedure may seem rather basic but should ensure trouble-free testing:

1. Check test leads for obvious defects.
2. Zero the instrument.
3. Select the correct range for the values anticipated. If in doubt, choose the highest range and gradually drop down.
4. Make a record of the test results, if necessary.
5. When a zero reading is expected and occurs (or, in the case of insulation resistance, an infinite reading), make a quick check on the test leads just to ensure that they are not open-circuited.
6. Return switches or selectors to the 'off' position.
7. Replace instruments and leads inside carrying case.

Initial inspection

Inspection and testing

Circumstances that require an initial verification

New installations or Additions or Alterations are the circumstances that require an initial verification.

General reasons for initial verification

1. To ensure that the equipment and accessories are to a relevant standard.
2. To prove compliance with BS 7671.
3. To ensure that the installation is not damaged, to impair safety.

Information required

The required information is the assessment of general characteristics as mentioned in sections 311, 312 and 313 of the IET Regulations 514.9.1 together with other information such as drawings, charts, etc.

Documentation required and to be completed

The document that is required as part of the verification process is the Electrical Installation Certificate signed or authenticated for the design and construction (could be the same person), and then for the inspection and test. A schedule of test results and an inspection schedule must accompany an Electrical Installation Certificate.

Sequence of tests

1. Continuity of all protective conductors.
2. Continuity of ring final circuit conductors.
3. Insulation resistance.
4. Protection by SELV, PELV or Electrical separation.
5. Protection by barriers and enclosures provided during erection.
6. Insulation of non-conducting floors and walls.
7. Polarity.
8. Earth electrode resistance.
9. Earth fault loop impedance.
10. Operation of RCDs.
11. Prospective fault current.
12. Phase sequence.
13. Functional testing.
14. Voltage drop.

Before any testing is carried out, a detailed physical inspection must be made to ensure that all equipments are to a relevant British or Harmonized European Standard, they are erected or installed in compliance with the IET Regulations, and that they are not damaged in such a way that it could cause danger. In order to comply with these requirements, the IET Regulations give a checklist (schedule of inspections) of items that, where relevant, should be inspected.

However, before such an inspection or a test is carried out, certain information **must** be available to the verifier. This information is the result of the Assessment of General Characteristics required by IET Regulations Part 3, Sections 311, 312 and 313, and drawings, charts and similar information related to the installation. It is at this point that most readers who work in the real world of electrical installation will be lying on the floor laughing hysterically.

Let us assume that the designer and installer of the installation are competent professionals, and all of the required documentation is available.

It is interesting to note that one of the items on the checklist **is** the presence of diagrams, instructions and similar information. If these are missing then there is a deviations from the IET Regulations.

Another item on the list is the verification of conductors for current-carrying capacity and voltage drop in accordance with the design. How on earth can this be verified without all the information? A 32 A Type B circuit breaker (cb) protecting a length of 4 mm^2 conductor may look reasonable, but there is no guarantee for it being completely correct. Let us look then at the general content of the checklist.

19

1.	Connection of conductors	Are terminations electrically and mechanically sound and has the insulation and sheathing been removed only to a minimum to allow satisfactory termination?
2.	Identification of conductors	Are conductors correctly identified in accordance with the IET Regulations?
3.	Routing of cables	Are cables installed in such a way that account is taken of external influences such as mechanical damage, corrosion, heat, etc.?
4.	Conductor selection	Are conductors selected for current-carrying capacity and voltage drop in accordance with the design?
5.	Connection of single pole devices	Are single pole protective and switching devices connected in the line conductor only?
6.	Accessories and equipment	Are all accessories and equipments correctly connected?
7.	Thermal effects	Are fire barriers present where required and protection against thermal effects provided?
8.	Protection against shock	What methods have been used to provide protection against electric shock?
9.	Mutual detrimental influence	Are wiring systems installed such that they can have no harmful effect on nonelectrical systems, and whether the systems of different currents or voltages are segregated where necessary?
10.	Isolation and switching	Are there correctly located and installed appropriate devices for isolation and switching?
11.	Undervoltage	Are there protective devices present at places where undervoltage might be a cause of concern?
12.	Protective devices	Are protective and monitoring devices correctly chosen and set to ensure fault protection and/or overcurrent?
13.	Labelling	Are all protective devices, switches (where necessary) and terminals correctly labelled?
14.	External influences	Have all items of equipment and protective measures been selected in accordance with the appropriate external influences?
15.	Access	Are all means of access to switchgear and equipment adequate?
16.	Notices and signs	Are danger notices and warning signs present?
17.	Diagrams	Are diagrams, instructions and similar information relating to the installation available?
18.	Erection methods	Have all wiring systems, accessories and equipment been selected and installed in accordance with the requirements of the IET Regulations, and are fixings for equipment adequate for the environment?

Therefore, we have now inspected all relevant items, and provided that there are no defects that may lead to a dangerous situation while testing, we can now start the actual testing procedure.

Testing continuity of protective conductors

All protective conductors, including main and supplementary protective bonding conductors must be tested for continuity using a low-resistance ohmmeter.

For main equipotential bonding there is no single fixed value of resistance above which the conductor would be deemed unsuitable. Each measured value, if indeed it is measurable for very short lengths, should be compared with the relevant value for the length and size of a particular conductor. Such values are shown in Table 19.3.

Table 19.3 Resistance (Ω) of copper conductors at 20°C

CSA (mm²)	Length (m)									
	5	10	15	20	25	30	35	40	45	50
1	0.09	0.18	0.27	0.36	0.45	0.54	0.63	0.72	0.82	0.9
1.5	0.06	0.12	0.18	0.24	0.3	0.36	0.43	0.48	0.55	0.6
2.5	0.04	0.07	0.11	0.15	0.19	0.22	0.26	0.30	0.33	0.37
4	0.023	0.05	0.07	0.09	0.12	0.14	0.16	0.18	0.21	0.23
6	0.02	0.03	0.05	0.06	0.08	0.09	0.11	0.13	0.14	0.16
10	0.01	0.02	0.03	0.04	0.05	0.06	0.063	0.07	0.08	0.09
16	0.006	0.01	0.02	0.023	0.03	0.034	0.04	0.05	0.05	0.06
25	0.004	0.007	0.01	0.015	0.02	0.022	0.026	0.03	0.033	0.04
35	0.003	0.005	0.008	0.01	0.013	0.016	0.019	0.02	0.024	0.03

Where a supplementary protective bonding conductor has been installed between **simultaneously accessible** exposed and extraneous conductive parts, because circuit disconnection times cannot be met, then the resistance (R) of the conductor must be equal to or less than $50/I_a$.

Hence, $R \leq 50/I_a$ where 50 is the voltage above which exposed metalwork should not rise, and I_a is the minimum current causing operation of the circuit protective device within 5 s.

For example, suppose a 45 A BS 3036 fuse protects a cooker circuit, the disconnection time for the circuit cannot be met, and hence a supplementary bonding conductor has been installed between the cooker case and an adjacent central heating radiator. The R of that conductor should not be greater than $50/I_a$, and I_a in this case is 145 A (see fig. 3.2B of the IET Regulations), that is,

$$50/145 = 0.34 \ \Omega$$

Then the question remains: is how do we conduct a test to establish continuity of main or supplementary bonding conductors? Quite simply we can do that by just connecting the leads from a low-resistance ohmmeter to the ends of the bonding conductor (Fig. 19.13). One end should be disconnected from its bonding clamp, or else the measurements may include the resistance of parallel paths of other earthed metalwork. Also remember to zero the instrument first or, if this facility is not available, record the resistance of the test leads so that this value can be subtracted from the test reading later on.

The continuity of circuit protective conductors (cpc's) may be established in the same way, but a second method is preferred, as the results of this second test indicate the value of ($R_1 + R_2$) for the circuit in question.

The test is conducted in the following manner:

1. Temporarily link together the line conductor and cpc of the circuit concerned in the distribution board (DB) or consumer unit.
2. Test between line and cpc at **each** outlet in the circuit. A reading indicates continuity.
3. Record the test result obtained at the furthest point in the circuit. This value is ($R_1 + R_2$) for the circuit.

Figure 19.14 illustrates the aforementioned method of conducting a test.

There may be some difficulty in determining the ($R_1 + R_2$) values of circuits in installations that comprise steel conduit and trunking, and/or steel wire armouring (s.w.a.) and mineral-insulated metal-sheath (m.i.m.s.) cables because of the parallel earth paths that are likely to exist. In these cases, continuity tests may have to be carried out at the installation stage before accessories are connected or terminations made off, as well as after completion.

Figure 19.13

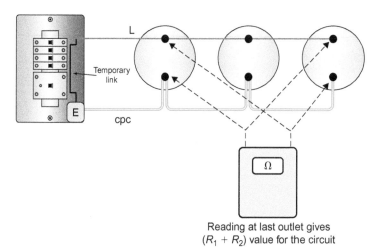

Figure 19.14

Although it is no longer considered a good working practice to use steel conduit or trunking as a protective conductor, it is permitted; and hence its continuity must be proved. The enclosure must be inspected along its length to ensure that it is sound and then the standard low-resistance test is performed.

Testing continuity of ring final circuit conductors

There are two main reasons for conducting this test:

1. To establish that interconnections in the ring do not exist.
2. To ensure that the cpc is continuous, and indicate the value of $(R_1 + R_2)$ for the ring.

What are interconnections in a ring circuit, and why is it important to locate them? Figure 19.15 shows a ring final circuit with an interconnection.

The most likely cause of the situation shown in Fig. 19.15 is where a DIY enthusiast has added sockets P, Q, R and S to an existing ring A, B, C, D, E and F.

Initially there is nothing wrong with this. The problem arises if a break occurs at, say, point Y, or the terminations fail in socket C or P. Then there would be four sockets all fed from the point X, which would then become a spur.

So, how do we identify such a situation with or without breaks at point 'Y'? A simple resistance test between the ends of the line, neutral or cpc's will only indicate that a circuit exists, and whether there are interconnections or not. The following test method is based on the philosophy that the resistance measured across any diameter of a perfect circle of conductor will always be of the same value (Fig. 19.16).

The perfect circle of conductor is achieved by cross-connecting the line and neutral legs of the ring (Fig. 19.17).

The test procedure is as follows:

1. Identify the opposite legs of the ring. This is quite easy with sheathed cables, but with singles, each conductor will have to be identified, probably by

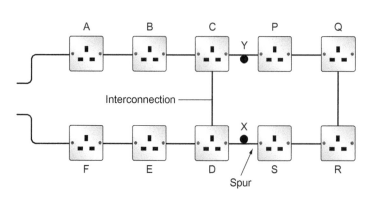

Figure 19.15 Ring circuit with interconnection.

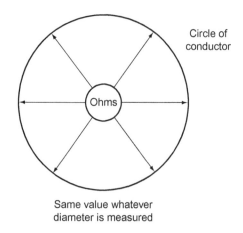

Figure 19.16 Perfect circle of conductor.

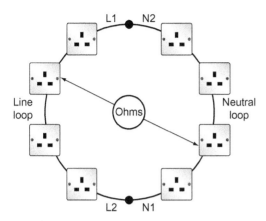

Figure 19.17 Perfect circle formed by ring circuit.

taking resistance measurements between each one and the closest socket outlet. This will give three high readings and three low readings thus establishing the opposite legs.

2. Take a resistance measurement between the ends of each conductor loop. Record this value.
3. Cross-connect the opposite ends of the line and neutral loops (Fig. 19.18).

Figure 19.18 L-N cross connection.

4. Measure between line and neutral terminals at each socket on the ring. The readings obtained should be, for a perfect ring, substantially the same. If an interconnection existed as shown in Fig. 19.15, then sockets A to F would all have similar readings, and those beyond the interconnection would have gradually increasing values to approximately the midpoint of the ring, then decreasing values back towards the interconnection. If a break had occurred at point Y then the readings from socket S would increase to a maximum at socket P. One or two high readings are likely to indicate either loose connections or spurs. A null reading, that is, an open-circuit indication is probably a reverse polarity, either line–cpc or neutral–cpc reversal. These faults would

clearly be rectified and the test at the suspect socket(s) repeated.

5. Repeat the above procedure, but in this case cross-connect the line and cpc loops (Fig. 19.19).

Figure 19.19 L-cpc cross connection.

In this instance, if the cable is of the flat twin type, the readings at each socket will increase very slightly and then decrease around the ring. This difference, caused by the difference in the size of the line and cpc, will not be significant enough to cause any concern. The measured value is very important and is $R_1 + R_2$ for the ring.

As before, loose connections, spurs and, in this case, L–N cross polarity, will be picked up.

The details in Table 19.4 are typical approximate ohmic values for a healthy 70 m ring final circuit wired in 2.5/1.5 flat twins and cpc cable. In this case the cpc will be approximately 1.67 × the L or N resistance.

As mentioned earlier, the null readings may indicate a reverse polarity. They could also indicate twisted conductors not in their terminal housing. The examples shown in Fig. 19.20 explain these situations.

Table 19.4 Typical ring circuit values

Initial measurements	L1–L2	N1–N2	cpc1–cpc2
	0.52	0.52	0.86
Reading at each socket	0.26	0.26	0.32–0.34
For spurs, each metre in length will add the following resistance to the above values	0.015	0.015	0.02

Testing insulation resistance

This is probably the most used and yet the most abused test among all. Affectionately known as 'meggering', an **insulation resistance test** is performed to ensure that the insulation of conductors, accessories and equipment is in a healthy condition,

Socket	L–N Reading	L–cpc Reading
A	OK	No reading
B	No reading	OK
C	No reading	No reading

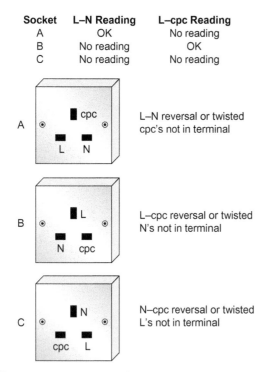

A — L–N reversal or twisted cpc's not in terminal

B — L–cpc reversal or twisted N's not in terminal

C — N–cpc reversal or twisted L's not in terminal

Figure 19.20 Possible faults.

Figure 19.21 Insulation resistance.

and will prevent dangerous leakage currents between conductors, as well as between conductors and earth. It also indicates whether any short circuits exist.

Insulation resistance is the resistance measured between conductors and is made up of countless millions of resistances in parallel (Fig. 19.21).

The more resistances there are in parallel, the **lower** the overall resistance would be, and as a consequence, the longer a cable is, the lower the insulation resistance is. In addition to the fact that almost all installation circuits are also wired in parallel, it becomes apparent that tests on large installations may give, if measured as a whole, pessimistically low values, even if there are no faults.

Under these circumstances, it is usual to break down such large installations into smaller sections, such as floor by floor, distribution circuit by distribution circuit, etc. This also helps to minimize disruption in the case of periodic testing. The test procedure is as follows:

1. Disconnect all items of equipment such as capacitors and indicator lamps as these are likely to give misleading results. Remove any items of equipment likely to be damaged by the test, such as dimmer switches, electronic timers, etc. Remove all lamps and accessories and disconnect fluorescent and discharge fittings. Ensure that the installation is disconnected from the supply, all fuses are in place, and cb's and switches are

in the 'on' position. In some instances it may be impracticable to remove lamps, etc. and in this case the local switch controlling such equipment may be left in the 'off' position.

2. Join together all live conductors of the supply and test between this joint and earth. Alternatively, test between each live conductor and earth in turn.

3. Test between line and neutral. For three-phase systems, join together all lines and test between this joint and the neutral. Then test between each of the lines. Alternatively, test between each of the live conductors in turn. Installations incorporating two-way lighting systems should be tested twice with the two-way switches in alternative positions.

Table 19.5 gives the test voltages and minimum values of insulation resistance for extra-low voltage (ELV) and LV systems.

Table 19.5 Minimum values of insulation resistance

System	Test Voltage (d.c.)	Minimum Insulation Resistance (MΩ)
SELV and PELV	250V	0.5
LV up to 500V	500V	1.0
Over 500V	1000V	1.0

If a value of less than 2 MΩ is recorded it may indicate a situation where a fault is developing, but as yet still complies with the minimum permissible value. In this case each circuit should be tested separately in order to locate the problem.

Example

An installation comprising six circuits has individual insulation resistances of 2.5, 8, 200, 200, 200 and 200 MΩ; and therefore the total insulation resistance will be

$$\frac{1}{R_t} = \frac{1}{2.5} + \frac{1}{8} + \frac{1}{200} + \frac{1}{200} + \frac{1}{200} + \frac{1}{200}$$
$$= 0.4 + 0.125 + 0.005 + 0.005 + 0.005 + 0.005$$
$$= 0.545$$
$$R_t = \frac{1}{0.545}$$
$$= 1.83 \text{ M}\Omega$$

This is clearly greater than the minimum insulation resistance of 1 MΩ but less than 2 MΩ, so the system could be considered satisfactory, but investigation should take place.

Special tests

The next three tests are special as they are not often required in the general type of installation. These tests also require special test equipment. Therefore, the requirements for these tests will only be briefly outlined in this chapter.

Protection by separation of circuits

When separated extra-low voltage (SELV) or protective extra low voltage (PELV) is used as a protective measure, then the separation from circuits of a higher voltage has to be verified by an insulation resistance test at a test voltage of 250 V and result in a minimum insulation resistance of 0.5 MΩ. If the circuit is at LV and supplied from, for example, a BS 3535 transformer, the test will be at 500 V with a minimum value of 0.555 MΩ.

Protection by barriers or enclosures during erection

If, on site, protection against direct contact is provided by fabricating an enclosure or erecting a barrier, it must be ensured that the enclosure can provide a degree of protection of at least IPXXB or IP2X. Readily accessible horizontal top surfaces should be at least IPXXD or IP4X.

An enclosure having a degree of protection IP2X can withstand the ingress of fingers and solid objects exceeding 12 mm diameter. IPXXB is protection against finger contact only. IPXXD or IP4X gives protection against wires and solid objects exceeding 1 mm in diameter.

The test for IPXXB or IP2X is conducted with a 'standard test finger', which is supplied at a test voltage not less than 40 V and no more than 50 V. One end of the finger is connected in series with a lamp and live parts in the enclosure. If the lamp does not light when the end of the finger is introduced into the enclosure, the protection is considered to be satisfactory.

The test for IPXXD or IP4X is conducted with a rigid 1 mm diameter wire with its end bent at right angles. Protection is ensured if the wire does not enter the enclosure.

Protection by non-conducting location

This is a rare location and demands special equipment to measure the insulation resistance between insulated floors and walls at various points.

Testing polarity

This simple test is just as important as all the other tests, but it is often overlooked. Many serious injuries and electrocutions could have been prevented if only polarity checks had been carried out.

The requirements are:

1. All fuses and single pole switches are in the line conductor.
2. The centre contact of an Edison screw-type lampholder is connected to the line conductor.
3. All socket outlets and similar accessories are correctly wired.

Although polarity is towards the end of the recommended test sequence, it would seem sensible, on lighting circuits, for example, to conduct this test at the same time as that for continuity of cpc's (Fig. 19.22).

As discussed earlier, polarity on ring final circuit conductors is achieved simply by conducting the ring circuit test. For radial socket outlet circuits, however, this is a little more difficult. The continuity of the cpc will have already been proved by linking line and cpc, and measuring between the same terminals at each socket. Whilst a line–cpc reversal would not have shown, a line–neutral reversal would, as there would have been no reading at the socket in question. This

Figure 19.22 Test for polarity on a lighting point.

would have been rectified, and therefore only line–cpc reversals need to be checked. This can be done by linking together line and neutral at the origin and testing between the same terminals at each socket. A line–cpc reversal will result in no reading at the socket in question.

When the supply is connected, it is important to check that the incoming supply is correct. This is done using an approved voltage indicator at the intake position or close to it.

Testing earth electrode resistance

In many rural areas, the supply system is TT and hence the general mass of earth is relied upon for a return path under earth fault conditions. Connection to earth is made by an electrode, usually of the rod type, and preferably installed as shown in Fig. 16.6 of Chapter 16.

In order to determine the resistance of the earth return path, it is necessary to measure the resistance that the electrode has with earth. If we were to make such measurements at increasingly longer distances from the electrode, we would notice an increase in resistance up to about 2.5–3 m from the rod, after which no further increase in resistance would be noticed (Fig. 16.4 of Chapter 16).

The maximum resistance recorded is the electrode resistance, and the area that extends the 2.5–3 m beyond the electrode is known as the earth electrode resistance area.

There are two methods of making the measurement, one using a proprietary instrument, and the other using a loop impedance tester.

Method 1: Protection by overcurrent device

This method is based on the principle of the potential divider (Fig. 19.23).

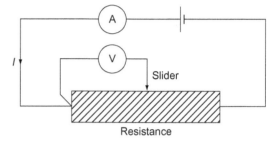

Figure 19.23 Potential divider.

By varying the position of the slider the resistance at any point may be calculated from $R = V/I$.

The earth electrode resistance test is conducted in a similar fashion, with the earth replacing the resistance and a potential electrode replacing the slider (Fig. 19.24). In Fig. 19.24, the earthing conductor to the electrode under test is temporarily disconnected.

The method of test is as follows:

1. Place the current electrode (C2) away from the electrode under test, approximately 10 times its length, that is, 30 m for a 3 m rod.
2. Place the potential electrode midway.
3. Connect test instrument as shown.
4. Record resistance value.
5. Move the potential electrode approximately 3 m on either side of the mid-position, and record these two readings.
6. Take an average of these three readings – this is the earth electrode resistance.

If three readings obtained from an earth electrode resistance test were 181, 185 and 179 Ω, then what would be the value of the electrode resistance and would the accuracy of the measurement be acceptable?

$$\text{Average value} = \frac{181 + 185 + 179}{3}$$
$$= 181.67 \, \Omega$$

Figure 19.24 Earth electrode resistance test.

For TT systems the result of this test will indicate compliance if the product of the electrode resistance and the operating current of the overcurrent device does not exceed 50 V.

Method 2: Protection by a RCD

In this case, an earth fault loop impedance test is carried out between the incoming line terminal and the electrode – a standard test for Z_e.

The value obtained is added to the cpc resistance of the protected circuits and this value is multiplied by the operating current of the RCD. The resulting

value should not exceed 50 V. If it does, then Method 1 should be used to check the actual value of the electrode resistance.

Testing earth fault loop impedance

Earth fault loop impedance has been discussed in Chapter 16, but let us remind ourselves of the component parts of the earth fault loop path (Fig. 19.25).

Starting at the point of fault, the things that need to be checked are as follows:

Figure 19.25 Earth fault loop path.

1. The cpc.
2. The main earthing conductor and earthing terminal.
3. The return path via the earth for TT systems, and the metallic return path in the case of TN–S or TN–C–S systems. In the latter case the metallic return is the Protective Earthed Neutral (PEN) conductor.
4. The earthed neutral of the supply transformer.
5. The transformer winding.
6. The line conductor back to the point of fault.

Overcurrent protective devices must, under earth fault conditions, disconnect fast enough to reduce the risk of electric shock. This is achieved if the actual value of the earth fault loop impedance does not exceed the tabulated maximum values given in the IET Regulations.

The purpose of the test, therefore, is to determine the actual value of the loop impedance (Z_s) for comparison with those maximum values, and it is conducted as follows:

1. Ensure that all main equipotential bonding is in place.
2. Connect the test instrument either by its BS 4363 plug, or the 'flying leads' to the line, and the neutral and earth terminals at the remote end of the circuit being under test. (If a neutral is not available, e.g. in the case of a three-phase motor, connect the neutral probe to earth.)
3. Press to test and record the value indicated.

It must be understood that this instrument reading is **not valid for direct comparison with the tabulated maximum values**, as account must be taken of the ambient temperature at the time of test, as well as the maximum conductor operating temperature, both of which will have an effect on conductor resistance. Hence, the ($R_1 + R_2$) could be greater at the time of fault than at the time of test.

Therefore, our measured value of Z_s must be corrected to allow these possible increases in temperature occurring at a later date. This requires actually measuring the ambient temperature and applying factors in a formula.

Clearly this method of correcting Z_s is time consuming and unlikely to be commonly used. Hence, a rule of thumb method may be applied, which simply requires that the measured value of Z_s does not exceed 0.8 of the appropriate tabulated value. Table 19.6 gives the 0.8 values of tabulated loop impedance for direct comparison with measured values.

In effect, a loop impedance test places a line to earth fault on the installation, and if an RCD is present it

may not be possible to conduct the test as the device will trip out each time the loop impedance tester button is pressed. Unless the instrument is of a type that has a built-in guard against such tripping, the value of Z_s will have to be determined from measured values of Z_e and ($R_1 + R_2$), and the 0.8 rule applied.

Important Note

Never short out an RCD in order to conduct this test.

As a loop impedance test creates a high earth fault current, albeit for a short space of time, some lower rated cb's may operate resulting in the same situation as with an RCD, and Z_s will have to be calculated. It is not really advisable to temporarily replace the present cb with one of a higher rating.

External loop impedance Z_e

The value of Z_e is measured at the intake position on the supply side and with all main equipotential bonding disconnected. Unless the installation can be isolated from the supply, this test should not be carried out, as a potential risk of shock exists with the supply on and the main bonding disconnected.

RCD/RCBO operation

Where RCDs/RCBOs are fitted, it is essential that they operate within set parameters. The RCD testers used are designed to do just this, and the basic tests required are as follows (Table 19.7):

1. Set the test instrument to the rating of the RCD.
2. Set the test instrument to half rated trip.
3. Operate the instrument and the RCD should not trip.
4. Set the instrument to deliver the full rated tripping current of the RCD.
5. Operate the instrument and the RCD should trip out in the required time.
6. For 30 mA RCDs used for additional protection, set the instrument to five times the tripping current.
7. Operate the instrument and the RCD should trip out in no more than 40 ms.

Where loop impedance values cannot be met, RCDs of an appropriate rating can be installed. Their rating can be determined from

$$I_n = 50/Z_s$$

where I_n is the rated operating current of the device, 50 is the touch voltage and Z_s is the measured loop impedance.

Table 19.6 Corrected Maximum Z_s Values for Comparison with Measured Values

PROTECTION RATING

PROTECTIVE DEVICE		5	6	10	15	16	20	25	30	32	40	45	50	60	63	80	100	125	160	200
BS EN 60898 and 61009 type **B**	0.4s & 5s		5.82	3.5		2.19	1.75	1.4		1.09	0.87		0.7		0.56	0.5	0.35	0.29		
BS EN 60898 and 61009 type **C**	0.4s & 5s		2.9	1.75		1.09	0.87	0.71		0.54	0.44		0.35		0.28	0.2	0.17	0.14		
BS EN 60898 and 61009 type **D**	0.4s		1.46	0.87		0.54	0.44	0.35		0.28	0.22		0.17		0.13	0.1	0.08	0.07		
	0.5s		2.91	1.75		1.09	0.87	0.69		0.55	0.44		0.35		0.27	0.22	0.17	0.14		
BS 3036 Semi enclosed	0.4s	7.22			1.94				0.83											
	5s	13.45			4.1		2.9		2			1.2		0.85			0.4			
BS 1362 Cartridge	0.4s	12.5			1.8															
	5s	17.63			2.9															
BS 88-2. Bolted type (**E**) and Clip-in type (**G**)	0.4s		6.24	3.71		1.95	1.35	1.03		0.79										
	5s		9.73	5.46		3.17	2.24	1.75		1.4	1.03		0.79		0.64	0.4	0.35	0.26	0.21	0.2
BS 88-3. Cartridge type (**C**)	0.4s	7.94				1.84	1.55			0.73										
	5s	11.62				2.14	2.57			1.24		0.79		0.55		0.4	0.3			

* 3A * 13A

19

Table 19.7

RCD Type	1/2 Rated	Full Trip Current
BS 4239 and BS 7288 sockets	No trip	<200 ms
BS 4239 with time delay	No trip	1/2 time delay 1 200 ms – time delay 1200 ms
BS EN 61009 or BS EN 61009 RCBO	No trip	300 ms
As above but Type S with time delay	No trip	130–500 ms

All RCDs have a built-in test facility in the form of a test button. Operating this test facility creates an artificial out of balance condition that causes the device to trip. This only checks the mechanics of the tripping operation, and not a substitute for the tests just discussed.

All other items of equipment such as switchgear, controlgear interlocks, etc., must be checked to ensure that they are correctly mounted and adjusted and that they function correctly.

Prospective fault current

This would normally be carried out at the same time as the measurement for Z_e using a PFC or PSCC tester. If this value cannot be measured it must be ascertained by either enquiry or calculation.

Phase sequence

For multi-phase circuits a check on the phase sequence must be made using a phase-rotation instrument similar to that shown in Fig. 19.11.

Functional testing

A check should be made to ensure that the operating mechanism of RCDs are functioning, using the integral test button.

Moreover, switchgear, assemblies, interlocks, etc., must be shown to have been erected correctly and function properly.

Voltage drop

A check on voltage drop should be made wherever required. This is not normally required during initial verification.

Periodic inspection

Periodic inspection and testing

Circumstances that require a periodic inspection and test

Test and inspection is due when there are insurance, mortgage, licensing reasons; change of use; change of ownership; after additions or alterations; after damage; change of loading and to assess compliance with current Regulations.

General reasons for a periodic inspection and test

1. To ensure the safety of persons and livestock.
2. To ensure protection of property from fire and heat.
3. To ensure that the installation is not damaged so as to impair safety.
4. To ensure that the installation is not defective and complies with the current Regulations.

General areas of investigation

Safety, wear and tear, corrosion, damage, overloading, age, external influences, suitability and effectiveness.

Documentation to be completed

Electrical Installation Condition Report (EICR), schedule of test results and an inspection schedule.

Sequence of tests

The sequence of tests conducted may be in any order that suits the circumstances.

This could be so simple. As it is, periodic inspection and testing tends to be complicated and frustrating. On the domestic scene, I doubt if any house owner actually decides to have a regular inspection. They say, 'If it works it must be OK'. It is usually only when there is a change of ownership that the mortgage companies insist on an electrical survey. The worst cases are, however, industry and commerce. Periodic inspections are requested, reluctantly, to satisfy insurers or an impending visit by the HSE. Even then it is usually the case that 'you can't turn that off' or 'why can't you just test this bit and then issue a certificate for the whole lot'. Under the rare circumstances when an inspection and test is genuinely requested it is difficult to convince the client that, as there are no drawings, or information about the installation, and that no switchgear is labelled, etc., you are going to be on site for a considerable time and at a considerable cost.

When there are no drawings or items of information, especially on a large installation, there may be a degree of exploratory work to be carried out in order to ensure safety whilst inspecting and testing. If it is felt that it may be unsafe to continue with the inspection and test, then drawings and information **must** be produced in order to avoid contravening the Health and Safety at Work Act, Section 6.

However, let us assume, as with the initial inspection, that the original installation was erected in accordance with the 16th edition, and that any alterations and/or additions have been faithfully recorded on the original documentation which is, of course, readily available.

A periodic inspection and test under these circumstances should be relatively easy, as little dismantling of the installation will be necessary, and the bulk of the work will only be inspection.

Inspection should be carried out with the supply disconnected as it may be necessary to gain access to wiring in enclosures, etc., and hence, with large installations it will probably need considerable liaison with the client to arrange convenient times for interruption of supplies to various parts of the installation.

This is also the case while testing protective conductors, as these must **never** be disconnected unless the supply can be isolated. This is particularly important for main equipotential bonding conductors that need to be disconnected in order to measure Z_e.

In general an inspection should reveal:

1. Any aspects of the installation that may impair the safety of persons and livestock against the effects of electric shock and burns.
2. That there are no installation defects that could give rise to heat and fire, and hence damage property.
3. That the installation is not damaged or deteriorated so as to impair safety.
4. That there is any defect or non-compliance with the IET Regulations identified, which may give rise to danger.

As was mentioned earlier, dismantling should be kept to a minimum and hence a certain amount of sampling will take place. This sampling would definitely increase if defects are detected.

From the testing point of view, not all of the tests carried out on the initial inspection may need to be applied. This decision depends on the condition of the installation.

The continuity of protective conductors is clearly important as is insulation resistance and loop impedance, but the question is whether polarity tests are necessary if the installation has remained undisturbed since the last inspection. The same applies to ring circuit continuity as the P–N test is applied to detect interconnections in the ring, which would not happen on their own.

It should be noted that if an installation is effectively supervised in normal use, then periodic inspection and testing can be replaced by regular maintenance by skilled persons. This would only apply to, say, factory installations where there are permanent maintenance staff.

Certification

Having completed all the inspection checks and carried out all the relevant tests, the only thing that remains is documenting all this information. This is done on electrical installation certificates, inspection schedules, condition reports, minor works certificates and any other documentation you wish to append to the foregoing. Examples of such documentation are shown in BS 7671 and in the IET Guidance Notes 3 on inspection and testing.

This documentation is vitally important. It has to be correct and signed or authenticated by a competent person. Electrical installation certificates and condition reports must be accompanied by a schedule of test results and an inspection schedule for them to be valid. It should be noted that three signatures are required on an electrical installation certificate, one in respect of the design, one for the construction and the third for the inspection and test. For larger installations there may be more than one designer, hence the certificate has space for two signatures, that is, of designer 1 and designer 2. It could be, of course, that for a very small company, one person signs all three parts. Whatever the case, the original must be given to the person ordering the work, and a duplicate retained by the contractor.

One important aspect of the electrical installation certificate is the recommended interval between inspections. This should be evaluated by the designer and will depend on the type of installation and its usage. In some cases the time interval is mandatory, especially where environments are subject to use by the public. IET Guidance Notes 3 give recommended maximum frequencies between inspections.

A condition report form is very similar in part to an electrical installation certificate in respect to the details of the installation, that is, maximum demand, type of earthing system, Z_e, etc. The rest of the form deals with the extent and limitations of the inspection and test, recommendations and a summary of the

19

installation. The record of the extent and limitations of the inspection is very important. The client or other third party must agree on exactly what parts of the installation will be covered by the report and those that will not. The interval until the next test is determined by the inspector.

With regard to the schedule of test results, test values should be recorded unadjusted; and any compensation for temperature, etc., will be made after the testing is completed.

Any alterations or additions to an installation will be subject to the issue of an electrical installation certificate, except where the addition is, say, a single point added to an existing circuit. In the later case, the work is subject to the issue of a minor works certificate.

Summarizing:

(a) The addition of points to existing circuits requires a Minor Works Certificate.

(b) A new installation or an addition or alteration that comprises new circuits requires an Electrical Installation Certificate.

(c) An existing installation requires a Periodic Inspection Report.

Note

(b) and (c) must be accompanied by a schedule of test results and an inspection schedule.

Inspection and testing

As the client/customer is to receive the originals of any certification, it is important that **all** relevant details are completed correctly. This ensures that future inspectors are aware of the installation details and test results that may indicate a slow progressive deterioration in some or all of the installations.

These certificates, etc., will also form part of a 'sellers pack' when a client wishes to sell a property.

The following is a general guide for completing the necessary documentation and should be read in conjunction with the examples given in BS 7671 and the IET On-site Guide.

Electrical installation certificate

1. **Details of client**
   Name: *Full name*
   Address: *Full address and postal code*
   Description: *Domestic, industrial, commercial*
   Extent: *What work has been carried out, for*

example, full re-wire, new shower circuit, etc. Tick a relevant box

2. **Designer/constructor/tester**
   Details of each or could be one person
   Note: Departures are not faults, they are systems/ equipment, etc., that are not detailed in BS 7671 but may be perfectly satisfactory

3. **Next test**
   When the next test should be carried out is decided by the designer

4. **Supply characteristics and earthing arrangements**
   Earthing system: *Tick relevant box (TT, TN–S, etc.)*
   Live conductors: *Tick relevant boxes*
   Nominal voltage: *Obtain from supplier, but usually 230 V single-phase U and U_0 but 400 V U and 230 U_0 for three phase*
   Frequency: *From supplier but usually 50*
   PFC: *From supplier or measured. Supplier usually gives 16 kA*
   Z_e: *From supplier or measurement. Supplier usually gives 0.8 Ω for TN–S, 0.35 Ω for TN–C–S and 21 Ω for TT systems*
   Main fuse: *Usually BS 1361, rating depends on maximum demand*

5. **Particulars of installation**
   Means of earthing: *Tick 'suppliers facility' for TN systems, 'earth electrode' for TT systems*
   Maximum Demand: *Value without diversity*
   Earth electrode: *Measured value or N/A*
   (a) Earthing and bonding:
   Conductors: *Actual sizes and material, usually copper*
   Main switch or Circuit breaker (could be separate units or part of a consumercontrol unit): *BS number; Rating, current and voltage; Location; 'not address', i.e. where is it located in the building; Fuse rating if in a switch-fuse, else N/A; RCD details only if used as a main switch*

6. **Comments on existing installation**
   Write down any defects found in other parts of the installation which may have been revealed during an addition or an alteration

7. **Schedules**
   Indicate the number of test and inspection 'schedules' that will accompany this certificate.

Electrical Installation Condition Report

1. **Details of client**
   Name: *Full name (could be a landlord, etc.)*
   Address: *Full address and post code (may be different to the installation address)*

Purpose: *e.g. Due date; change of owner/tenant; change of use, etc.*

2. Details of installation

Occupier: *Could be the client or a tenant*

Installation: *Could be the whole or part (give details)*

Address: *Full and post code*

Description: *Tick relevant box*

Age: *If not known, say so, or make an educated guess*

Alterations: *Tick relevant box and insert age where known*

Last inspection: *Insert date or 'not known'*

Records: *Tick relevant box*

3. Extent and limitations

Full details of what is being tested (extent) and what is not (limitations)

If there is not enough space on the form add extra sheets

4. Next inspection

Filled in by inspector and signed, etc., under declaration

5. Supply details

As per an Electrical Installation Certificate

6. Observations

Tick the relevant box, if work is required, record details and enter relevant code in space on right-hand side

7. Summary

Comment on overall condition. Only common sense and experience can determine whether satisfactory or unsatisfactory

8. Schedules

Attach completed schedules of inspections and test results.

Minor electrical installation works certificate

Only to be used when simple additions or alterations are made, *not when a new circuit is added.*

1. Description: *Full description of work*

Address: *Full address*

Date: *Date when work was carried out*

Departures: *These are not faults, they are systems/ equipment, etc., that are not detailed in BS 7671 but may be perfectly satisfactory (this is usually N/A)*

2. Installation details

Earthing: *Tick a relevant box*

Method of fault protection: *99% of the time this will be EEBADS. Other methods should be recorded*

Protective device: *Enter type and rating. For*

example, BS EN 60898 cb Type B, 20 A

Comments: *Note any defects/faults/omissions in other parts of the installation seen while conducting the minor works*

3. Tests

Earth continuity: *Measure and then tick in box if OK*

Insulation resistance: *Standard tests and results*

EFLI (Z_s): *Standard tests and results*

Polarity: *Standard tests, then tick in box if OK*

RCD: *Standard tests, record operating and current time*

4. Declaration

Name, address, signature, etc.

Schedule of test results (as per BS 7671)

1. Contractor: *Full name of tester*

2. Date: *Date of test*

3. Signature: *Signature of tester*

4. Method of fault protection: *99% EEBADS but could be SELV, etc.*

5. Vulnerable equipment: *Dimmers, electronic timers, CH controllers, etc., i.e. anything electronic*

6. Address: *Full, or if in a large installation, the location of a particular DB*

7. Earthing: *Tick the relevant box*

8. Z_e at origin: *Measured value*

9. PFC: *Record the highest value, i.e. PEFC or PSCC (should be the same for TN–C–S)*

10. Confirmation of supply polarity: *Tick in box*

11. Instruments: *Record serial numbers of each instrument, or one number for a composite instrument*

12. Description: *Suggest initial or periodic or whatever part of the installation is involved, e.g. Initial verification on a new shower circuit*

13. kVA rating: *Taken from the device (difficult when there are different devices in an installation). Nothing to stop adding sheets to this form!*

14. Type and rating: *For example, BS EN 60898 cb Type B, 32 A, or BS 88 40 A, etc.*

15. Wiring conductors: *Size of Live and cpc, e.g. 2.5 mm^2/1.5 mm^2*

16. Test results: *Fill in all measured values ($R_1 + R_2$), etc. Tick box if ring P–N is OK. If any test does not appear on the sheet, e.g. $5 \times I\Delta_n$, write the results in the remarks column.*

19

Schedule of inspections (as per BS 7671)

Do not leave boxes incomplete.

N/A in a box if it is not relevant

✓ in a box if it has been inspected and is OK

✗ in a box if it has been inspected and is incorrect.

Fault finding

This is not an exact science as faults in electrical systems can be many, varied and difficult to locate.

However, we can only state the main symptoms of electrical faults as follows:

▶ Loss of supply
▶ Fire
▶ Shock.

Table 19.8 indicates such symptoms, their possible common causes and the action to be taken. Column 2 of the same table illustrates, in general terms, the possible causes of faults. Table 19.9 summarizes these in more detail.

Table 19.8 Typical faults and diagnosis

Symptom	Possible Common Cause	Diagnosis	Action
Complete loss of supply	1. Fault on suppliers (DNO) main cable/equipment 2. Fault on service cable 3. Main fuse or cb operated 4. Main DB switch OFF 5. Main RCD operated	1. Check adjacent properties are also OFF 2. Check adjacent properties are ON 3. Check adjacent properties are ON 4. Visual check 5. Visual check	1. Contact DNO 2. Contact DNO 3. Contact DNO 4. Switch back on 5. Reset, if it trips, then switch off all cb's, reset and turn on each cb until one causes the main RCD to operate. This is the likely faulty circuit
Loss of supply to a circuit	1. Circuit fuse or cb operated 2. Conductor broken or out of terminal	1. Visual check 2. Check fuse/cb are OK	1. Replace or reset as operation may be due to an overload. If protection still operates, do NOT reset until fault has been found, usually by carrying out insulation tests 2. Locate fault by carrying out visual check/continuity tests
Fire/burning	1. Overloaded cable 2. Damaged insulation 3. Water in fittings/accessories	1, 2 and 3 visual check and smell	1, 2 and 3 turn off supply to circuit(s), investigate fuse and cable sizes, check for water ingress, damaged insulation, visually and using an insulation resistance tester
Electric shock	1. Exposed live parts 2. Insulation breakdown 3. Earthing and bonding inadequate 4. Appliances incorrectly wired or damaged and with inappropriate fusing 5. Incorrect polarity in accessories	Use of an approved voltage indicator between exposed and/or extraneous conductive parts	Turn off supply to circuit(s), check visually for covers missing, etc. Carry out insulation resistance and polarity tests on circuits and cables, and establish that all earthing and bonding is in place and that all protective devices are suitable for disconnection times

Table 19.9 Summary of faults and causes

General Cause	Details
Insulation breakdown	1. Damage by installer 2. Damage by other trades Damage by user (misuse, nails in walls, etc.) 3. Overloading
Fuse, cb or RCD operating instantly, as circuit is switched on	1. Short circuit caused by (a) damaged insulation (b) crossed polarity at terminations (c) water penetration in JB's, seals, glands, etc. 2. Faulty appliances
Fuse or cb operating regularly after a period of time	1. Overload caused by too many loads on a circuit, or machinery stalling or with too much mechanical load 2. Slight water penetration or general dampness
Fuse or cb operates with no apparent fault	Transient overvoltage caused by switching surges, motor starting, etc.

Many faults are easily located, while many remain unnoticed. Therefore for all cases observe the following general procedure whenever possible:

1. Determine the nature/symptom of the problem.
2. Ask client/personnel for their recollections, as to how, when and where the problem occurred (this can save a lot of time).
3. Carry out relevant visual and instrument checks to locate the fault.
4. Rectify if possible.
5. Retest.
6. Reinstate system.

19

CHAPTER 20

Basic Electronics Technology

Most of us in the world of electrical installation work are familiar with values such as 230 V, 3 kW or 60 A. We tend to view the terms and quantities used in electronics with a certain unease, but the relationship between ohms, volts and amperes in the world of electronics is no different from that in electrical installation work.

To feel more at home with electronics, we first look at some of the many components used in electronics.

Electronics components

Resistors

The ohmic value of 230 V appliances rated at 60 W, 1000 W, 3 kW, etc. should be familiar to us by now, but the values of resistors used in electronics are many and varied. To readily identify all the different values, a colour code is used. The same colour code is also used for capacitor values.

Resistor (or Capacitor) Colour Code

Colour	Value
Black	0
Brown	1
Red	2
Orange	3
Yellow	4
Green	5
Blue	6
Violet	7
Grey	8
White	9

Tolerance Colour Code

Colour	Percentage
Brown	1
Red	2
Gold	5
Silver	10
None	20

Each resistor carries a series of coloured bands to indicate its value and tolerance:

Three bands for a resistor with tolerance of 20%
Four bands for a resistor with tolerance between 10% and 2%
Five bands for a resistor with tolerance of 1%

These bands are interpreted as shown in Fig. 20.1. Figure 20.2 shows two examples of how to 'read' the colour coding to ascertain the value and tolerance.

Sometimes a resistor code will use numbers and letters rather than colours. The letters used are as follows:

R, K and M indicate multiples of 1, 1000 and 1 000 000 Ω, respectively.
F, G, J, K, and M indicate tolerances of 1%, 2%, 5%, 10% and 20%, respectively.

The R, K and M code can also be used for decimal points, for example,

3.3 kΩ may be shown as 3K3 Ω.
0.2 Ω may be shown as R2 Ω.

Other examples are as follows:

1. 2K2G indicates a value of 2.2 kΩ (2200 Ω) ± 2%.
2. 6M8J indicates a value of 6.8 MΩ (6 800 000 Ω) ± 5%.

Electrical Installation Work. 978-1-138-84927-3 © Brian Scaddan. Published by Taylor & Francis. All rights reserved.

Figure 20.1 Coding bands.

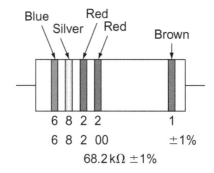

Figure 20.2 Colour values decoded.

3. 33KF indicates a value of 33.0 kΩ (33 000 Ω) ± 1%.

4. 470RM indicates a value of 470.0 Ω ± 20%.

Resistor types

There are three types of resistors in common use: carbon, wire wound and carbon preset or variable.

The wire-wound type is usually chosen where high voltage is present. It is also more accurate than the carbon variety. The preset type, as shown in

Figure 20.3 Preset resistor.

Fig. 20.3, is used in the simple metal detectors used by electricians and to adjust EXIT and ENTRY times in alarm panels.

Capacitors

There are a number of ways of marking a capacitor with its value. The most common, apart from actually writing the value on the capacitor, is to use the same colour code as for resistors (Fig. 20.4). Remember, small capacitors are usually in the picofarad (pF) range (1 nanofarad (nF) = 1000 pF).

One other method commonly used is the three-digit method, in which the third digit gives the number of zeros that follow the first two digits, to give the value in pF. For example, 104 is the code for a capacitor of 100 000 pF or 100 nF.

Yellow	1st number
Violet	2nd number
Orange	Zeros
Black	Tolerance
Red	Voltage

Tolerance		Voltage	
Green	5%	Brown	100 V
White	10%	Red	250 V
Black	20%	Yellow	400 V

47 000 pF or 47 nF 20% 250 V

Figure 20.4 Capacitor colour coding.

Capacitor types

The many types of capacitors in use range from waxed paper and foil, electrolytic, polyester, mica and ceramic, to air – all of which have different applications depending on frequency, voltage, supply (alternating current or direct current), losses, etc.

Inductors and transformers

Inductors and transformers have already been discussed in Chapter 4. The types of inductors and transformers used in electronics work on the same principles. Probably the main difference is the use of ferrite as a core for inductors that are used to tune for radio frequencies. Ferrite is a much more efficient material than iron as the hysteresis losses are much less.

Semi-conductors

These devices are neither strictly conductors nor insulators but, under certain circumstances, can become either.

Silicon is the most common semi-conductor material, and the addition of impurities such as aluminium or arsenic creates the circumstances under which it will conduct or insulate. For example, if we add aluminium to a sample of silicon, it becomes a 'p'-type material whereas adding arsenic to the sample makes it an 'n'-type material.

The junction diode

If we now take a sample of 'p'-type and a sample of 'n'-type silicon and join them together, we will have a 'junction diode'. By connecting a positive charge

Figure 20.6 Junction diode.

to the 'p'-type sample and a negative charge to the 'n'-type sample, the whole assembly will act as a conductor. Reversing the connections will result in the arrangement acting as an insulator (Fig. 20.5). Hence, it is used commonly as a means of rectification.

The symbol for the diode is shown in Fig. 20.6. The diode will conduct when current flow is in the direction of the arrow.

Figure 20.7 shows how the forward and reverse current for a silicon diode varies with the applied voltage. At point 'X' in the graph, the reverse voltage is so great that the diode breaks down and conducts. This value is called the reverse breakdown voltage.

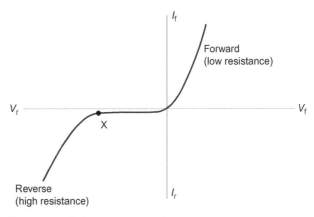

Figure 20.7 Reverse breakdown voltage.

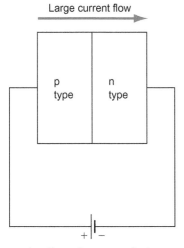

Junction acts as a conductor

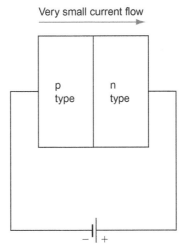

Junction acts as an insulator

Figure 20.5 Semi-conductor.

Rectification

Most electricity supply systems are alternating current (a.c.) and since many appliances require a direct current (d.c.) supply, it is necessary to change a.c. to d.c. This change is called **rectification**.

Figure 20.8a illustrates how a diode or group of diodes can be used to rectify an a.c. supply. It can be seen that the rectified d.c. output is not true d.c., for which the waveform would be a straight line, but has something of a pulsating nature. This type of output is usually quite acceptable for most purposes in electrical installation work. Should a more refined or smoothed output be required, the addition of capacitance and inductance (Fig. 20.9) can provide this.

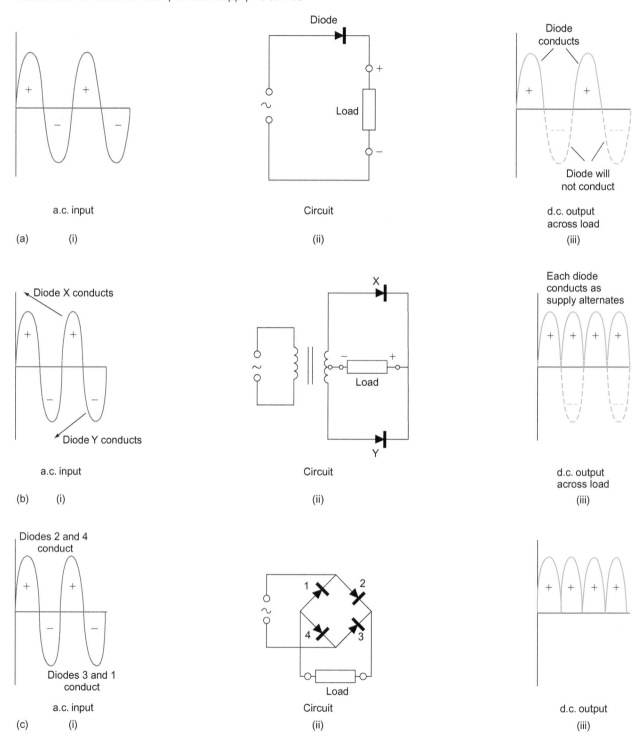

Figure 20.8 (a) Half-wave rectification, (b) full-wave rectification from transformer supply and (c) bridge-type full-wave rectification.

Figure 20.9 Smoothing circuit.

Rectifier output

Since the output from a full-wave rectifier is a series of sinusoidal pulses, the average value of this output is given by

Average value = maximum or peak value × 0.637

and for half-wave

$$\text{Average value} = \frac{\text{maximum or peak value} \times 0.637}{2}$$

Note

When selecting a diode for a particular duty, ensure that it is capable of operating at the peak voltage. For example, consider a circuit that is required to operate at 12 V d.c. and a 230/12 V transformer is to be used in conjunction with diodes.

The 12 V output from the transformer is 12 V r.m.s. and therefore has a peak value of

$$\frac{12}{0.7071} = 16.97 \text{ V}$$

The diode must be able to cope with this peak voltage.

Example

Calculate the average value of a full-wave rectified d.c. output if the a.c. input is 16 V.

$$\text{Peak value of a.c. input} = \frac{16}{0.7071} = 22.61 \text{ V}$$
$$\text{Average value of d.c. output} = 22.61 \times 0.637 = 14.4 \text{ V}$$

Applications

1. The d.c. machines (supply).
2. Bell and call systems.
3. Battery charging.
4. Emergency lighting circuits.

Thyristors or silicon-controlled rectifiers

A thyristor is a four-layer p–n–p–n device with three connections (Fig. 20.10).

Under normal circumstances, with positive on the anode and negative on the cathode, the thyristor will not conduct. If, however, a large enough positive firing potential is applied to the 'gate' connection, the thyristor will conduct and will continue to do so even if the signal on the gate is removed. It will cease to conduct when the anode potential falls below that of the cathode. The device, like the diode, will not conduct at all in the reverse direction.

Let us now consider what happens when a thyristor is wired in an a.c. circuit as shown in Fig. 20.11. Resistor R_1 ensures that the minimum gate potential required for firing is maintained and resistor R_3 ensures that the gate potential does not rise to a level that could cause damage to the gate circuit. Variable resistor R_2 enables various gate potentials to be selected between maximum and minimum, and it will also be seen that,

Figure 20.10 Thyristor.

Figure 20.11 Thyristor circuit.

20

as the circuit is resistive, the applied voltage and the gate voltage are in phase (Fig. 20.12).

It will be seen from Fig. 20.13 that if the gate voltage is adjusted to a maximum, the gate will fire at point A. When it is at a minimum it fires at point B.

The current flowing in a resistive circuit is also in phase with the voltage and in this case it will flow only when the thyristor is conducting. Since this conduction takes place only when the thyristor is triggered by the gate, the amount of current flowing in any positive half-cycle can be controlled by the gate potential (Fig. 20.13a and b).

This control of the amount of current flowing in each half-cycle can be used to control the speed of small motors such as those used in food mixers and hand drills. A simple circuit is shown in Fig. 20.14.

More complicated circuitry is now in use, utilizing thyristors, to control the speed of induction motors – something that in the past proved very difficult.

Transistor

The transistor is basically a semi-conductor, which can act as a switch and an amplifier. It is manufactured from silicon just as diodes are manufactured from silicon and could be thought of as two diodes 'back-to-back' (Fig. 20.15a and b).

The connections to a transistor are known as the emitter, collector and base, and the symbols for transistors are as shown in Fig. 20.16a and b.

To identify transistor connections, turn the device so that the connection leads are facing you and with the

Figure 20.12   Thyristor wareforms.

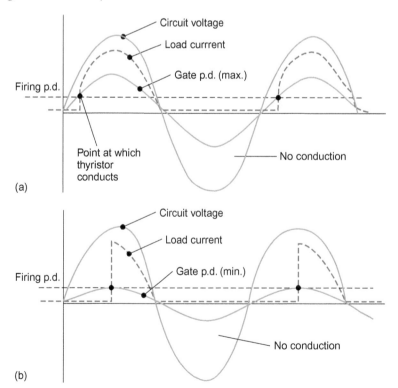

(a)

(b)

Figure 20.13 (a) Maximum gate potential and (b) minimum gate potential.

Figure 20.14 Basic speed control circuit for a motor.

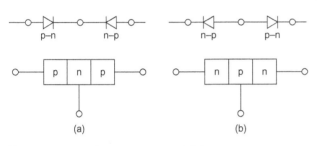

Figure 20.15 (a) p–n–p type and (b) n–p–n.

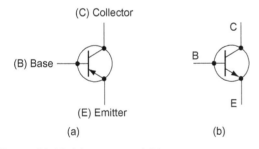

Figure 20.16 (a) p–n–p and (b) n–p–n.

Figure 20.17 Transistor.

tag at the bottom left, the connections are as shown in Fig. 20.17.

There are two types of transistors: the field effect transistor (FET) and the bi-polar-type transistor. FETs are extremely small and are used mainly in integrated circuits (ICs), whereas the bi-polar-type transistor, being much larger, is used extensively as a stand-alone component in circuitry.

Basis transistor action

When a small voltage is applied to the base, it switches the transistor on, and allows current to flow from the collector to the emitter and vice versa. If this current is passed through a resistor, a voltage will be developed across that resistor, which can be many times greater than the input voltage to the base. Hence we have amplification.

Other commonly used components

	Zener diode Zener diode is used in circuits to give voltage control or stabilization.
	Triac Triacs are often used 'back-to-back' to provide a smoother and more efficient thyristor effect. They will be found in dimmer switches.
	Diac Diacs are used in conjunction with triacs as a triggering device.
	Thermistor Thermistor is used as a means of sensing temperature change. It is commonly found embedded in motor windings to detect overheating.
	Light emitting diode Light emitting diode (LED) is simply a semi-conductor signal lamp.
	Light-sensitive diode or photodiode Light-sensitive diode (LSD) is used to activate a circuit in response to light.
	Light-dependent resistor The light-dependent resistor (LDR) is similar to the LSD but is able to handle much larger currents. It is used as a switching device in such areas as street lighting, etc.
	Integrated circuit The IC or **chip** as it is often called is a minute electronic circuit comprising many hundreds of thousands of microscopic components.

Heat sinks

In many instances, semi-conductors will become hot when in operation and would be damaged if this heat were not dissipated. Embedding the device in the centre of a large plate or series of plates helps dissipate the heat to the surrounding air (Fig. 20.18).

Figure 20.18 Heat sink.

20

281

Electronics diagrams

Three main types of diagram are used:

1. **Block diagrams** (Fig. 20.19) are used to give a general indication of a complete system.
2. **Circuit diagrams** (Fig. 20.20), as in installation work, indicate how a circuit **works**.
3. **Layout diagrams** (Fig. 20.21) are similar to wiring diagrams in installation work; they show how a circuit should be **wired** and components are shown in their correct locations.

Note

Cross-reference should be possible between circuit and layout diagrams by means of pin numbers of a positional reference system.

Electronics assembly

Unlike installation circuits, electronic circuitry is almost entirely constructed using soldered joints. The formation of such joints is critical to ensure healthy circuit performance.

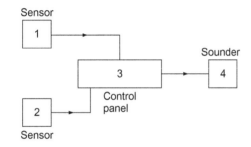

Figure 20.19 Block diagram of an intruder alarm system.

Soldering

A soldered joint comprises the surfaces to be joined and a material (solder), which is an alloy of 40% tin and 60% lead, melted onto the surfaces. To aid the soldering process, a flux is used and usually this is incorporated in the solder.

Cleanliness is vital to ensure a good soldered joint – cleanliness of not only the surfaces to be joined but also the 'bit' of the soldering iron.

Remember that excessive solder will make a poor joint and that the iron should be at the correct temperature. Many modern irons have built-in temperature controls. Figure 20.22 shows some effects of soldering faults.

Heat shunts

As heat can damage electronic components, it is important to ensure that too much heat does not reach the component during the soldering process. Heat shunts are used for this purpose and can simply consist of a pair of pliers or a crocodile clip attached to the component lead. This

Figure 20.21 Layout diagram of an amplifier circuit.

Figure 20.20 Simple transistor amplifier circuit.

Figure 20.22 Soldered joints.

helps to dissipate the heat before it can reach the component itself (Fig. 20.23).

De-soldering

It may be necessary on occasion to remove a component from a circuit. The pro-cess is similar to that of soldering; the joint is flooded with new hot solder and, as a heat balance is achieved, the old solder softens. When the whole lot is fluid, the solder is sucked away from the joint using either a copper braid or a specially designed solder-sucker.

Handling electronic circuits

Many electronic components are very sensitive to small electrostatic charges and are often damaged when subjected to them. To avoid such damage, work benches and the immediate floor area are supplied with anti-static mats and the operator has an earth-tag around his/her wrist to ensure that there is a discharge of static.

Figure 20.23 Heat shunts.

Self-Assessment Questions

1. Write down the value of the resistors with the following colour codes:
   (a) Orange–silver–red–silver
   (b) Red–white–green–red
   (c) Green–green–black–gold–brown
   (d) Brown–red–gold–silver.
2. Give the colour code for the following values of resistor:
   (a) 8.7 K 1%
   (b) 15K8J
   (c) 100RG
   (d) 350PK 20%.
3. Indicate the values of the following colour-coded capacitors:
   (a) Brown–silver–gold–green–brown
   (b) Red–violet–gold–white–yellow
   (c) Violet–green–gold–black–brown
   (d) Brown–black–yellow–white–red.
4. Write down the colour code for the following capacitors:
   (a) 25 nF 20% 400 V
   (b) 53 nF 10% 100 V
   (c) 15 nF 5% 250 V
   (d) 200 nF 20% 250 V
5. Explain with the aid of diagrams the difference between a heat sink and a heat shunt.
6. Draw the BS 3939 symbol for the following components:
   (a) An LED
   (b) A p–n–p transistor
   (c) An iron-cored inductor
   (d) A variable capacitor.
7. With the aid of drawings explain the difference between block, circuit and layout diagrams.

20

Answers to Self-Assessment Questions

CHAPTER 1

Indices

1. 8^4
2. The whole numbers are **not** the same
3. 1
4. $10^1 \times 10^{-1}$; 10^0
5. 1 and 3^0
6. 1
7. (a) 1 (b) 10 (c) 25

Algebra

1. (a) $4X$
   (b) $5F$
   (c) $8Y + 4X$
   (d) $2M^2$
   (e) $6P^3$
   (f) 12
   (g) $2A$
   (h) X

2. (a) $X = P + Q - Y$
   (b) $X = A + D - F$
   (c) $X = L - P - W - Q$
   (d) $X = 2$
   (e) $X = \dfrac{P \cdot D}{M}$
   (f) $X = \dfrac{A}{W}$
   (g) $X = \dfrac{H \cdot K}{2B}$
   (h) $X = \dfrac{M \cdot Y}{A \cdot B \cdot C}$
   (i) $X = \dfrac{W}{(A+B)}$
   (j) $X = \dfrac{R(M+N)}{2P}$

Pythagoras and trigonometry

1. A right-angled triangle
2. $H = \sqrt{B^2 + P^2}$
3. 19.416
4. 9.8
5. 18.33
6. Trigonometry
7. (a) $\sin\phi = \dfrac{P}{H}$; (b) $\cos\phi = \dfrac{B}{H}$; (c) $\tan\phi = \dfrac{P}{B}$
8. 9.46; 8.58ϕ
9. 51.31°; 12.49
10. 66.8°; 15.23

CHAPTER 2

1. +ve and −ve; protons are +ve, electrons are −ve
2. The number of electrons orbiting the nucleus of an atom
3. Those electrons which leave their orbits and wander through the molecular structure of the material
4. (a) A conductor has many random electrons, an insulator has very few
   (b) Conductor: copper, aluminium, silver, tungsten, steel, gold, etc. Insulator: PVC, rubber, mica, wood, glass, paxolin, etc.
5. (a) Amperes; (b) volts; (c) ohms; (d) coulombs; (e) $\mu\Omega$ mm
6. 160 m
7. 4.0 mm²
8. 0.6 Ω
9. 64 Ω
10. (a) Current halves
    (b) Current doubles
    (c) Current remains the same
11. Nothing
12. The current in a circuit is directly proportional to the voltage and inversely proportional to the resistance, at constant temperature
13. (a) 24 Ω; (b) 6.25 A; (c) 24 V
14. 2200 Ω
15. Alternating current and direct current
16. Extra low: 50 V and below, between conductors, and between conductors and earth
    Low: between 50 and 1000 V, conductor to conductor, and 50 and 600 V conductor to earth

17.

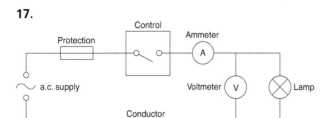

18.

$$R = \frac{V}{I}$$
$$= \frac{12}{0.5}$$
$$= 24\,\Omega$$

CHAPTER 3

1. $97\,\Omega$
2. $1440\,\Omega$
3. $21.33\,\Omega$
4. (a) $0.272\,\Omega$
   (b) $38.672\,\Omega$
5. (a) 3% of nominal for lighting 5% of nominal for power
   (b) 9.2 V; yes
6. 8 A; 12 A; 4 A; 2 A; 26 A
7. (a) $17.6\,\Omega$, 13 A
   (b) $88.16\,\Omega$, 2.6 A
   (c) $52.9\,\Omega$; 4.57 A
   (d) $1322.5\,\Omega$; 0.174 A
8. 19.84 W; 9 W; 5.63 W
9. $6\,\Omega$; 2 A; 3 A; 24 V; 12 V; 12 V
10. £274.35
11. 3 kW
12. 3.13 h
13. 86%

CHAPTER 4

1. $40\,\text{cm}^2$
2. 60 T
3. 12 N
4. $5\,\Omega$
9. 46.98 A; 35.36 A
13. 30 turns
14. 1 H
15. 0.08 S; 20 A; 2 A
16. 5 A

17. 50 Hz
18. 240 V
19. 240 V
20. 20 A; 30 A; 36.05 A
21. 240 V; 144 V
22. 0.6
23. 330

CHAPTER 5

1. 7.3 mC
2. 1.92 mC
3. 10 µF
4. 50 µF; 11.5 mC; 4.6 mC
5. $80\,\text{cm}^2$
6. 50 Hz
7. 144 V; 192 V; 240 V
9. 9.2 A; 4.6 A; 10.28 A

CHAPTER 6

1. 6.25 A; 86.5 µF
2. 110 V; 0.95 leading
3. 69 µF
4. 5.89 µF
5. 7.8 A at 0.4 lagging
6. 10.8 kVA; 0.82 lagging

CHAPTER 7

3. 10 A; 230 V
4. 0.97 lagging
5. 5.77 A; 110 V
6. 0.829
7. 6 kW; 6.9 kVA
8. 22 A

CHAPTER 8

2. (b) 434 V; 390.6 V; 46.2 A
3. 28 A
4. (b) 8.33 revs/s
6. (b) 2.76%
9. 75.22 A, 462 µF
10. 118.7 Nm

CHAPTER 9

4. (b) $0.516\,\Omega$; 1.55 V

CHAPTER 10

8. 136 lx; 85.65 lx
9. (a) 64; (b) 45
10. (a) 44; (b) 41.25 A
14. 1.27 A; 0.94 lagging

CHAPTER 16

1. 151.8 Ω
2. 18 m
3. 0.5 Ω
4. No
5. Yes. Change the cb to a Type 'B'

CHAPTER 20

1. (a) 3.8 k 10%; (b) 29.5 k 5%;
   (c) 550 k 1%; (d) 12 k 10%
2. (a) Silver–violet–red–brown
   (b) Brown–green–silver–red–green
   (c) Brown–black–black–silver
   (d) Gold–green–black–gold
3. (a) 18 nF 5% 100 V
   (b) 27 nF 10% 400 V
   (c) 75 nF 20% 100 V
   (d) 100 nF 10% 250 V
4. (a) Brown–green–gold–black–yellow
   (b) Green–red–gold–white–brown
   (c) Brown–green–gold–green–red
   (d) Red–black–yellow–black–red

Appendix

General Formulae and Information

Multiples of units

Prefix	Value	Symbol
tera	10^{12}	T
giga	10^{9}	G
mega	10^{6}	M
kilo	10^{3}	k
centi	10^{-2}	c
milli	10^{-3}	m
micro	10^{-6}	μ
nano	10^{-9}	n
pico	10^{-12}	p

Typical examples

1 million ohms = 1 megohm = $1M\Omega$

1 thousand watts = 1 kilowatt = 1kW

1 thousandth of an ampere = 1 milliampere = 1mA

1 millionth of a volt = 1 microvolt = $1\mu V$

1 million millionth of a farad = 1 picofarad = 1pF

Note

For example: 10^{-6} may be shown as $\dfrac{1}{10^{6}}$ and $\dfrac{1}{10^{-6}}$ may be shown as 10^{6}

So $x \times 10^{-6}$ may be shown as $\dfrac{x}{10^{6}}$ and vice versa and $\dfrac{x}{10^{-6}}$ may be shown as $x \times 10^{-6}$ and vice versa

Systeme International (SI units)

The base units are as follows.

Length	metre	m
Mass	kilogramme	kg
Time	second	s
Electric current	ampere	A
Luminous intensity	candela	cd
Thermodynamic temperature	kelvin	K
Amount of substance	mole	mol

Length

The units in general use and their conversions are as follows.

Millimetre (mm);
centimetre (cm);
metre (m);
kilometre (km).

To obtain	multiply	by
mm	cm	10^{1}
	m	10^{3}
	km	10^{6}
cm	mm	10^{-1}
	m	10^{2}
	km	10^{5}
m	mm	10^{-3}
	cm	10^{-2}
	km	10^{3}
km	mm	10^{-6}
	cm	10^{-5}
	m	10^{-3}

Area

square millimetre (mm^2);
square centimetre (cm^2);
square metre (m^2);
square kilometre (km^2)
also, $1km^2$ = 100 hectares (ha)

To obtain	multiply	by
mm^2	cm^2	10^{2}
	m^2	10^{6}
	km^2	10^{12}
cm^2	mm^2	10^{-2}
	m^2	10^{4}
	km^2	10^{10}
m^2	mm^2	10^{-6}
	cm^2	10^{-4}
	km^2	10^{6}
km^2	mm^2	10^{-12}
	cm^2	10^{-10}
	m^2	10^{-6}

Volume

cubic millimetre (mm^3);
cubic centimetre (cm^3);
cubic metre (m^3)

To obtain	multiply	by
mm³	cm³ m³	10^3 10^9
cm³	mm³ m³	10^{-3} 10^6
m³	mm³ cm³	10^{-9} 10^{-6}

Capacity

millilitres (ml);
centilitre (cl); litre (l);
also, l litre of water has a mass of l kg.

To obtain	multiply	by
ml	cl l	10^1 10^3
cl	ml l	10^{-1} 10^2
l	ml cl	10^{-3} 10^{-2}

Mass

milligramme (mg);
gramme (g);
kilogramme (kg); tonne (t).

To obtain	multiply	by
mg	g kg t	10^3 10^6 10^9
g	mg kg t	10^{-3} 10^3 10^6
kg	mg g t	10^{-6} 10^{-3} 10^3
t	mg g kg	10^{-9} 10^{-6} 10^{-3}

Temperature

Kelvin (K) = °C + 273.15
Celsius (°C) = K − 273.15

Celsius (°C) $= \dfrac{5}{9}$ (°F − 32)

Fahrenheit (°F) $= \left(\dfrac{9 \times °C}{5}\right) + 32$

Boiling paint of water at sea level = 100°C or 212°F
Freezing paint of water at sea level = 0°C or 32°F
Normal body temperature = 98.4°F or 36.8°C

Areas of figures and solids

Square $= a \times a = a^2$

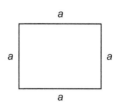

Rectangle $= h \times l$

Parallelogram $= h \times l$

Cube $= 6a^2$

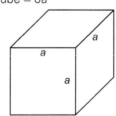

Cuboid $= 2hb + 2hl + 2bl$
$\qquad$ or $2(hb + hl + bl)$

Circle $= \pi r^2$ or $\dfrac{\pi d^2}{4}$ ($\pi = 3.1416$)

Circumference $= 2\pi r$ or πd

Ellipse = πxy
Perimeter = $\pi(x + y)$

Sphere = $4\pi r^2$

Cone:
  open ended = πrl
  solid = $\pi rl + \pi r^2$
  or $\pi r(l + r)$

Cylinder:

  hollow = $2\pi rh$
  one ended = $2\pi rh + \pi r^2$
  or $\pi r(2h + r)$
  solid = $2\pi rh + 2\pi r^2$
  or $2\pi r(h + r)$

Volumes of solids

Cube = $a \times a \times a = a^3$

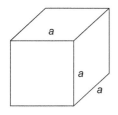

Cuboid = $h \times b \times l$

Sphere = $\dfrac{4}{3}\pi r^3$

Cone = $\dfrac{1}{3}\pi r^2 h$

Cylinder = $\pi r^2 h$

Trigonometrical ratios

For a right-angled triangle:

$$\sin\theta = \frac{B}{A}, \quad \cos\theta = \frac{C}{A}$$

$$\tan\theta = \frac{B}{C}$$

$$\frac{\sin\theta}{\cos\theta} = \tan\theta$$

$\sin^{-1}$ means 'The angle whose sine is ....'.

So, $\sin^{-1} 0.5$ means, the angle whose sine is 0.5; also $\cos^{-1} 0.5$ means, the angle whose cosine is 0.5; and $\tan^{-1} 0.5$ means, the angle whose tangent is 0.5.

The theorem of Pythagoras

For a right-angled triangle:
'The square on the hypotenuse (A) is equal to the sum of the squares on the other two sides.'
Hence:

$$A^2 = B^2 + C^2$$
$$\therefore A = \sqrt{(B^2 + C^2)}$$
$$\text{or } A = (B^2 + C^2)^{1/2}$$

$x^{1/2}$ means $\sqrt{x}$

Basic mechanical formulae

$$\begin{array}{lll} \text{Force } (F) \text{ in} & = & \text{mass } (m) & \times & \text{acceleration } (a) \\ \text{newtons} & & \text{in kg} & & \text{in m/s}^2 \end{array}$$

$$\begin{array}{lll} \text{Work done in} & = & \text{force } (F) & \times & \text{distance (m) meters} \\ \text{joules} & & \text{in newtons} & & \end{array}$$

$$\begin{array}{lll} \text{Load force } (F) & = & \text{mass } (m) & \times & \text{acceleration due to} \\ \text{in newtons} & & \text{in kg} & & \text{gravity (g) where} \\ & & & & g = 9.81 \text{ m/s}^2 \end{array}$$

Simple leavers

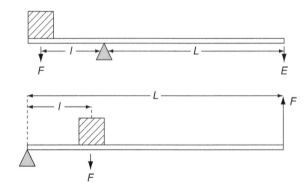

At balance $F \times l = E \times L$.

Basic Electrical formulae

Ohm's law

$$R = \frac{V}{I} \text{ ohms}$$

where: I = current in amperes
$\quad\quad\quad V$ = voltage in volts
$\quad\quad\quad R$ = resistance in ohms

Resistivity

$$R = \frac{\rho \cdot l}{a}$$

where: ρ = resistivity in ohm metres
$\quad\quad\quad l$ = length in metres
$\quad\quad\quad a$ = cross-sectional area in square metres

Temperature coefficient

For rise in temperature from 0°C to t°C

$$R_f = R_o(1 + \alpha t) \text{ ohms}$$

where: R_f = final resistance
$\quad\quad\quad R_o$ = resistance at 0°C
$\quad\quad\quad \alpha$ = temperature coefficient
$\quad\quad\quad t°C$ = change in temperature

For change from t_1 to t_2

$$\frac{R_1}{R_2} = \frac{(1 + \alpha t_1)}{(1 + \alpha t_2)}$$

Resistances in series

$$R_{total} = R_1 + R_2 + R_3 \text{ etc.}$$

Resistances in parallel

$$\frac{1}{R_{total}} = \frac{1}{R_1} + \frac{1}{R_2} + \frac{1}{R_3} + \text{etc.}$$

Power in resistive circuits

$$P = I \cdot V \text{ watts}$$
$$\text{or } P = I^2 \cdot R \text{ watts}$$
$$\text{or } P = \frac{V^2}{R} \text{ watts}$$

Electrical heating

Heat energy = mass × change in temperature
$\quad\quad\quad\quad\quad\quad$ × specific heat capacity (joules)

$$\text{kWh (output)} = \frac{\text{Mass} \times \text{change in temperature} \times \text{SII}}{3\,600\,000}$$

Quantity of electricity and charge on a capacitor

$Q = I \cdot t$ coulombs

where: Q = charge
I = current
t = time in seconds

Charge is also
$Q = C \cdot V$ coulombs

where: C = capacitance in farads
V = voltage across plates

Electric force of field strength

$$E = \frac{V}{d} \text{ volts/m}$$

where: V = voltage across plates
d = distance between plates in metres

Electric flux density

$$D = \frac{Q}{a} \text{ coulombs/m}^2$$

where Q = charge
a = area of plates in square metres

Absolute permittivity

$$\varepsilon = \frac{D}{E} \text{ farads/metre}$$

where E = electric force
D = electric flux density
or $\varepsilon = \varepsilon_o \varepsilon_r$

where: ε_r = relative permittivity of dielectric
ε_o = permittivity of free space
$= 8.85 \times 10^{-12}$ F/m

also $C = \dfrac{\varepsilon_o \varepsilon_r A(n-1)}{d}$ farads

where: C = capacitance of capacitor
n = number of parallel plates
d = distance between plates in metres
A = area of each plate in square metres

Energy stored in a capacitor

$W = \frac{1}{2} \cdot C \cdot V^2$ joules

where: C = capacitance in farads
V = voltage across plates

Capacitors in parallel

$C_{\text{total}} = C_1 + C_2 + C_3$ etc.

Charge on each capacitor:
$Q_1 = C_1 \cdot V : Q_2 = C_2 \cdot V : Q_3 \cdot V$: etc.

Capacitors in series

$$\frac{1}{C_{\text{total}}} = \frac{1}{C_1} + \frac{1}{C_2} + \frac{1}{C_3} \text{ etc.}$$

Charge on each capacitor is the same as the total charge:

$Q_1 = C \cdot V : Q = C_1 \cdot V_1 : Q = C_2 \cdot V_2$: etc.

Time constant

$$\tau = C \cdot R$$

where: C = capacitance in farads
R = resistance in ohms

Magneto motive force (m.m.f.)

$F = NI$ ampere turns
where: F = m.m.f.
N = no. of turns
I = current in amperes

Magnetizing force

$$H = \frac{NI}{l} \text{ ampere turns/metre}$$

where: l = length of magnetic circuit

Absolute permeability

$$\mu = \frac{B}{H}$$

where: B = flux density in teslas
H = magnetizing force in ampere-turn/metre

or $\mu = \mu_0 \cdot \mu_r$

where: μ_0 = permeability of free

space $= 4\pi \times 10^{-7}$ H/m

μ_r = relative permeability of magnetic material

Magnetic flux

$$\Phi = \frac{F}{S}$$

where: Φ = flux in webers
F = m.m.f.
S = reluctance

also $S = \dfrac{l}{\mu A}$

for reluctance in series:

$$S = S_1 + S_2 + S_3$$

Magnetic flux density

$$B = \frac{\Phi}{A} \text{ teslas}$$

where Φ = flux in webers

A = area at right angles to field

Force on a conductor

$F = B \cdot l \cdot I$ newtons

where: B = flux density

l = length of conductor in metres

I = current

E.m.f. induced by conductor

$E = B \cdot l \cdot v$ volts

where: B = flux density

l = length of conductor in metres

v = velocity in metres/second

For a conductor moving at an angle, then:

$E = B \cdot l \cdot v \cdot \sin \theta$

where: θ = angle in degrees

Induced e.m.f. due to flux change

$$E = \frac{(\Phi_2 - \Phi_1)}{t} \cdot N \text{ volts}$$

where: ϕ = flux in webers

N = no. of turns

t = time in seconds

Self inductance

$$E = \frac{-L(I_2 - I_1)}{t} \text{ volts}$$

Note: The minus sign denotes the e.m.f. is a back e.m.f.

$$\text{also } L = \frac{N \cdot \Phi}{I} \text{ henrys}$$

where: E = induced e.m.f.

L = self inductance in henrys

I = current

t = time in seconds

Mutual inductance

$$E = \frac{-M(I_2 - I_1)}{t} \text{ volts}$$

$$M = \frac{(\Phi_2 - \Phi_1)}{(I_2 - I_1)} \cdot N \text{ henrys}$$

Energy stored in a magnetic field

$W = \frac{1}{2} \cdot L \cdot I^2$ joules

Time constant

$$\tau = \frac{L}{R}$$

Electrical Machines

D.c.motors

$$E = \frac{2 \cdot p \cdot \Phi \cdot n \cdot z}{c}$$

where: E = back e.m.f.

p = no. of pairs of poles

ϕ = flux per pole

n = speed in rev/s

z = no. of armature conductors

$c = 2_p$ for lap wound and 2 for wave wound

also $E = V - I_a \cdot R_a$

where: V = supply voltage

I_a = armature current

R_a = armature resistance

Torque equation

$$\frac{T_1}{T_2} = \frac{I_{a1} \cdot \Phi_1}{I_{a2} \cdot \Phi_2}$$

D.c generators

$$E = \frac{2 \cdot p \cdot \Phi \cdot n \cdot z}{c}$$

and $E = V + I_a R_a$

$$E_1 = \frac{n_1 \cdot \Phi_1}{n_2 \cdot \Phi_2}$$

$$E \propto n \cdot \Phi$$

where: E = generated e.m.f.

Mechanical output

$P = 2 \cdot \pi \cdot n \cdot T$ watts

where: n = speed in rev/s

T = torque in newton metres

Induction motors

$F = n \cdot p$

where: f = frequency in hertz

n = speed in rev/s

p = no. of pairs of poles

Slip

$$\text{Percentage slip } (s) = \frac{n_s - n_\tau}{n_s} \times 100$$

$$\text{Per unit slip} (s) = \frac{n_s - n_\tau}{s}$$

where: n_s = synchronous speed

n_τ = rotor speed

Slip speed $(s) = n_s \, n_\tau$

Slip frequency $f_s = sf$

where: f = supply frequency

s = slip

A.C. Circuits

R.m.s. value = 0.707 × maximum value
Average value = 0.636 × maximum value

Inductive reactance

$X_1 = 2 . \pi . f . L$ ohms

$V = I . X_L$ volts

where: f = frequency in hertz

L = inductance in henrys

Inductance (*L*) and resistance (*R*) in series

$V_{X_L} = I . X_L$ ohms

$V_R = I . R$

$V = V_{X_L} + V_R$ by phasors only

or $V^2 = V_{X_L}^2 + V_R^2$ if components are assumed pure

Total opposition = impedance = Z ohms

$Z = \sqrt{(R^2 + X_L^2)}$ ohms

or $Z = \dfrac{V}{I}$

L and *R* in parallel

$V = I_2 . X_L$

or $V = I_1 . R$

$Z = \dfrac{V}{I}$

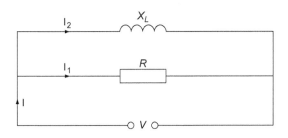

$I = I_1 + I_2$ by phasors or active and reactive components only

or $I^2 = I_1^2 + I_2^2$ if components are assumed pure

Power factor

$$\text{Power factor } (PF) = \cos\theta$$

$$\text{or} = \frac{R}{Z}$$

Capacitive reactance

$$X_C = \frac{1}{2 . \pi . f . C} \text{ ohms}$$

where f = frequency in hertz

C = capacitance in farads

$V = I . X_C$

Capacitance and resistance in series

$V_{X_C} = I . X_C$

$V_R = I . R$

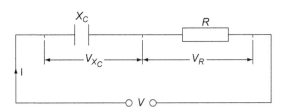

$V = V_{X_C} + V_R$ by phasors only

or $V^2 = V_R^2 + V_{X_C}^2$ if components are assumed pure

Impedance

$$Z = \sqrt{(R^2 + X_C^2)} \text{ ohms}$$

$$\text{or } Z = \frac{V}{I}$$

Capacitance and resistance in parallel

$$V = I_2 . X_c$$
$$\text{or } V = I_1 . R$$
$$\text{or } V = I . Z$$

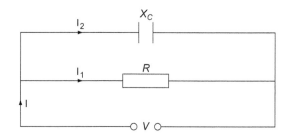

$I = I_1 + I_2$ by phasors, or active and reactive components only

or $I^2 = I_1^2 + I_2^2$ if components are assumed pure

Inductance, capacitance and resistance in series

$$V_{X_L} = I . X_L$$
$$V_{Xc} = I . X_C$$
$$V_R = I . R$$
$$V = I . Z$$

$V = V_{X_L} + V_{X_c} + V_R$ by phasors or active and reactive components only

or $V^2 = V_R^2 + (V_{X_L} - V_{X_c})^2$ if components are assumed pure

Impedance

$$Z = \sqrt{\left[R^2 + \left(X_L - X_C \right)^2 \right]} \text{ ohms}$$

$$\text{or } Z = \frac{V}{I}$$

Resonance

$$f_o = \frac{1}{2\pi \sqrt{(L.C)}} \text{ hertz}$$

Also at resonance
$$X_L = X_C$$
$$\text{so } Z = \sqrt{R^2} = R \text{ ohms}$$

Voltage magnification, dynamic impedance or *Q* factor

$$Q = \frac{1}{R} \sqrt{\left(\frac{L}{C} \right)}$$

Power

Single phase

Power triangle:

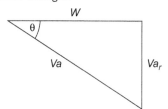

$$\text{Power factor} = \cos \theta = \frac{W}{Va}$$

$$\text{hence } W = Va \cos \theta$$
$$\text{and } Va = Va . \sin \theta$$

Three phase

Star connected:

$$V_L = \sqrt{(3)} . V_p$$
$$I_L = I_p$$

where: V_L = line voltage
V_p = phase voltage
I_L = line current
I_p = phase current

Delta connection:

$$V_L = V_P$$
$$I_L = \sqrt{(3)} . I_P$$

For star or delta connection:

$$\text{Total } Va = \sqrt{(3)} . V_L . I_L$$
$$\text{and, total watts} = \sqrt{(3)} . V_L . I_L . \cos \theta$$

Delta ← → star conversions

To convert (1) to (2)

$$R_1 = \frac{R_{AB} \times R_{CA}}{R_{AB} + R_{BC} + R_{CA}}$$

$$R_2 = \frac{R_{AB} \times R_{BC}}{R_{AB} + R_{BC} + R_{CA}}$$

$$R_3 = \frac{R_{BC} \times R_{CA}}{R_{AB} + R_{BC} + R_{CA}}$$

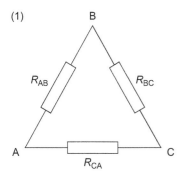

(1)

To convert (2) to (1)

$$R_{AB} = R_1 + R_2 + \frac{(R_1 \times R_2)}{R_3}$$

$$R_{BC} = R_2 + R_3 + \frac{(R_2 \times R_2)}{R_1}$$

$$R_{CA} = R_3 + R_1 + \frac{(R_3 \times R_1)}{R_2}$$

(2)

Transformers

$$\frac{V_p}{V_s} = \frac{N_p}{N_s} = \frac{I_s}{I_p}$$

Regulation

$$\text{Percentage} \times \text{regulation} = \frac{E_s - V_s}{E_s} \times 100$$

where: E_s = no load voltage
V_s = on load voltage

Losses

Hysteresis loss $\propto$ frequency

Eddy current $\propto \Phi_{max}^2 \times f^2 \times$ (lamination thickness)^2

Efficiency

$$\text{Percentage efficiency} = \frac{\text{output}}{\text{input}} \times 100$$

Rectifiers

Half wave:

$$\text{Average value} = \frac{0.636 \times \text{max.value}}{2}$$

Wheatstone Bridge

$$\frac{P}{Q} = \frac{R}{X}$$

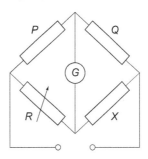

Cells and Batteries

Internal e.m.f.

$E = V + Ir$

where: E = e.m.f. of cell
V = terminal p.d of cell
I = current drawn
r = internal resistance of cell

or $E = V + I(R + r)$
where: R = resisteance of connected load

Ampere hour (Ah) efficiency

$$\text{Ah efficiency}(\%) = \frac{\text{Ah output}}{\text{Ah input}} \times 100$$

Watt hour (Wh) efficiency

$$\text{Wh efficiency}(\%) = \frac{\text{Wh output}}{\text{Wh input}} \times 100$$

Charging current

$$I = \frac{V - E}{R + r}$$

where: V = supply voltage

E = e.m.f. of cell

R = resistance of series resistor and leads

r = internal resistance of cell

Illumination

Luminous flux

$$F = \frac{E \times a}{MF \times CU}$$

where: E = illuminance in lux or lm/m²

a = area in m²

MF = maintenance factor

CU = coefficient of utilization

Illuminance

$$E = \frac{I}{d^2}$$

where: I = luminous intensity in candelas

d = vertical distance from source in metres

also $E = \frac{I \cos^3 \theta}{d^2}$

where: θ = some angle from the vertical in degrees

Index

Index

Index